PROGRESS IN BIOORGANIC CHEMISTRY

PROGRESS IN
BIOORGANIC CHEMISTRY
VOLUME ONE

Edited by

E. T. KAISER

Departments of Chemistry and Biochemistry
University of Chicago

F. J. KÉZDY

Department of Biochemistry
University of Chicago

WILEY-INTERSCIENCE,

A DIVISION OF JOHN WILEY & SONS, INC.

NEW YORK · LONDON · SYDNEY · TORONTO

Cat for
Chem.

Library of Congress Catalogue Card Number: 75   142715

ISBN 0   471   45485   0

Printed in the United States of America.

10 9 8 7 6 5 4 3 2 1

2/15/72

CONTRIBUTORS

JOSEPH E. COLEMAN, *Department of Molecular Biophysics and Biochemistry, Yale University, New Haven, Connecticut*

A. R. FERSHT, *Medical Research Council Laboratory for Molecular Biology, Cambridge, England*

GORDON A. HAMILTON, *Department of Chemistry, The Pennsylvania State University, University Park, Pennsylvania*

A. J. KIRBY, *University Chemical Laboratory, Cambridge, England*

1837

PREFACE TO THE SERIES

Bioorganic chemistry is a new discipline emerging from the interaction of biochemistry and physical organic chemistry. Its origins can be traced to the enzymologists whose curiosity was not satisfied with the purification and the superficial characterization of an enzyme, to the physical organic chemists who had the conviction that the elementary steps of biological reactions are identical to those observed in organic chemistry, and to the physical and organic chemists who wished to understand and to imitate *in vitro* the unequaled catalytic power and specificity exhibited by living organisms. As with all interdisciplinary sciences, bioorganic chemistry uses many of the methods and techniques of the disciplines from which it is derived; many of its protagonists qualify themselves as physical organic chemists, enzymologists, biochemists, or kineticists. It is, however, a new science of its own by the criterion of having developed its own goals, concepts, and methods.

The principal goal of bioorganic chemistry can be defined as the understanding of biological reactions at the level of organic reaction mechanisms, that is, the identification of the basic parameters which govern these reactions, the formulation of quantitative theories describing them, and the elucidation of the relationships between the reactivity and the structures of the molecules participating in the process. This definition is narrower than one which some scientists would give. They might prefer to include areas such as medicinal chemistry, for example, as part of the field of bioorganic chemistry. Accordingly, the goals which they would cite would differ from those which we have considered. We do not seek here to argue or to defend our concept of what constitutes bioorganic chemistry, but within the framework of our definition we believe that there is a real distinction between much work in present day medicinal chemistry and that in the bioorganic field. In our conception of bioorganic chemistry the emphasis is on mechanism.

The theoretical formulation of the understanding of biochemical and, therefore, enzyme-catalyzed reactions has required the elaboration of new concepts, such as multifunctional catalysis, stereospecificity by three-point attachment, and control of reactivity by conformational changes. Many of the new concepts will not survive; they will be redefined, discarded, or reevaluated as fortunately always happens in science. But the trend is clearly apparent — these new concepts are providing us with efficient tools of great power which can be used to describe and discuss enzymatic reactions.

As to the methods of bioorganic chemistry, they are conceptually the same as those for the study of any chemical reaction; they include analytical and physical techniques. However, the complexity of the reacting molecules has resulted in methods which are new and unique in their ability to probe the chemistry of a functional group surrounded by a multitude of very similar groups or a chemical event accompanied by a host of satellite reactions. The discovery of numerous methods involving active-site directed reagents, "reporter molecules," and chromophoric substrates illustrates the usefulness and the elegance of the new science.

The future of bioorganic chemistry appears very promising, and the fields to cover in the future are immense and unexplored. The earliest work has concentrated on the understanding of general acid–general base catalyzed reactions, hydrolytic reactions, and the role of proteins in enzymatic catalysis. The mechanism of enzymatic catalysis by most coenzymes is very far from being well described, and the very prominent role of metal ions in catalysis is only beginning to emerge. Other important problems, such as surface catalysis at biological membranes, transport mechanisms, the process by which ribosome-catalyzed reactions occur, and the reactivity of RNA and DNA molecules, are at an early stage of development or are completely unexplored.

As a result, a rapid growth of bioorganic chemistry is desirable and is currently underway, as evidenced by the large number of papers published on the subject. An unfortunate result of this rapid expansion is the scarcity of comprehensive treatments of bioorganic chemistry. The rapid progress in this field makes it likely that large portions of any comprehensive textbook will become obsolete soon, although the student of bioorganic chemistry may still learn some of the basic concepts of the subject from them. Because of many factors, it would be possible to revise textbooks only at infrequent intervals. For this reason the format of presenting comprehensive treatments of limited subjects seemed more appropriate to us. It would provide the investigators, interested readers, and students with a thorough and critical evaluation of those aspects of bioorganic chemistry where definite and substantial progress has been achieved.

It is the hope of the Editors of this volume to be able to respond to the need for up-to-date comprehensive treatments of important topics in bioorganic chemistry. In attempting to do so we would like to provide treatments of bioorganic subjects which will be general enough to retain the attention of most workers in the field and which will be at a level beyond that of a usual review article or literature survey. Since many aspects of bioorganic chemistry are still in the process of evolution, we also would like to provide a forum where the authors can express challenging new ideas and present stimulating and, frequently, controversial discussions. For this reason we hope to give the authors somewhat more latitude than is customary in this kind of publication, while still retaining the requirement of scientific sobriety.

In this first volume of the series on *Progress in Bioorganic Chemistry* several topics which are essential to the understanding of enzymatic reaction mechanisms are discussed. We expect that further volumes will appear at one- to two-year intervals, and we will welcome suggestions as to suitable topics for future volumes.

Finally, we wish to express our gratitude to the authors who contributed to the present volume and to Mrs. Hanna Posner for her help.

E. T. KAISER
F. J. KÉZDY

Chicago, Illinois
January 1971

CONTENTS

PROGRESS IN BIOORGANIC CHEMISTRY

INTRAMOLECULAR CATALYSIS

A. J. KIRBY

University Chemical Laboratory, Cambridge, England

A. R. FERSHT

Medical Research Council Laboratory for Molecular Biology, Cambridge, England

1

1 INTRODUCTION

The relevance of intramolecular catalysis for the study of enzyme mechanism rests on one simple, basic assumption: reactions between functional groups held in close proximity on a single molecule may be valid models for reactions between the same groups in the enzyme–substrate complex. More than ten years of exploratory work in the area have led to the identification of a wide range of intramolecular reactions involving the functional groups available in amino acid side chains, and it is possible to come to working conclusions about the scope and limitations of the approach. In this chapter we attempt to use these conclusions to identify the areas that seem likely to be the most productive. Our discussion includes the most important advances since the systematic and authoritative review of the subject by Bruice and Benkovic[1].

It is clear that proximity per se cannot in general account for the efficiency of either enzymic or intramolecular catalysis. Koshland[2, 3] has used a very simple model to calculate the rate enhancements expected if the sole function of an enzyme is to bring substrates and catalysts into close, properly orientated proximity. The predicted accelerations are far too low to account for the efficiency of enzymic catalysis (although much larger rate enhancements are predicted if a sufficiently large number of functional groups are involved in the catalysis). Implicit in this model is the assumption that the effects of proximity are no more than statistical. This would mean, as pointed out in particular by Jencks[4] that there should be a limiting value for the rate enhancement expected for an intramolecular reaction. This would be 55 M in aqueous solution, corresponding to the intermolecular reaction in which every substrate molecule is completely surrounded by molecules of reactant. The rate enhancements observed in intramolecular reactions are commonly much greater than this, and clearly some factor or factors as yet unconsidered must be involved.

There is little direct evidence as to the nature of these factors, but since they will certainly be involved in enzymic catalysis also it is a matter of some importance to identify and examine them. There are a number of plausible candidates. For example, one possibility is steric inhibition of resonance. π-Delocalization can be interrupted by rotation about a single bond with partial π-character. Huisgen and Ott[5] have shown that small-ring lactones, held necessarily in the *cis* configuration, are more rapidly hydrolyzed than open chain esters, which exist in the thermodynamically more stable *trans* configuration. Some delocalization is possible in both these cases, since the ester group remains planar; and considerably larger effects may be expected if a system is forced into a geometry where delocalization is completely interrupted.

Steric inhibition of resonance is an attractive possibility for substrates like esters and amides, in which the activation energy for nucleophilic attack could be considerably reduced by conformational changes during binding, but it is not applicable to all substrates. If there are entirely general factors operating in enzymic catalysis, we must therefore look elsewhere. Two particularly attractive candidates are local solvation effects and steric compression.

An integral part of any intermolecular reaction in solution is the mutual penetration by the reactants of their separate solvation shells. It is not a simple matter to estimate how large a proportion of the total free energy of activation for a reaction might be accounted for by this process, but there is evidence that it may be a large factor. Small anions with high charge densities, such as fluoride, chloride, and hydroxide ions, are strongly solvated in water and other hydroxylic solvents by a hydrogen-bonding mechanism. In dipolar aprotic solvents, such as dimethyl sulfoxide, these anions are very weakly solvated and have much higher activities. The rates of many S_N2 reactions of small anions, for example, are 10^6–10^7 times as fast in dimethyl sulfoxide as in hydroxylic solvents [6], and similar effects are observed on the alkaline hydrolysis of esters [7, 8].

When reacting groups are already close in the ground state, as they are in intramolecular reactions, their solvation shells cannot be entirely independent. Depending on the molecular geometry of the particular reaction, mutual penetration of the solvation shells by the two groups will be generally well under way in the ground state. In some cases, it will be complete, with the two groups in direct contact. This extreme situation may be common in intramolecular reactions between adjacent groups on conformationally rigid molecules, and here information from molecular models may be useful. For example, Jencks[9] points out that the carboxylate group of the highly reactive bicyclic compound 3,6-endoxo-Δ^4-tetrahydrophthalate is too close to the ester group which it attacks for solvent to come between; and the same is probably true of the maleic acid derivatives discussed on page 28. In such circumstances reactions would be expected to be substantially accelerated, and this factor might account for some part of the discrepancy between the rate enhancements observed for intramolecular reactions and those expected on proximity grounds alone.*

With the exclusion of the last intervening molecule of solvent, the remaining barriers to reaction are van der Waal's repulsion and the electronic activation energy involved in making and breaking covalent bonds. In intermolecular reactions and in intramolecular reactions between the same groups on conformationally mobile molecules, the remaining activation energy is the irreducible minimum required to overcome these barriers. But in intramolecular reactions between groups on conformationally rigid molecules, and perhaps in enzymic reactions also, the situation may arise where some progress towards surmounting these barriers has occurred in the ground state. This could happen if groups were forced into sufficiently close proximity, by what may loosely be termed steric compression. The remaining activation energy rapidly becomes a very sensitive function of distance along the reaction coordinate in the early part of many reactions, and interactions of this sort could lead to large rate accelerations. A possible example is described on page 28. In the special case of reactions between groups bearing like charges, electrostatic repulsion may also be overcome in this way, and a possible instance of this effect is found in the hydrolysis of phosphate diesters catalyzed by the ionized carboxyl group (p. 47).

Ideas like those described above are most easily tested by studying simple intramolecular reactions. No single system is likely to be a useful model for more than a fraction of the mechanism of an enzymic reaction, but it is possible to draw firm conclusions about the separate parts of an enzymic mechanism by studying several different models. It should also be possible to design simple polyfunctional systems with predictable properties and catalytic efficiencies comparable to those of enzymes.

If one approaches enzyme mechanism by studying intramolecular catalysis, the problem falls naturally into two parts: (1) why intramolecular

---

*The first experimental search for a differential effect of solvation on reactivity in intramolecular and intermolecular catalysis (by the ionized carboxyl group) proved negative (T. C. Bruice and A. Turner, *J. Amer. Chem. Soc.*, 92, 3422 (1970).

reactions go so fast and (2) why enzymic reactions go so much faster. Neither part of the problem is near to solution at the present time, but a deeper understanding of the first part is a necessary preliminary to a solution of the second. In particular, when we understand the factors that make intramolecular reactions go so fast, we will be in a position to design systems of greater catalytic efficiency. And not until we know how efficient intramolecular catalysis can be can we know how much intrinsically more efficient is enzymic catalysis.

Thus an area of increasing importance is the study of the relationship between structure and efficiency in intramolecular catalysis. For example, we do not know *a priori* whether two groups held *ortho* to each other on a benzene ring react with each other particularly efficiently or particularly slowly. The bonds that hold groups together on the same molecule may at the same time be holding them apart. But this lack of flexibility, generally the great disadvantage of the simple intramolecular reaction as a model for an enzymic reaction, is its great strength, when the model can be shown to be true. For then it becomes possible to specify mechanism and geometry with sufficient precision to define a possible part of the enzymic mechanism. So it is essential to study the interactions between functional groups in a range of environments before drawing any general conclusions about the effects of proximity on reactivity.

2 INTRAMOLECULAR CATALYSIS BY THE CARBOXYL GROUP

Intramolecular catalysis by the carboxyl group has attracted particular attention in recent years. The group is generally ionized under physiological conditions, but the undissociated form may be present in significant amounts, especially in hydrophobic regions of macromolecules where ionization can be suppressed. Catalytically the carboxyl group is the most versatile available to enzymes. In the undissociated form it can act as a general acid, as it does in the hydrolysis of ortho esters[10] and in intramolecular reactions with certain acetals (see p. 31). As the anion it can act as a nucleophile or a general base catalyst; and numerous examples, both intermolecular and intramolecular, have been recognized in ester hydrolysis reactions. Several are discussed in the next section, and a comprehensive review is available[11]. The anion can also apparently stabilize the conjugate acid of a very weak oxygen base in aqueous solution, as in the hydrolysis of salicyl phosphate (see p. 44), and can act as a carrier of acyl groups in series nucleophilic catalysis, because the group itself becomes susceptible to nucleophilic attack when it has successfully carried out a nucleophilic displacement (see pp. 15–16):

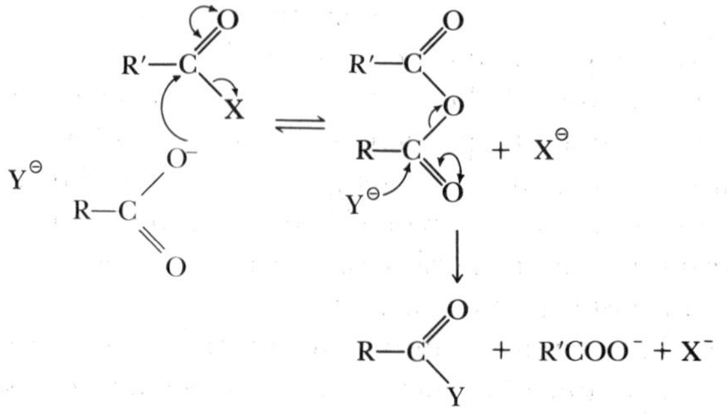

$$R'-C\overset{O}{\underset{X}{\overset{\|}{<}}} \quad \overset{O}{\underset{R-C}{\overset{\|}{<}}} \rightleftharpoons \quad \begin{matrix} R'-C\overset{O}{\underset{O}{\overset{\|}{<}}} \\ R-C\overset{O}{\underset{Y^{\ominus}}{<}} \end{matrix} \quad + \quad X^{\ominus}$$

$$\downarrow$$

$$R-C\overset{O}{\underset{Y}{\overset{\|}{<}}} \quad + \quad R'COO^- + X^- \tag{1}*$$

2.1 Catalysis of Ester Hydrolysis

Ester hydrolysis has always been a favorite proving ground for kinetic and mechanistic theories. Most of our present ideas about catalysis in aqueous solution stem from work with carboxylic esters, and recent studies on intramolecular catalysis continue to use them in preference to other substrates.

2.1.1   The Hydrolysis of Aspirin. Aspirin hydrolysis is often described as the classic example of intramolecular catalysis[12]. The aspirin anion (**1**) is hydrolyzed in a pH-independent reaction nearly 180 times as fast as expected from a linear free energy relationship and from comparison with the rate of hydrolysis of acetyl methyl salicylate[13–15]. Hydrolysis

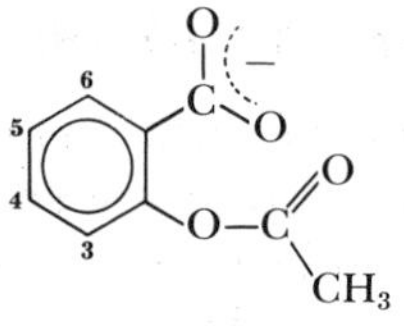

1

clearly involves intramolecular catalysis by the ionized carboxyl group, but the mechanism of this catalysis has been securely established only recently[13]. The work is discussed in some detail here, since it illustrates a number of points of general importance for reactions of this type and depends on a detailed understanding of catalysis in aqueous solution that has only become available in recent years.

Three reasonable mechanisms can be written for the catalyzed hydrolysis of aspirin involving the anion and a molecule of water only, or the kinetic equivalent, the undissociated acid and a hydroxide ion[13]. These cannot be distinguished from a knowledge of the concentration depend-

*Double-headed arrows will be used throughout this chapter to indicate when intermediates are formed with the generation of electric charge and then decompose.

Nucleophilic catalysis

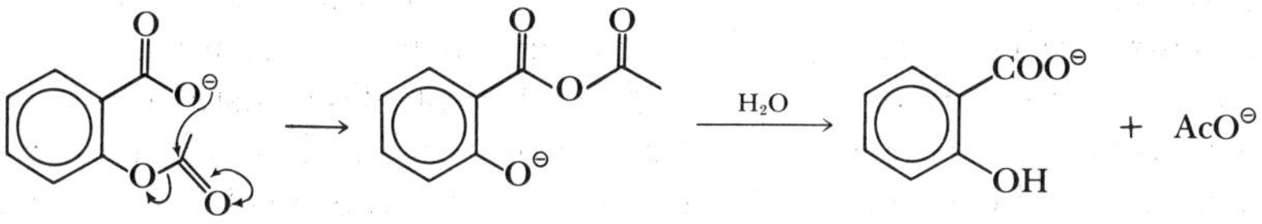

Mechanism A

General base catalysis

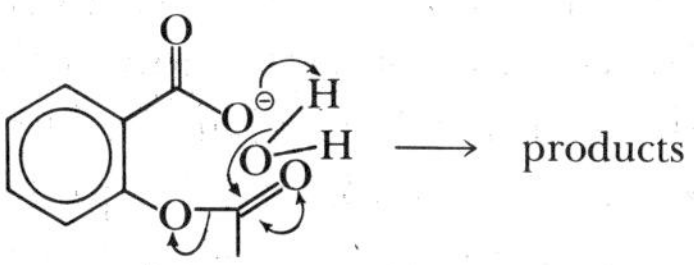

Mechanism B

Specific base–general acid catalysis

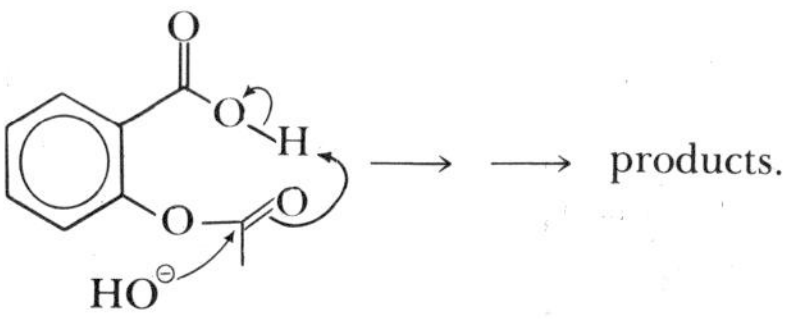

Mechanism C

ence of the reaction. For a number of years the reaction was believed to involve nucleophilic catalysis (Mechanism A) because hydrolysis in ^{18}O-enriched water led to incorporation of labeled oxygen into the salicylic acid produced[16], as would be expected if the mixed anhydride were an intermediate. But when the experiment was repeated, using more highly enriched $H_2^{18}O$, no incorporation could be detected[13]. This evidence reinforced doubts expressed about the nucleophilic mechanism by Garrett[17] (who found that solvolysis was faster in mixed alcohol–water solvents) and indeed made the nucleophilic mechanism most unlikely.

In an attempt to resolve the ambiguities arising from the kinetic equivalence of the various mechanisms, Fersht and Kirby[18] studied a series of aspirins with substituents in the 4- and 5-positions. They used a modification of the Hammett equation, due to Jaffé,[19] which assumes that the effects of substituents on the reacting carboxyl and ester groups are independent (eq. 2).

$$\log k/k_0 = \sigma_1\rho_1 + \sigma_2\rho_2 \qquad (2)$$

From the linear plot of $1/\sigma_1 \log k/k_0$ against σ_2/σ_1 (σ_1 and σ_2 are σ_m or σ_p, appropriate), they obtained values of ρ_1 and ρ_2, the Hammett reaction constants for the attacking carboxylate and the leaving group.

The algebraic signs of the ρ-values do not allow distinctions to be drawn between mechanisms. The rate is increased by electron withdrawal from the O—$COCH_3$ group whatever the mechanism. In the nucleophilic and the general base catalysis mechanisms, electron-withdrawing substituents reduce the reactivity of the COO^- group. In the specific base–general acid catalysis mechanism, electron withdrawal increases the catalytic strength of the —COOH group, but also increases its acidity and thus reduces the concentration of the catalytic form. This factor can outweigh the increased reactivity, so that the effects of substituents on all three mechanisms are likely to be qualitatively the same.

This is a point of some importance, since it has led to confusion in the past. For example, the fact that the hydrolysis of salicylamide is faster than that of the 5-nitro compound led Bruice and Tanner[20] to prefer the general base catalysis mechanism (2) to the kinetically equivalent

2   3

specific base–general acid catalysis mechanism (3). It is true that the introduction of the nitro group *para* to the phenoxide oxygen decreases its reactivity. But this argument neglects the effect of substitution on the pK_a of the phenolic group, as explained above and as pointed out originally by Capon and Ghosh[21]. Thus pairs of mechanisms of this type cannot be distinguished by the qualitative effects of substitution; nor does the deuterium solvent isotope effect allow a choice[22].

The *magnitudes* of the ρ-values obtained by Fersht and Kirby[18] for the effects of substituents on intramolecular catalysis of aspirin hydrolysis did allow a tentative choice between the three possible mechanisms. For the effects of substituents on the leaving group $\rho = 0.96$, and for the carboxylate group $\rho = -0.52$. Comparison with ρ-values for intermolecular reactions of known mechanism led to the conclusion that only in the case of the general base catalysis mechanism are both ρ-values of the expected magnitude[18].

The nucleophilic mechanism (Mechanism A), involving rate-determining anhydride formation, can be ruled out on numerous counts. The entropy of activation ($\Delta S\ddagger = -22.5$ eu) is very low for a unimolecular reaction [13]. The deuterium solvent isotope effect $(k_H/k_D = 2.2)$ is very high for nucleophilic catalysis; and the observed catalysis of the attack of protic nucleophiles is not predicted [13, 14]. But the lack of any incorporation of oxygen-18 from the enriched solvent is conclusive [13]. Mechanism C, involving general acid catalyzed attack of hydroxide ion, was shown to be unlikely because of the calculated rate constants necessary [13] and because (general acid) catalysis of the attack of other oxyanion nucleophiles is not observed in the pH region concerned [13].

All the available evidence is consistent with mechanism B, involving general base catalysis by the ionized carboxyl group of the attack of water. And as the data for all the mono-substituted aspirins are correlated by a single linear free-energy relationship and no incorporation of ^{18}O from the enriched solvent is observed with the most reactive or the least reactive ester, it may be assumed that the same mechanism is involved in the hydrolysis of all the monosubstituted compounds. In the case of aspirin, intermolecular general base catalysis by oxyanions is observed, and from the Bronsted plot correlating the data it can be estimated that the "effective" concentration [23] of the neighboring carboxylate group of aspirin is $13\ M$ [13].

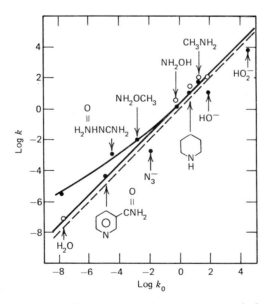

Figure 1   A comparison [14] of the second order rate constants, k, for the reactions of nucleophiles with (●) the aspirin anion and (○) p-carboxyphenyl acetate anion with the second order rate constants, k_0, for the corresponding reactions with phenyl acetate. The straight lines are drawn with slope 1.0, and the dashed line represents $k = k_0$.

St. Pierre and Jencks[14] showed that the (aminolysis) reaction of aspirin anion with semicarbazide is also subject to intramolecular catalysis (Fig. 1), the effective concentration of the ionized carboxyl group in this case being 28 M.

2.1.2 Nucleophilic versus General Base Catalysis.

The hydrolysis of substituted-phenyl acetate esters is catalyzed by acetate anion, but the mechanism of catalysis changes from nucleophilic catalysis, observed for esters with very good leaving groups, such as the dinitrophenyl acetates, to general base catalysis for esters, such as phenyl acetate itself, of less strongly acidic phenols[24]. For esters with leaving groups of intermediate basicity, such as p-nitrophenyl acetate, both mechanisms account for a significant proportion of the observed catalysis.

Entirely analogous behavior is observed in the hydrolysis of substituted aspirins catalyzed by the neighboring carboxylate group. Monosubstituted derivatives, including the 3-nitro and 5-nitro compounds, are hydrolyzed by the general base catalysis mechanism [13], but the hydrolysis of 3,5-dinitro-aspirin (4) involves nucleophilic catalysis, as shown by the formation of large amounts of methyl 3,5-dinitrosalicylate (6) on solvolysis in 50% aqueous methanol and the incorporation of ^{18}O into the 3,5-dinitrosalicylate formed on hydrolysis in enriched water (Scheme 1).

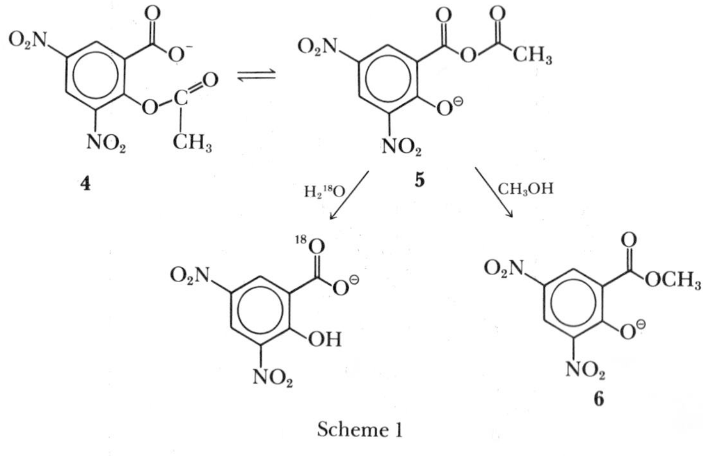

Scheme 1

Although the product distribution shows clearly that this reaction must involve nucleophilic catalysis, its kinetic parameters are closely similar to those for the hydrolysis of aspirin itself, which are characteristic of a reaction involving general base catalysis. For example, $\Delta S\ddagger = -20.6$ eu., $k_H/k_D = 2.0$, and solvolysis is faster in aqueous methanol. Fersht and Kirby conclude that the rate-determining step of the reaction is the hydrolysis of the mixed anhydride intermediate **5** and that this involves intramolecular general base catalysis by the phenolate oxygen, as in **7**. [This mechanism is thought to be involved in the hydrolysis of salicylate ester anions **8** (see p. 68)]. The concentrations of the starting material and the anhydride intermediate **7** are related by a simple

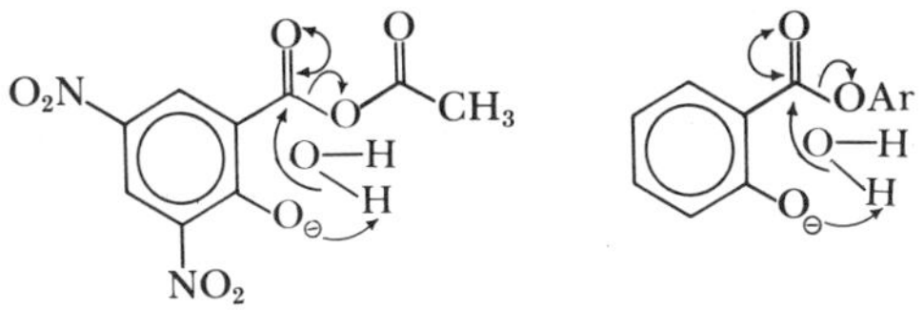

7   8

equilibrium constant, so any reaction of the anhydride anion depends on the concentration of 3,5-dinitroaspirin anion. Thus the apparent pK_a obtained from the pH–rate profile is that of the starting material [25].

2.1.3 Exocyclic versus Endocyclic Displacement. Intramolecular nucleophilic catalysis is the well established mechanism of the rapid hydrolysis of the anions of the monophenyl esters of phthalic, maleic, and 3,6-endoxo-Δ^4-tetrahydrophthalic acids [26, 27]. The rate-determining step in each of these reactions is the hydrolysis of the intermediate cyclic anhydride, which accumulates and is readily detected (eq. 3). Cyclic

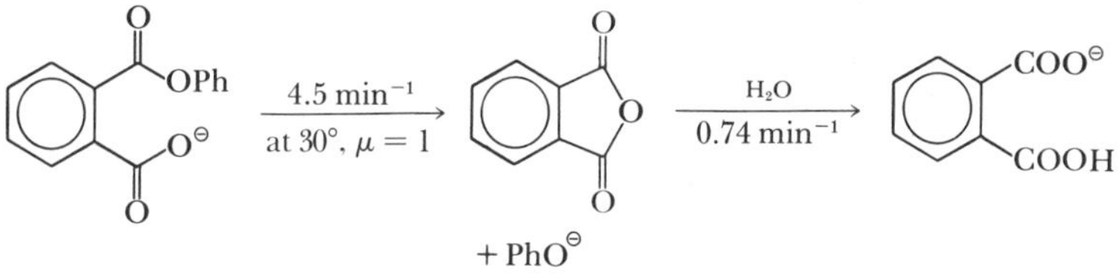

anhydrides are also probable intermediates in the hydrolyses of monophenyl glutarates and succinates [28], but in these cases ring closure is much slower than the subsequent hydrolysis of the anhydride, and the evidence is indirect. The rates of these reactions, and thus the efficiency of catalysis by the ionized carboxyl group, are very much greater than those observed for aspirin and its derivatives. Monophenyl phthalate is hydrolyzed over 10^4 times as fast as aspirin, and nucleophilic catalysis must be at least 10^6 times more effective. This observation leads to the generalization, which will be rationalized below, that reactivity in intramolecular nucleophilic catalysis is controlled not only by the relative basicities of nucleophile and leaving group, as it is in intermolecular catalysis, but also, and in particular, by whether or not the leaving group remains attached to the intermediate formed — in other words, whether displacement is endocyclic or exocyclic.

In the nucleophilic mechanism for the hydrolysis of aspirin (Scheme 2), the reacylation of the leaving group (k_2) is a rapid, intramolecular

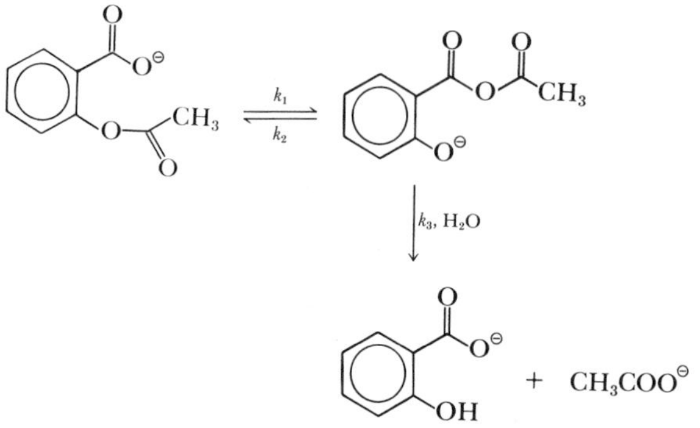

Scheme 2

reaction. St. Pierre and Jencks showed that $0.01\,M$ phenol in $0.01\,M$ phosphate buffer at pH 7.3 reacts with $10^{-3}\,M$ acetic anhydride to give a 44% yield of phenyl acetate[14]. In the intramolecular reaction the effective concentration of the phenolate anion will be very large, certainly greater than $10\,M$. Clearly then in Scheme 2 $k_2 \gg k_3$, and the hydrolysis of the anhydride will be rate determining. The overall rate constant, assuming a low, steady-state concentration of anhydride, is

$$k_{\mathrm{obs}} = k_1 k_3/(k_2 + k_3) \tag{4}$$

which, when $k_2 \gg k_3$, simplifies to

$$k_{\mathrm{obs}} = k_1/k_2 \times k_3 = k_3 \times K_e \tag{5}$$

where K_e is the equilibrium constant for the formation of the anhydride anion from the aspirin anion in Scheme 2. An order of magnitude calculation of K_e can be made as follows. From their free energies of hydrolysis[29–31] we know that simple acyclic anhydrides [and acyl tertiary ammonium ions also (146)] are less stable than phenyl acetate by some 8–10 kcal mole^{-1}. This corresponds to an equilibrium constant for the formation of a neutral anhydride from an ester in the region of 10^{-6}. The aspirin *anion* is also more stable than the anhydride anion by virtue of its much lower basicity (at least 5 pK units). So the value of K_e (some 10^{-6} for the protonated form of aspirin) is of the order of 10^{-11} for the anions shown in Scheme 2.

This very large adverse factor in the rate constant for hydrolysis by the nucleophilic mechanism is specific for reactions involving endocyclic displacement. In exocyclic displacements, by contrast, the disappearance of starting material depends primarily on the rate constant for the displacement reaction (eq. 6). Under typical conditions the reacylation

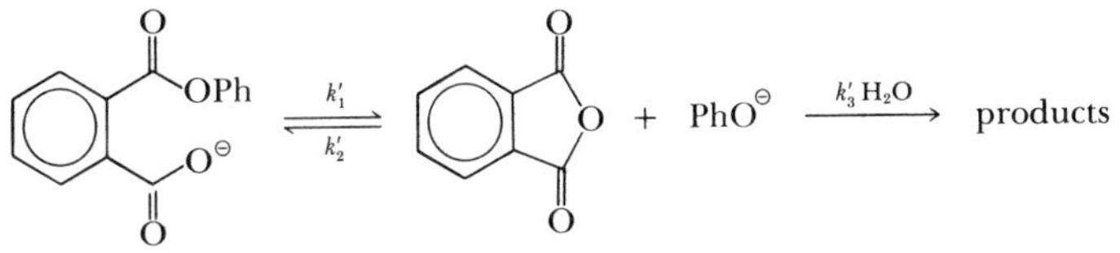

$$\text{(6)}$$

of the leaving group, now an intermolecular reaction, is not kinetically significant,* so that the overall rate constant for the disappearance of starting material is k_1'. The relative efficiencies of the endocyclic and exocyclic mechanisms depend on the relative magnitudes of k_3 (Scheme 2) and k_1', as well as on K_e. But normally the decisive factor is this highly unfavorable equilibrium constant, so that the much lower efficiency of an intramolecular nucleophilic catalysis involving an endocyclic displacement is basically a result of a rapid and highly unfavorable equilibrium of the type shown in Scheme 2.

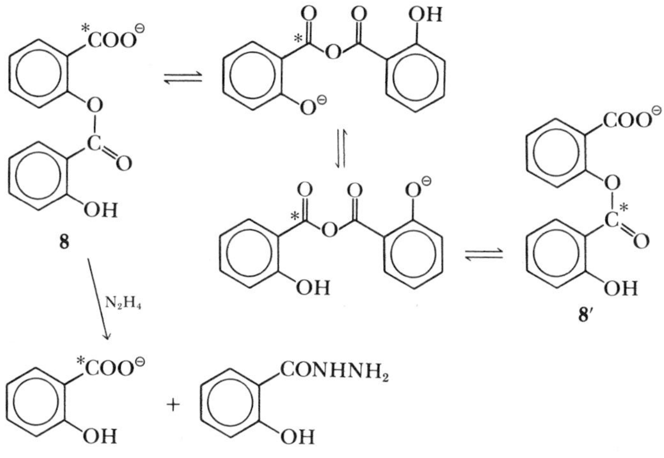

Scheme 3

*The back reaction of intermediate with leaving group to regenerate starting materials is, of course, not prohibited. It can be detected in favorable cases in intermolecular reactions involving nucleophilic catalysis of hydrolysis and can cause large effects on the rates of such reactions[32, 33].

Independent evidence for the rapid preequilibrium postulated for the aspirin reaction has been found by Kemp and Thibault[34]. Working with salicyl salicylate (8), labeled with carbon-14 in the free carboxyl group, and using hydrazine to trap unrearranged starting material, they were able to measure the rate of formation of 8′ from 8. The rate constant for this reaction is 2.6 min^{-1} at 30° and ionic strength 0.5 (NaCl), under conditions where the rate constant for hydrolysis is only 1.73×10^{-4} min^{-1} (Scheme 3).

The foundation of Kemp's work must surely have been laid in some earlier work with Woodward[35], in which an analog of the acetic salicylic mixed anhydride rearranges to an analog of aspirin, as part of the reaction of the N-ethylbenzisoxazolium cation with acetate ion (Scheme 4).

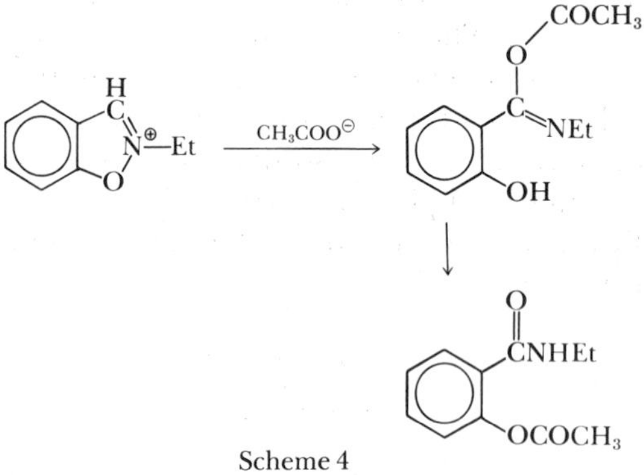

Scheme 4

Although the imino anhydride cannot be detected directly, indirect evidence is obtained from the reaction of the N-ethylbenzisoxazolium cation with the hippurate anion, in which some 2-phenyloxazol-5-one is formed as a side product. This is most easily rationalized by postulating the imino anhydride intermediate 9.

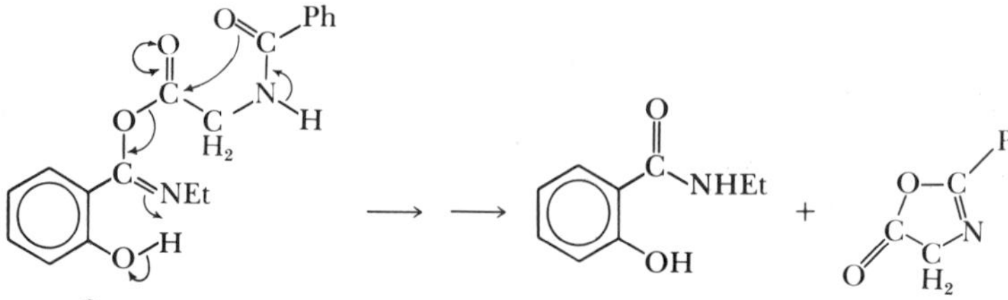

$$(7)$$

A similar rearrangement has been suggested by Higuchi and coworkers[36] to account for the formation of acetate esters when acetic anhydride is added to solutions of certain hydroxy acid anions. For example, $0.02\,M$ acetic anhydride reacts with $0.5\,M$ salicylate in aqueous solution at 25° and pH 5.1 to give a 22.6% yield of aspirin.

Higuchi and his co-workers favor initial formation of a mixed anhydride, with subsequent rapid intramolecular acyl transfer to give the ester (Scheme 5).

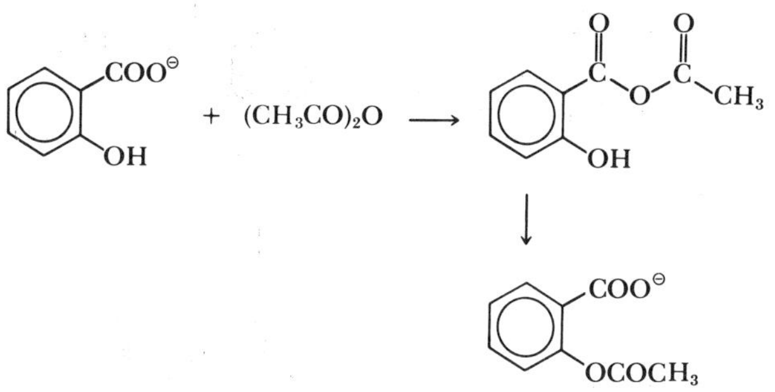

Scheme 5

The hydrolyses of a number of analogs of aspirin appear to involve the same mechanism as that of aspirin itself. These include the esters of thiolsalicylic acid, and a recent reinvestigation[37] of the hydrolysis of benzoyl thiosalicylate showed that general base catalysis is the likely mechanism in this case also. The ester is formed as an intermediate in the hydrolysis of p-nitrophenyl benzoate catalyzed by thiolsalicylate, and no oxygen-18 is incorporated into thiolsalicylic acid when the reaction is carried out in enriched water (Scheme 6).

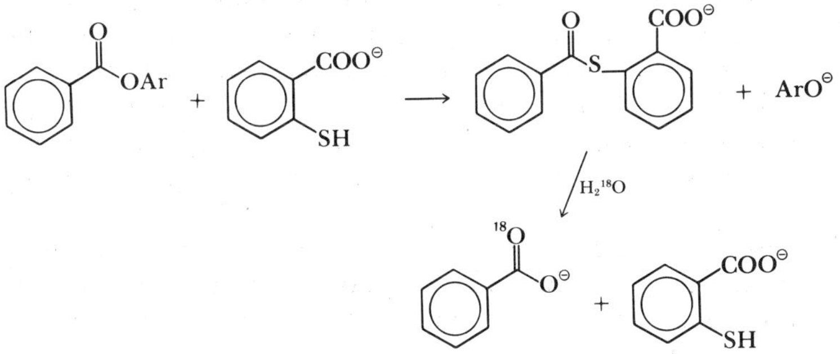

Scheme 6

2.1.4 Series Nucleophilic Catalysis.
To test the conclusions described in the previous sections, Fersht and Kirby[38] examined the hydrolysis of 3-acetoxyphthalate (10), a molecule which is a derivative of both salicylic and phthalic acids. If there is a rapid equilibrium between aspirin

anion and the anion of salicylic–acetic anhydride (Scheme 2 above), then the corresponding mixed anhydride (**11 ⇌ 12**) from 3-acetoxyphthalate might be expected to be hydrolyzed very rapidly, with participation of the second carboxyl group, as in **12** (Scheme 7).

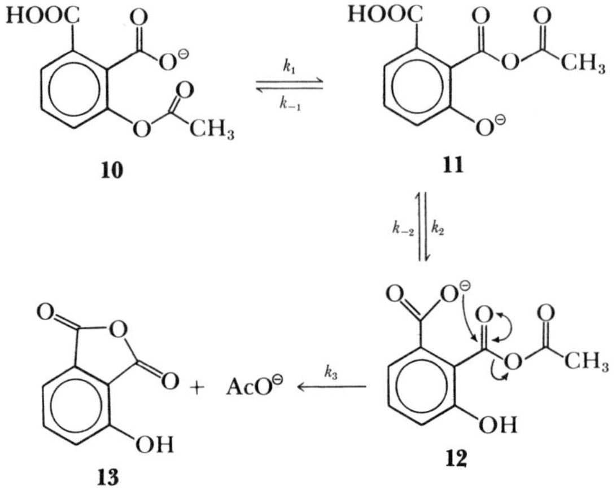

Scheme 7

The rate of deacetylation of the monoanion of 3-acetoxyphthalate is in fact over 6000 times greater than that of the aspirin anion and over 10^6 times greater than the rate of the uncatalyzed hydrolysis of aspirin methyl ester[38]. The mechanism of hydrolysis is that shown in Scheme 7, as shown by the isolation of 3-hydroxyphthalic anhydride (**13**) from the reaction mixture[38]. The equilibrium constant k_2/k_{-2} is very large, of the order of 10^5, so that reversion to products (k_{-1}) is slowed by this factor. Under these circumstances intramolecular nucleophilic attack on the salicyloyl end of the anhydride by the carboxylate group of **12** (k_3) outweighs nucleophilic attack on the acetyl end by the small proportion of phenoxide anion present (k_{-1}). The rate-determining step is therefore k_1.

The pH–rate profile for the hydrolysis of 3-acetoxyphthalate is shown in Figure 2. The hydrolysis of the dianion is much slower than that of the monoanion, but it is still 13 times as fast as that of the aspirin anion under the same conditions and still involves the nucleophilic route (Scheme 8). In this case the kinetically important species of the anhydride intermediate is the dianion **14**. In **14** the phenoxide and carboxylate groups compete directly to attack the anhydride, and attack by carboxylate is apparently faster by about one order of magnitude. This is in spite of its enormously lower basicity, and thus nucleophilicity, towards the carboxyl group, and means that the exocyclic displacement

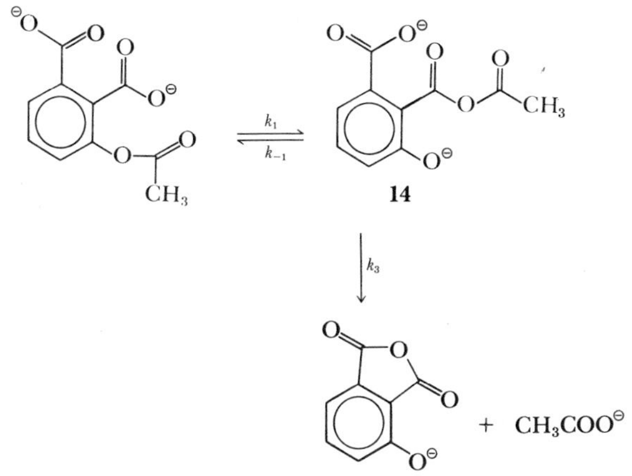

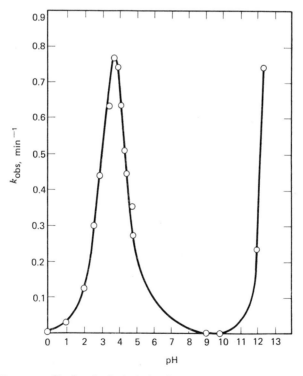

Scheme 8

Figure 2   pH–rate profile for the hydrolysis of 3-acetoxyphthalic acid, at 25° and ionic strength 1.0[34].

of acetate is favored over the endocyclic displacement of phthalate anion by aryloxide (k_{-1}) by a factor of up to 10^5–10^6.

This result suggests that it should be possible to enhance the efficiency of intramolecular catalysis in other reactions involving acyl transfers by way of endocyclic nucleophilic displacements. As long as a carboxyl group is acylated as part of the reaction, an exocyclic displacement of the acyl group as the carboxylate anion can be brought about by adding a suitable nucleophilic center. In principle, the acyl group can be transferred via a nucleophilic carrier or carriers from an unreactive to a reactive site where the exocyclic displacement occurs. Fersht and Kirby [38] suggest that this type of process, which they call series nucleophilic catalysis, may be of some general importance.

2.1.5 Catalysis by the COOH Group. Intramolecular general acid catalysis of ester hydrolysis is a fairly common phenomenon in the sense that there are esters, such as methyl hydrogen phthalate [27], for which the hydrolysis reaction depends on the concentration of the undissociated or acid form of the molecule. Mechanistically these reactions do not, as a rule, appear to involve classical general acid catalysis of the attack of water, and for esters at least other mechanisms are more efficient.

The hydrolysis of acetyl 3,5-dinitrosalicylic acid (**15**) is 28 times as fast as that of the anion [15] although the hydrolysis of the anion is itself accelerated by a factor of about 50 by intramolecular catalysis. The first order rate constant for the hydrolysis of **15** (at 39° and ionic strength 1.0) is 0.75 min^{-1}, over 200 times greater than that for the hydrolysis of its methyl ester, **16**. Catalysis also occurs with monosubstituted aspirins,

15   16

although the rate enhancement is less marked and to some extent is obscured by the more efficient general base catalysis of hydrolysis of the anions (Fig. 3). Aspirin itself is hydrolyzed some 16 times as fast as its methyl ester at 39° [13, 15] and some 19 times as fast at 25° [14]. Furthermore, considerable enhancements of the rates of attack of various added nucleophiles on aspirin acid are observed (see Table 1). This behavior is consistent with intramolecular general acid catalysis of attack by water and other nucleophiles (**17**), but other evidence rules out this mechanism. In particular, solvolysis of 3,5-dinitro-aspirin acid in 50% methanol–water gives a 12% yield of methyl 3,5-dinitrosalicylate [15], as expected

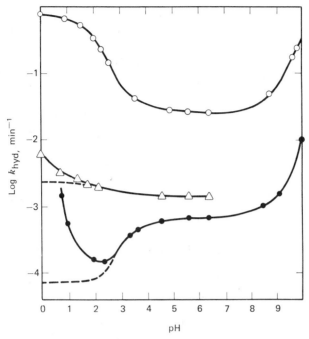

Figure 3 A comparison[15] of the pH–rate profiles for hydrolysis of (●) aspirin and its (△) 5-nitro and (○) 3,5-dinitro derivatives, at 39° and ionic strength 1.0. The broken lines represent the spontaneous hydrolysis of the acid (neutral) forms and have been corrected for external acid catalysis.

TABLE 1 SECOND ORDER RATE CONSTANTS FOR THE REACTIONS OF NUCLEOPHILES WITH *p*-CARBOXYPHENYL ACETATE (*p*-CPA), ASPIRIN (*o*-CPA), METHYL ASPIRIN (*o*-CPAM), AND PHENYL ACETATE(PA) AT 25°, IONIC STRENGTH 1.0 *M* [14]

| Nucleophile | pK'_a | k, $M^{-1}\,min^{-1}$ | | | | |
		p-CPA$^-$	*o*-CPA$^-$	*o*-CPA	*o*-CPAM	PA
Water	−1.7	5.3×10^{-8}	2.8×10^{-6}	3.4×10^{-7}	1.8×10^{-8}	2×10^{-8}
Nicotinamide	3.56	—	5.3×10^{-5}	1.9×10^{-2}		1.2×10^{-5}
Semicarbazide	3.84	—	1.2×10^{-3}	7.0×10^{-2}	4.5×10^{-4}	4.3×10^{-5}
Azide ion	4.45	—	$ca.2 \times 10^{-3}$	3.6×10^{-1}	—	1.4×10^{-2}
Methoxyamine	4.73	—	1.0×10^{-2}	3.3×10^{-1}	—	1.5×10^{-3}
Hydroxylamine	6.06	4.2	1.3	1.7×10	3.0	7.0×10^{-1}
Methylamine	11.0	1.5×10^2	4.2×10	—	1.4×10^2	1.7×10
Piperidine	11.4	3.7×10	1.2×10	—	5.2×10	4.3
Hydroperoxide ion	11.6	—	5.6×10^3	2.0×10^5	3.5×10^4	3.2×10^4
Hydroxide ion	15.8	1.3×10^2	1.3×10	—	9.5×10	7.6×10

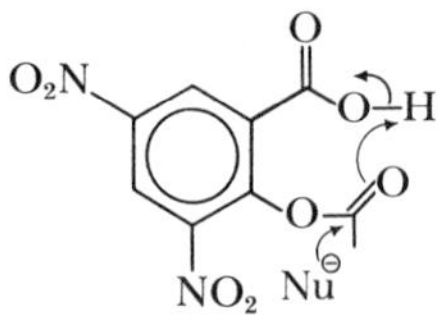

17

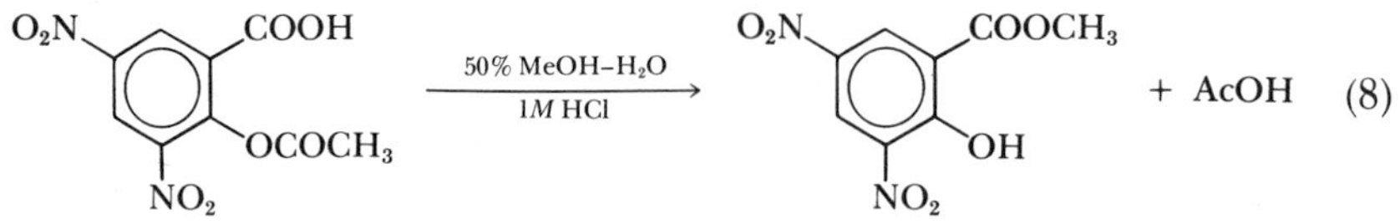

$$\xrightarrow[\text{1}M\text{ HCl}]{\text{50\% MeOH–H}_2\text{O}}$$

+ AcOH   (8)

if an anhydride intermediate were involved (eq. 8). The methyl ether
(**18**) of the presumed anhydride gave a similar yield of the corresponding
methyl ester on solvolysis in aqueous methanol under the same conditions[15] (eq. 9). Fersht and Kirby consider that the nucleophilic

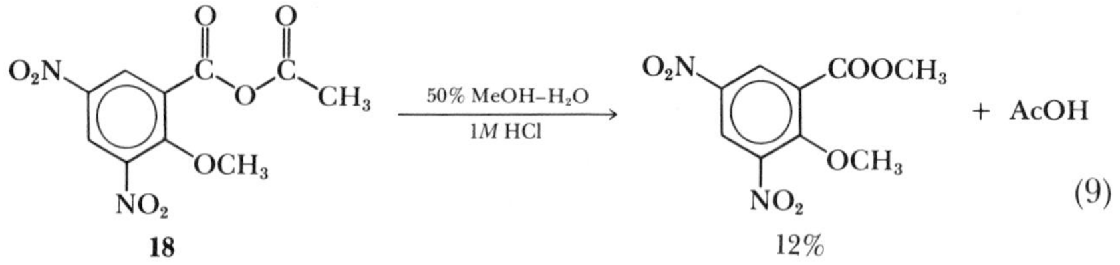

$$\xrightarrow[\text{1}M\text{ HCl}]{\text{50\% MeOH–H}_2\text{O}}$$

+ AcOH

18   12%   (9)

mechanism accounts for the observed catalysis of the reactions of monosubstituted aspirin acids also[15]. This mechanism involves the same
equilibrium between aspirin and mixed anhydride that is too unfavorable
to lead to catalysis in the anion reaction (Scheme 2); but the position of
equilibrium is shifted towards the anhydride in the protonated form,
because the dissociation constant of the phenol group of the mixed anhydride (**19**), K_P, is smaller than that of the original acetylsalicylic acid
(K_A). Thus, since $K_2 = K_1 \times K_A/K_P$ (Scheme 9), the equilibrium concentration of anhydride is much higher at low pH, and the nucleophilic
route is preferred. This argument contains a principle of general importance. In two reactions, the hydrolysis of aspirin acids and the hydrolysis of 3-acetoxyphthalate monoanion, the addition of a proton to the
system has been shown to favor intramolecular nucleophilic catalysis by
a large factor. A large part of this factor is simply the ratio of the dissociation constants of the nucleophile and the leaving group. This always
favors the reaction of the protonated form when the leaving group is
more basic than the carboxylate anion in endocyclic displacements. But
the argument can also be extended to exocyclic displacements if it is

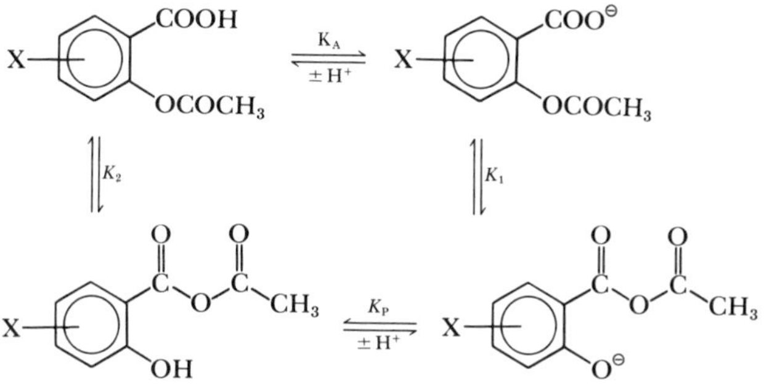

Scheme 9

assumed that these involve a tetrahedral intermediate. In the hydrolysis of an ester like methyl hydrogen phthalate[27] the back reaction of the leaving group with the anhydride is not kinetically significant. The rate-determining step for nucleophilic catalysis by the ionized carboxyl group is the breakdown of the tetrahedral intermediate **20** (Scheme 10), formed

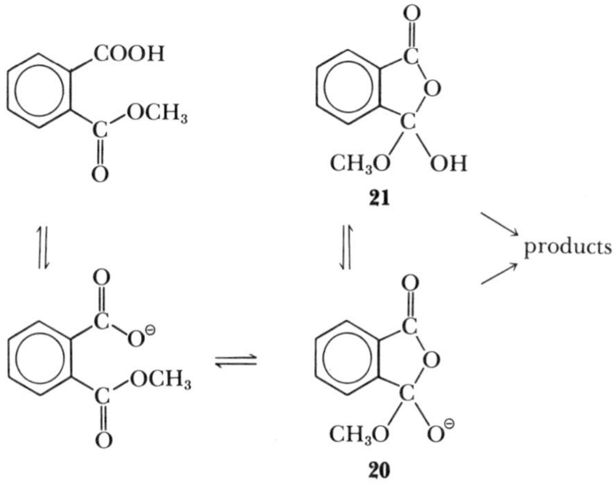

Scheme 10

in a rapid preequilibrium. This intermediate is stabilized by protonation, and the equilibrium is favored by a factor given by the ratio of the dissociation constants of the starting material and the tetrahedral intermediate **21**. Thus, if **21** is a full intermediate and the rate-determining step is the same, the protonated carboxyl group may be expected to be a more effective nucleophilic catalyst than the anion. This is of course the case for reactions in which very poor leaving groups must be displaced from esters and amides. So although this treatment is oversimplified, there is good reason to suppose that the simple nucleophilic mechanism (Scheme 10) can account for the observed catalysis in these systems.

Intramolecular general acid catalysis is involved in the bifunctional catalysis mechanism **22**, suggested by Morawetz and Oreskes[39] to account for the rapid hydrolysis of the monoanion of succinyl salicylate.

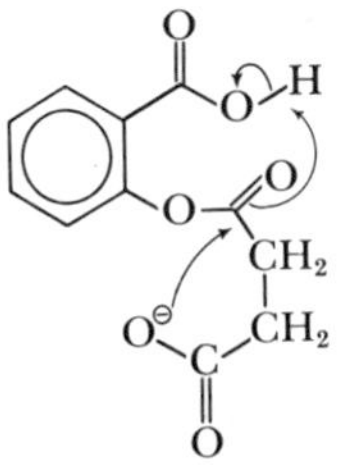

22

The pH–rate profile for the hydrolysis of this compound is strikingly similar to that of 3-acetoxyphthalate[38] (Fig. 2), which cannot hydrolyze by bifunctional catalysis. At 25° the monoanion is hydrolyzed with a rate constant of about 5 min^{-1} (the first pK_a is 3.62). This is just five times as fast as the corresponding reaction of 3-acetoxyphthalate (pK_a = 3.11), and since a similar mechanism, involving series nucleophilic catalysis, is available for the hydrolysis of succinyl salicylate (Scheme 11), it no longer seems necessary to invoke the intriguing bifunctional catalysis mechanism[38].

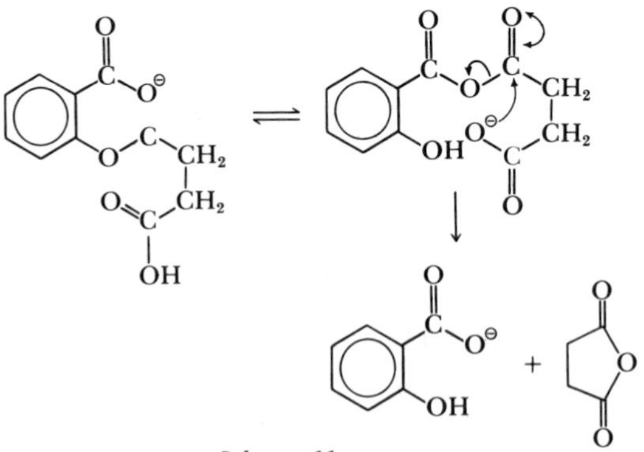

Scheme 11

2.2 Catalysis of Amide Hydrolysis

The undissociated carboxyl group is an efficient catalyst for the hydrolysis of a neighboring amide function[40]. The first detailed investigation of this reaction involved the hydrolysis of phthalamic acid[41, 42]. Bender found that this amide is hydrolyzed 10^5–10^6 times as fast as p-carboxybenzamide and proposed a mechanism[42] involving concerted electrophilic–nucleophilic catalysis (**23**). Evidence from an isotopic label-

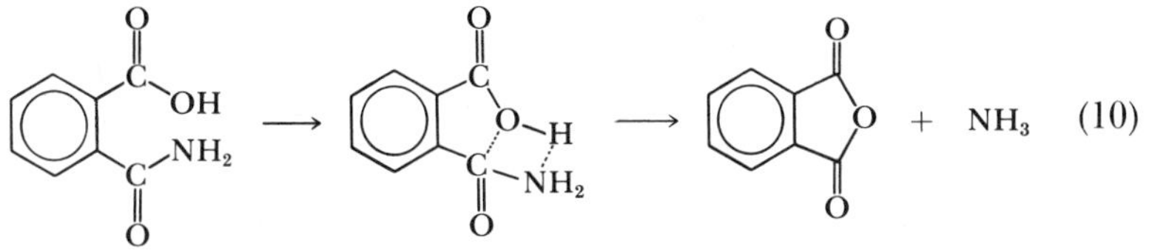

23

ing experiment was consistent with the presence of phthalic anhydride as an intermediate, although this was not formed in detectable concentrations.

The nucleophilic addition of the carboxyl group to the amide carbonyl group requires the plane of the amide group to be rotated out of the plane of the benzene ring and would be impossible in the case of o-carboxyphthalimide (**24**). As predicted, intramolecular catalysis of the hydrolysis of **24** is a much less efficient and mechanistically quite different

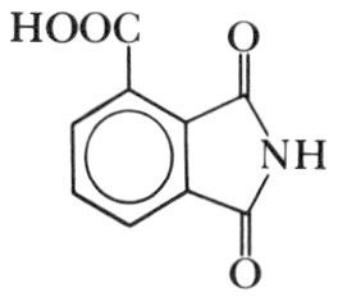

24 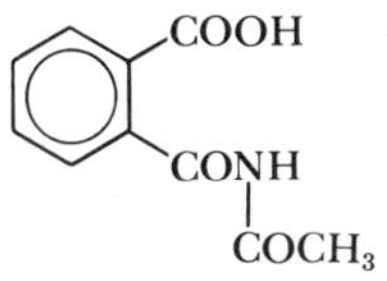   **25**

reaction[43]. Catalysis follows a bell-shaped pH–rate profile, and the maximum rate enhancement, compared with phthalimide, is little more than an order of magnitude (at 100° and pH 2.9). Also, the reaction shows a considerable solvent deuterium isotope effect, $k_H/k_D = 2.87$ at 100°. These data can be explained if the hydrolysis of o-carboxyphthalimide involves general acid or base catalysis[43]. Such a drastic change in mechanism, if it is brought about solely by the different geometry of **24**, supports Bender's conclusions about the mechanism of hydrolysis of phthalamic acid. But an imide is not a good model for an amide. For example, Shafer and co-workers[44] have shown that the hydrolysis of N-acetylphthalamic acid (**25**) is also much less susceptible to catalysis by the neighboring carboxyl group than is that of phthalamic acid, and it is clear that the basicity of the leaving group may be as important as the geometry of the system in reactions of this sort.

Bender's electrophilic–nucleophilic mechanism (**23**) for intramolecular catalysis of the hydrolysis of phthalamic acid involves a proton transfer as part of the rate-determining step and is not readily reconciled with his observation that hydrolysis is some 50% *faster* in D_2O at 24.8°[41]. The simplest alternative mechanism[42], in which the proton transfer and the nucleophilic attack are separate steps (Scheme 12) is at once chemically more straightforward and kinetically more complex, and thus at least less obviously inconsistent with the solvent deuterium isotope effect.

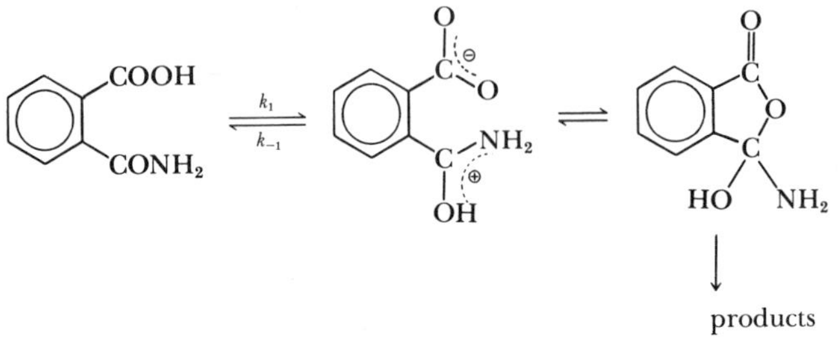

Scheme 12

The initial proton transfer is unlikely to be rate determining: if k_{-1} is not less than $10^{-11} \sec^{-1}$[45], k_1 will be at least $10^4 \sec^{-1}$, since the dissociation constant of a benzoic acid is some 10^7 times smaller than that of a protonated benzamide. The attack of COO^- on an adjacent protonated amide group will also be very fast, so that the breakdown of the tetrahedral intermediate is the likely slow step of this reaction (and of analogous reactions in which the nucleophilic addition step is sufficiently fast). The breakdown of the tetrahedral intermediate might be expected to involve the mechanism shown in equation 11[46].

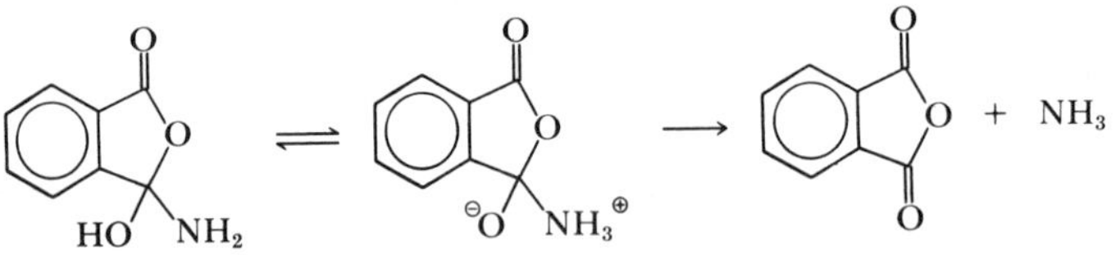

Recent work on intramolecular catalysis of amide hydrolysis by the carboxyl group has been particularly concerned with the effects of structure on reactivity. In the mechanism discussed above, and in most other possible mechanisms, changes in structure are most likely to affect reactivity through their effect on the equilibrium constant for ring closure rather than on the rate constant for the rate-determining step itself. There should therefore be a correlation between the efficiency of intramolecular catalysis of amide hydrolysis, via the anhydride, and the equilibrium constant for the formation of the anhydride from the free acid. Such a correlation would be useful, because the latter reaction has been studied in some detail and a good deal of information is available.

It is well known that the equilibrium constant for the formation of the cyclic anhydride from an aliphatic dicarboxylic acid is profoundly affected by the pattern of substitution on the intervening carbon chain. In general, alkyl substitution favors the cyclic anhydride (the Thorpe-Ingold effect [47]; for a general discussion see Eliel [48]). For example, tetramethylsuccinic anhydride is formed when the acid is steam distilled, and $\alpha,\alpha,\alpha',\alpha'$-tetramethyladipic anhydride is (at any rate kinetically) stable even at alkaline pH. Higuchi and Eberson [49] have observed similar effects in intramolecular catalysis of the hydrolysis of the corresponding half-amides of such dicarboxylic acids. The various methyl-substituted succinanilic acids show pH–rate profiles for hydrolysis of the general type shown in Figure 4. The rate constant for hydrolysis of the neutral species increases steadily as the number of methyl substituents is increased until that for tetramethylsuccinanilic acid (26) at 25° is some 1200 times as fast as that of the unsubstituted compound. Succinanilic acid is itself hydrolyzed with carboxyl participation in the pH range 1.5–2.5, more than an order of magnitude faster than acetanilide. At lower pH

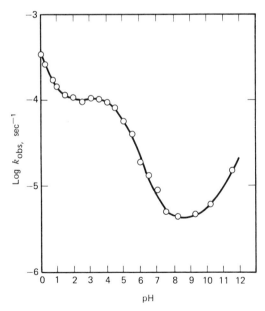

Figure 4   pH–rate profile for the hydrolysis of tetramethylsuccinanilic acid (26) at 25.5°. The pH-independent region between pH 1.5 and 3.5 represents the hydrolysis of the neutral compound [49].

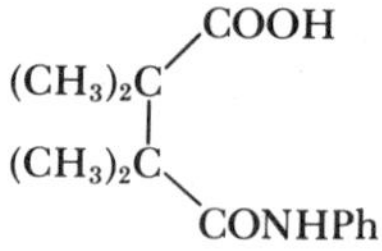

$$(CH_3)_2C \overset{\displaystyle COOH}{\underset{\displaystyle (CH_3)_2C \diagdown CONHPh}{|}}$$

26

specific acid catalysis is observed. The second order rate constant for this reaction is actually smaller than that for acetanilide hydrolysis[49] and presumably represents the normal external acid catalyzed reaction. The corresponding reaction of tetramethylsuccinanilic acid, on the other hand, is over 2000 times faster[49] and clearly involves intramolecular catalysis.*

The effects of alkyl substitution illustrated above have been investigated in most detail in intramolecular reactions involving the formation of anhydrides from ester acids, such as the monophenyl succinates and glutarates studied by Bruice and others[28, 40]. The general conclusion reached by Bruice is that alkyl substitution decreases the population of extended chain conformations of these compounds[26, 50], thus bringing the functional groups together for more of the time. Two relevant observations are consistent with this interpretation. Much larger effects are observed if the groups are held together on conformationally rigid molecules; examples are the aryl esters of maleic acid (**27**) and 3,6-endoxo-Δ^4-tetrahydrophthalate (**28**). Also, the enthalpies of activation

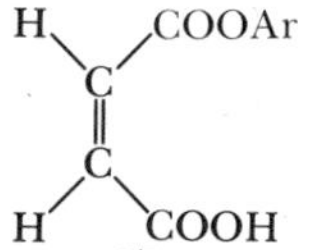

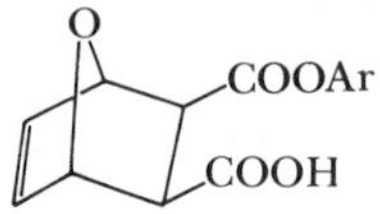

for the reactions of the *p*-bromophenyl half-esters of a group of 3- and 3,3-mono- and disubstituted glutaric acids are independent of the nature of the substituents, so that their effects are reflected solely in the entropy of activation[51].

Data of this sort are available for only one series of amides, but these suggest strongly that a second, quite distinct effect exists. The rate enhancement caused by the substitution of one, two, or three methyl groups into succinanilic acid is reflected in more favorable entropy and enthalpy terms, with no well defined trend apparent. But the very rapid

---

*Near neutral pH tetramethylsuccinanilic acid is converted, reversibly, to the imide[49], and a similar reaction is observed with *N*-methylphthalamic acid[44]. These reactions must involve attack by the amide on the carboxyl group, but little is known about the mechanisms involved, except that they must be complex[44].

hydrolysis of tetramethylsuccinanilic acid (**26**) is the result of a sharp decrease in $\Delta H\ddagger$, while $\Delta S\ddagger$ is considerably *less* favorable than for the hydrolysis of the unsubstituted compound[49]. A simple explanation for this result is that the molecule in the ground state is constrained by hydrophobic interactions into a conformation close to that of the transition state, with the adjacent pairs of methyl groups partially eclipsed. As the eclipsing process proceeds the carboxyl and amide groups must be forced into closer proximity, possibly to the point where van der Waals' forces become important and certainly to the point where no molecule of solvent remains between them. One or both of these effects could explain the larger decrease in the enthalpy term for the reaction. The smaller, partially compensating decrease in the entropy term suggests that the transition state is relatively more strongly solvated than the ground state for this compound, which it would be if desolvation of the ground state occurred at this stage, as described above, or if steric hindrance to solvation were relatively more important in the ground state than in the transition state.

At this point a cautionary note should be sounded. The rationalizations made in the two previous paragraphs are plausible explanations of highly complex processes, but it is unlikely that the simple conformational explanation of the alkyl substituent effect can be the whole story because the rate enhancements observed are greater than the limit predicted for intramolecular catalysis on the basis of proximity alone (see p. 2). Also the interpretation of enthalpies and entropies of activation in solution is always difficult because changes in solvent structure may have greater effects than do changes in the substrate. Rationalizations in this situation stand or fall on whether they lead to useful predictions.

One of the most reactive half-esters studied by Bruice and Pandit[26] is that of maleic acid. The maleamic acids, obtained by reacting amines with maleic anhydride, are also rapidly hydrolyzed in the free acid form, and the reaction has been used in a procedure for the reversible blocking of protein amino groups[52]. Maleic acid derivatives are conformationally rigid, with the carboxyl groups already eclipsed. Thus, if the differences caused by alkyl substitution in succinic acid derivatives are due solely to conformational effects, similar differences would not be expected in the reactions of maleic acid derivatives.

In fact, the substitution of methyl groups into maleamic acids enhances the rate of hydrolysis even more than it does in the case of the succinanilic acids. Dixon and Perham[53] have used citraconic anhydride in place of maleic anhydride to block amino groups reversibly, with advantage, because the removal of the blocking group is faster. The compounds obtained from amino acids and dimethylmaleic anhydride are more labile still. The effects of alkyl substitution on the rate of hydrolysis of N-methylmaleamic acid are illustrated in Figure 5, and the data are given in Table 2[54]. The rate constant for the hydrolysis of the most reactive compound, the dimethylmaleic acid derivative, is estimated from the observed part of the pH–rate profile. Below pH

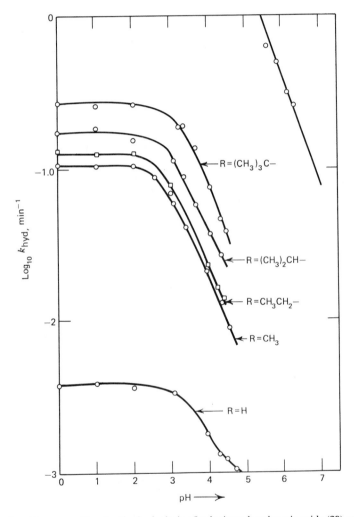

Figure 5   pH–rate profiles for the hydrolysis of substituted maleamic acids (**29**) at 39° and ionic strength 1.0. The labeled curves are for N-methyl monoalkylmaleamic acids. The straight line (top right) is the lower end of the pH–rate profile for the hydrolysis of N-propyl dimethylmaleamic acid (**30**) [54].

5.6 the hydrolysis of the cyclic anhydride becomes rate determining, so that this amide has the remarkable property of being converted rapidly to a more stable anhydride in this region of pH [54] (eq. 12).

TABLE 2 RATES OF HYDROLYSIS OF SUBSTITUTED N-METHYLMALEAMIC ACIDS (29) AT 39° AND IONIC STRENGTH 1.0 [54]

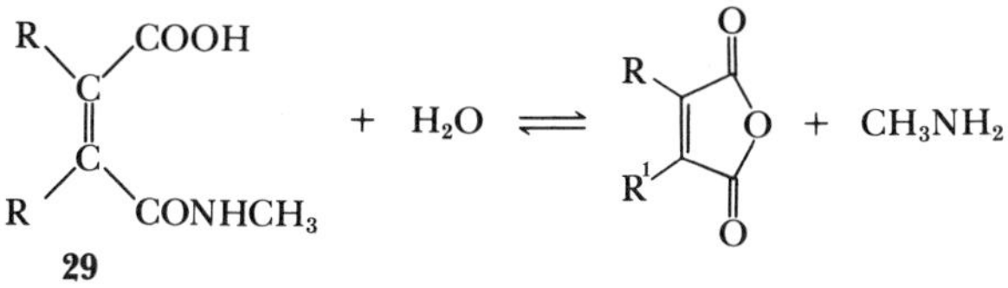

Compound 29	k_{obs},[a] min^{-1}	k/k_0[b]
R = R' = H	3.91×10^{-3}	1
R = H; R' = CH$_3$		26
R = H; R' = C$_2$H$_5$		33
R = H; R' = (CH$_3$)$_2$CH		44
R = H; R' = (CH$_3$)$_3$C		69
R = R' = CH$_3$		16000
R + R' = —(CH$_2$)$_4$—		700–800
R + R' = —(CH$_2$)$_3$— (31)		6×10^{-5}

[a]Mean of values at pH 0, 1, and 2.
[b]Rate relative to that of unsubstituted compound.

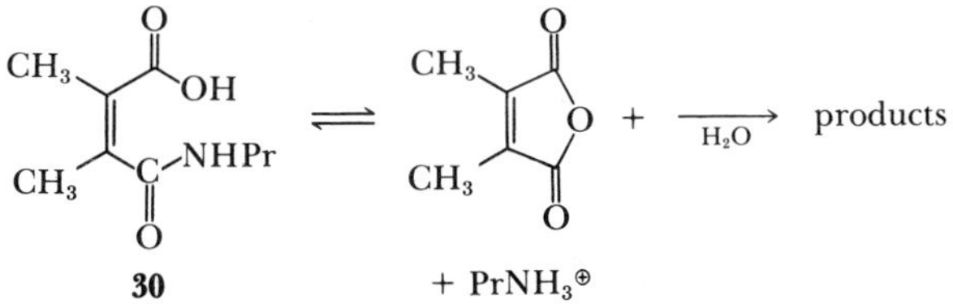

$$\hspace{8cm} (12)$$

Of the possible explanations of this effect of alkyl substitution, only one appears to be consistent with all the evidence so far. This is that the substituents act to buttress each other and force the reacting groups into closer proximity, thus making progress towards overcoming van der Waals' repulsion in the ground state (see p. 4). Only accurate X-ray structural studies can provide conclusive evidence for this effect, but one recent result provides important corroborative evidence. If steric compression of the sort described is involved, it would be expected to be relieved if the two alkyl substituents were held back in a ring. The cyclohexene-1,2-dicarboxylic acid derivative is indeed nearly an order of magnitude less reactive than the dimethylmaleamic acid; but the effect on reactivity becomes dramatic when the size of the ring is reduced by just one carbon. Although the hydrolysis of the dimethylmaleamic acid **30** is some 10^8–10^9 times as fast as that of N-methylacetamide at pH 3 and 39°, the hydrolysis of the N-methylamide of cyclopentene-1,2-dicarboxylic acid (**31**) shows little, if any, rate enhancement[54]. That such relatively minor structural changes should have such profound effects on reactivity has important implications for our thinking about enzymic catalysis.

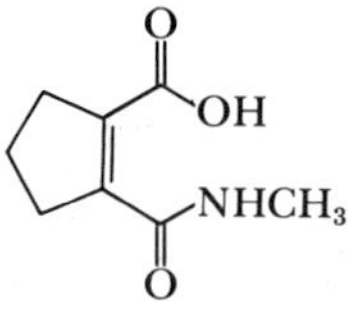

31

2.3 The Hydrolysis of Acetals

Whereas the enzyme chymotrypsin was intensively studied and the nature of the catalytically active residues was partially inferred[55] before the elucidation of the tertiary structure by Blow and his co-workers[56], lysozyme has the inverse history. The crystallographic studies by Phillips and his colleagues[57–62] made possible and stimulated serious mechanistic studies.

Lysozyme catalyzes the hydrolysis of the (1–4)-glycosidic linkages in certain N-acetylglucosamine derivatives of the general form **32**[63]. It

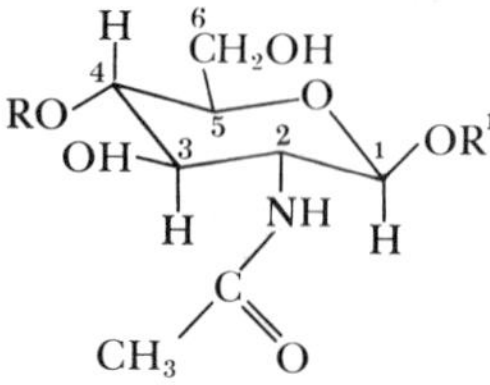

32

is inferred from chemical[64] and X-ray diffraction[57, 60, 62] studies of enzyme-inhibitor complexes that the only groups catalytically involved in the rate-determining step of the reaction are the carboxyl groups of an aspartate (Asp 52) and a glutamate residue (Glu 35). The former is thought to operate on the acetamide side of the (1–4) linkage of the glycoside (**32**), probably in the ionized form, while the latter is involved on the aglycone or leaving group side of the glycoside linkage, probably in the undissociated form. It appears further that the pyranose ring is distorted into a half-chair conformation, the ring being compressed towards the configuration expected for a carbonium center, with sp^2 hybridization, at C_1. Some controversy exists as to whether the enzymic reaction goes by way of a carbonium ion intermediate or involves concerted nucleophilic displacement at C_1. The point is complicated further by the possibility that the neighboring acetamido group at C_2 may act as a nucleophile in the N-acetylglucosamine derivatives. An excellent, comprehensive, review dealing with both enzymatic and chemical processes involved in acetal hydrolysis is given by Capon[65]. Shorter reviews of work on lysozyme have been written by Chipman and Sharon[66] and by Jollés[67].

2.3.1 Intramolecular General Acid Catalysis. Acetal hydrolysis is a classic example of a reaction involving specific acid catalysis[10]. Only recently has intermolecular general acid catalysis been shown to occur in this reaction. Earlier work is reviewed by Cordes[10].

Two types of mechanisms can account for general acid catalysis in this system. The first involves the conjugate acid of the acetal as a full intermediate (Scheme 13). This scheme is that generally accepted for

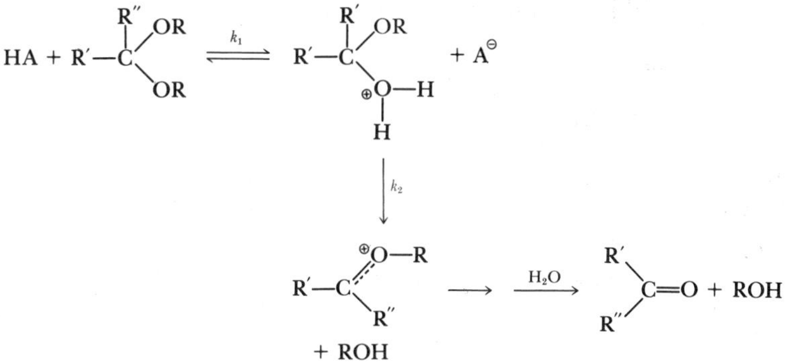

Scheme 13

specific acid catalysis of hydrolysis when protonation of the acetal is a rapid preequilibrium process ($k_1 > k_2$). If $k_2 > k_1$, on the other hand, general acid catalysis is observed.

Alternatively, the cleavage of the carbon–oxygen bond may be concerted with the protonation of the acetal (Scheme 14), as it is believed to be, for example, in the general acid catalyzed hydrolysis of orthoesters[10].

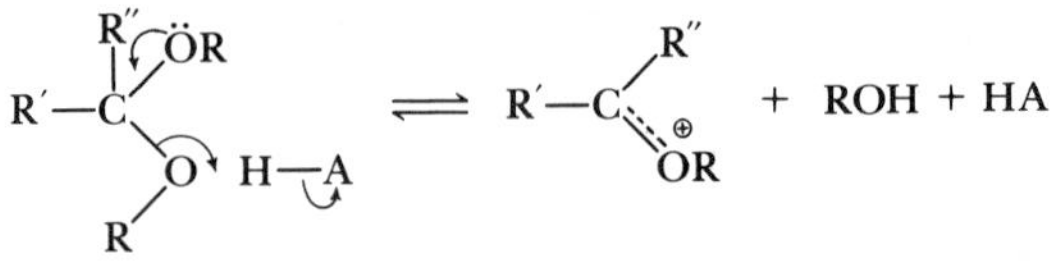

Scheme 14

The first striking example of intermolecular general acid catalysis of acetal hydrolysis was found by Anderson and Capon[69, 71] in the hydrolysis of benzaldehyde aryl methyl acetals (eq. 13). The structure

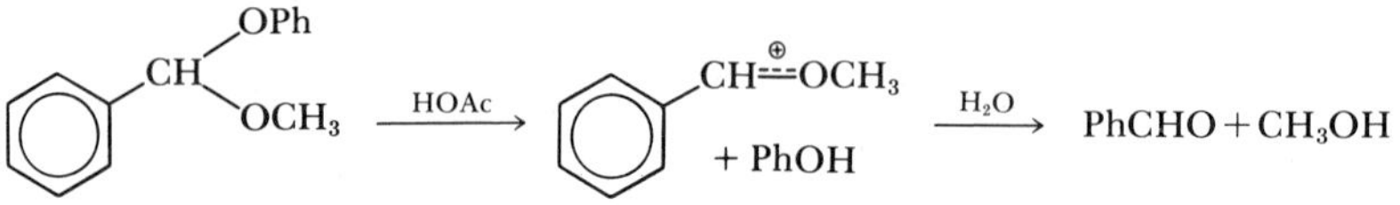

$$PhCHO + CH_3OH \quad (13)$$

of the acetals was designed to favor general acid catalysis: the methoxy benzylcarbonium ion is particularly stable and the phenoxy group is both weakly basic and a good leaving group when protonated. These factors act to increase k_2 relative to k_1 (Scheme 13) or to favor Scheme 14. The authors prefer the latter mechanism, with proton transfer and carbon–oxygen bond breaking concerted.

Subsequently Anderson and Fife[72] demonstrated catalysis by weak general acids of the hydrolysis of tropone diethyl ketal. Here too a particularly stable carbonium ion is involved (eq. 14).

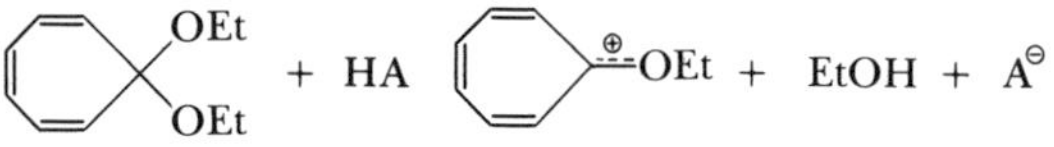

Intramolecular general acid catalysis of acetal hydrolysis was established before these intermolecular reactions were discovered. It is not a simple matter to prove a mechanism involving general species catalysis because no intermediates are involved and the approach must necessarily be to try to rule out the other possible mechanisms. This is illustrated by the careful investigation of the hydrolysis of 2-methoxymethoxy-benzoic acid (**33**) by Capon and his co-workers[74]. At pH 4.08 the observed pseudo first order rate constant for hydrolysis is 300 times as great as that for 4-methoxymethoxybenzoic acid (**33a**), and 600 times as great as that estimated for the methyl ester (**33b**).

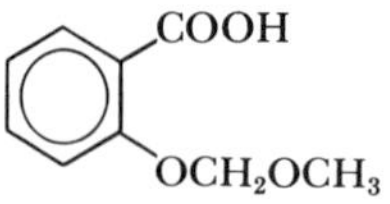

       33 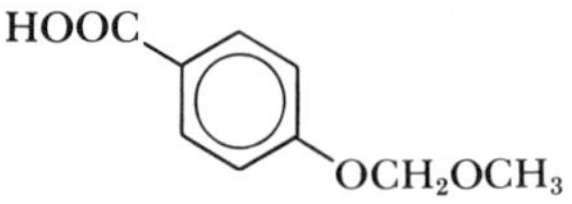   **33a** 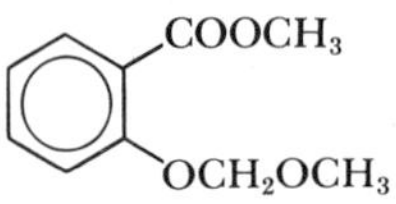   **33b**

Three types of mechanism could account for the observed catalysis: (1) intramolecular nucleophilic catalysis (Mechanism A), (2) intramolecular general acid catalysis (Mechanism B), and (3) specific acid catalyzed hydrolysis of the anion (Mechanism C).

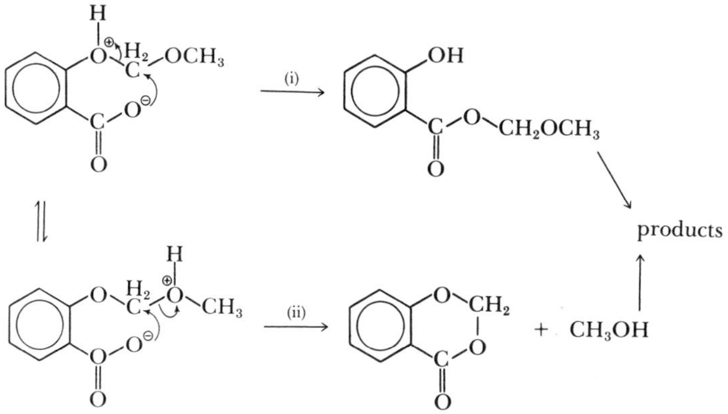

Mechanism A

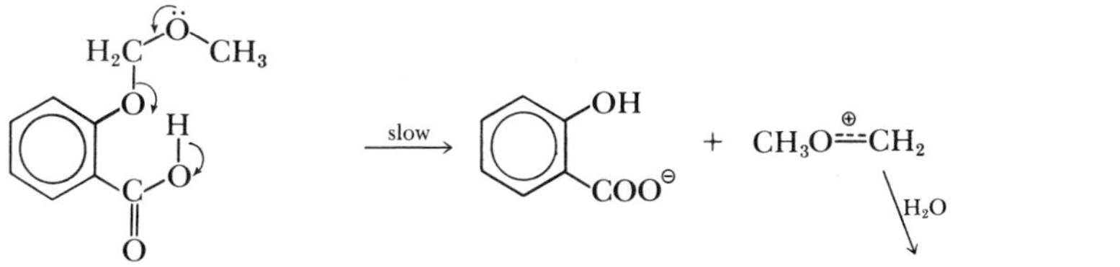

Mechanism B

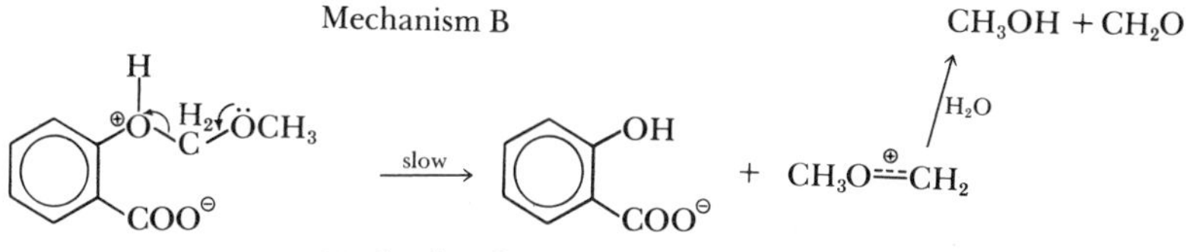

Mechanism C

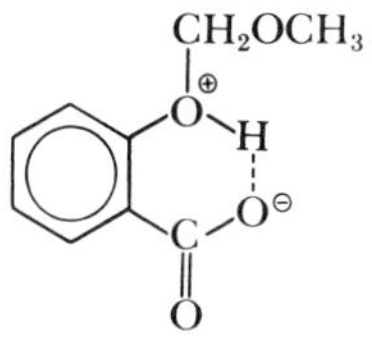

34

The hydrolysis of the undissociated acid is too fast, by two orders of magnitude, to represent simple specific acid catalyzed hydrolysis of the anion; although the enhanced rate might be due to specific acid catalysis if the field effect of the ionized carboxyl group stabilizes and thus increases the equilibrium concentration of the protonated intermediate (**34**), as it is thought to do in the hydrolysis of the salicyl phosphate dianion (see p. 44). Kinetic parameters such as the deuterium solvent isotope effect are not very informative in this situation [75, 76]. Capon et al. [74] prefer the mechanism involving classical general acid catalysis, with proton transfer in the rate-determining step, on two counts. They consider that the fact that 2-nitro-methoxy-methoxybenzene is little more reactive than the 4-nitro compound suggests that the field effect is not important; and the mechanism is known in intermolecular catalysis.

In an earlier study Bender and Silver [77] showed that plateaus in pH–rate profiles are not necessarily evidence for intramolecular catalysis. An examination of the pH–rate profile (Fig. 6) for the hydrolysis of 2-(2-hydroxy-5-nitrophenyl)-1,3-dioxane (**35**) shows an enhanced rate of hydrolysis of the undissociated phenol. Bender and Silver amply

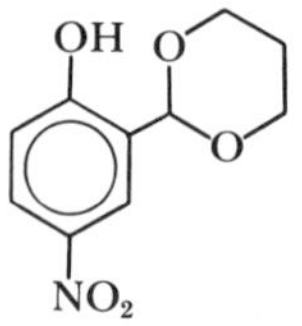

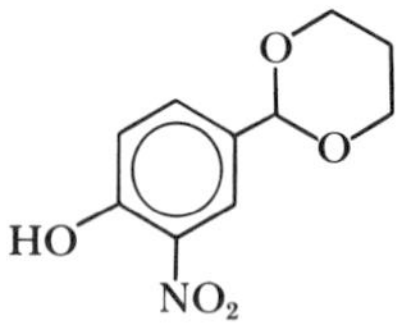

demonstrated that this apparent general acid catalysis is, in fact, simply the specific acid catalyzed hydrolysis of the anion. The change in σ-value of the OH group on ionization increases the basicity of the acetal oxygen atoms and increases the equilibrium constant for the pre-equilibrium protonation.

Similar conclusions were reached [76] for the hydrolysis of a number of other cyclic acetals, with general formulas **37–40**. The pH–rate

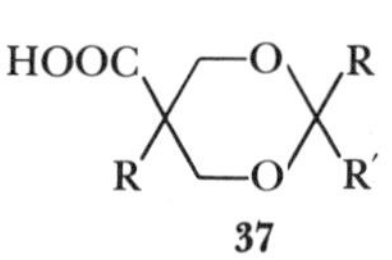

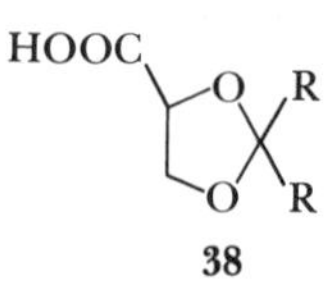

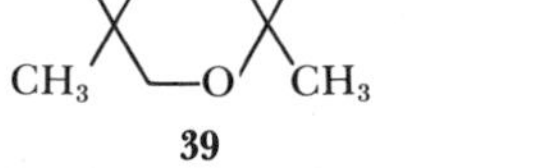

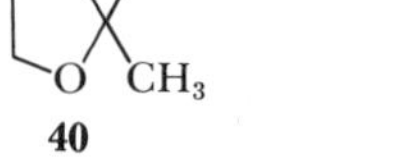

profiles for hydrolysis obtained in this systematic study were similar to that shown in Figure 6, consistent apparently with intramolecular catalysis by the carboxyl group. But the negative slopes of $\rho\sigma$ plots

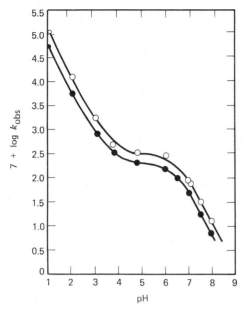

Figure 6   A comparison [77] of the pH–rate profiles in 10% acetonitrile–water at 25° for the hydrolysis of (●) 2-(2-hydroxy-5-nitrophenyl)-1,3-dioxane (**35**) and (○) 2-(4-hydroxy-5-nitrophenyl)-1,3-dioxane (**36**).

clearly showed that the mechanism was specific acid catalysis of the hydrolysis of the conjugate base (carboxylate) form of the substrate. Negative ρ-values were also obtained for the specific acid catalyzed hydrolysis of aryl-β-glucosides ($\rho = -0.7$), aryloxytetrahydropyrans ($\rho = -0.9$) [68] and 2-aryl-1,3-dioxanes [77]. The decrease in σ-value for the carboxyl group when it ionizes to COO^-, combined with the negative ρ-values, causes the apparent, but not real, catalysis. Clearly, in these cases the protonation step is more sensitive to electronic effects than the bond-breaking step.

Intermolecular general acid catalysis by acetic acid of the hydrolysis of benzaldehyde aryl methyl acetals (see p. 32) is characterized by a positive ρ-value of 0.89 [71]. Specific acid catalysis of hydrolysis is more sensitive to effects on the protonation equilibrium and thus differs in the expected direction from general acid catalysis, for which sensitivity to bond breaking is more important than the sensitivity to proton transfer.

The various compounds examined in the search for intramolecular general acid catalysis of acetal hydrolysis are collected in Table 3. The high rates of hydrolysis of compounds 1–5 of the table (see, for example,

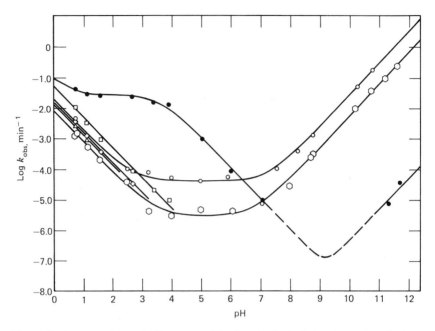

Figure 7 A comparison of pH–rate profiles for the hydrolysis of substituted phenyl
β-D-glucopyranosides, at 78.2° and ionic strength 0.3: (◊) phenyl; (△) l-naphthyl; (○)
o-nitrophenyl; (◊) p-nitrophenyl; (□) o-methoxyphenyl; (●) o-carboxyphenyl[63].

Fig. 7), suggest that catalysis is occurring and, by analogy within the
series, that the mechanism is general acid catalysis. Capon and his co-
workers [74] suggest that intramolecular general acid catalysis also occurs
in the hydrolysis of compound 6, in which the carboxyl group is considered as being in the aglycone. If the carboxyl group is in the glycone,
as in compound 13 in the table, catalysis does not occur [82].

A stereochemical explanation of why catalysis occurs in some but not
others of these compounds has been suggested by Capon and co-workers
[74]. [The fact that those that do not show catalysis are almost all alkyl
acetals and are thus less susceptible to general acid catalysis than aryl
acetals (see p. 32) cannot be a general reason, because the hydrolyses
of compounds 4–6 are catalyzed.] With the 2-carboxyphenyl acetals
the favored conformation of the carboxyl group is in the plane of the
benzene ring. This is also the most favorable conformation for hydrogen-
bonding to the acetal oxygen, and thus probably for intramolecular
general acid catalysis also. With acetals 7–9, on the other hand, the
conformation required for hydrogen bonding is not favored thermodynamically. For example, compound 7 would need to adopt the

conformation **41**, with the carboxyl group not only axial, which might not be too unfavorable, but lying over the dioxane ring. With compound 8 hydrogen bonding involves a five-membered ring and is thus even less favorable.

Piszkiewicz and Bruice also investigated the hydrolysis of *o*- and *p*-nitrophenyl 2-acetamido-2-deoxyglucopyranosides (**42**) and *o*- and *p*-nitrophenyl glucosides (**43**)[83]. For the β-isomers, which have the 2-acetamido or 2-hydroxy group *trans* to the nitrophenolate leaving group, spontaneous hydrolysis reactions are readily observed, with rates independent of pH in the range pH 2–10 (**42**), or 3–6 for the 2-hydroxy compound, which is less reactive by a factor of about 300. The α-anomers do not show spontaneous hydrolysis reactions.

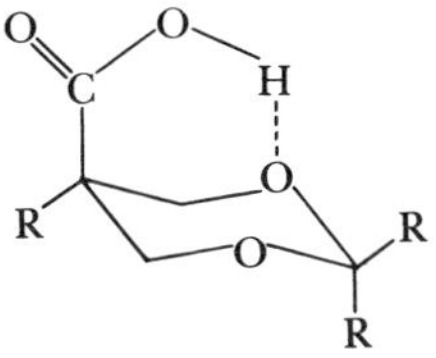

41

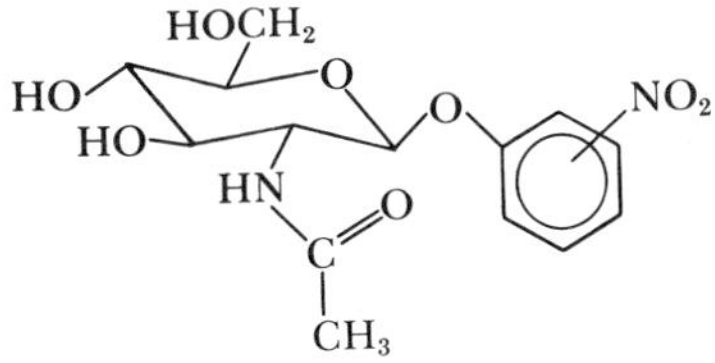

42(β)

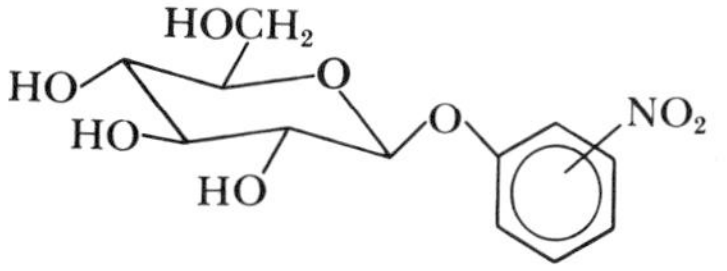

43(β)

The results are illustrated by the respective pH–rate profiles (Figs. 8 and 9). Intramolecular catalysis by the acetamido group is much more efficient than catalysis by the hydroxy group for the β-anomers. The requirement for the catalyzing and leaving groups to be *trans* to each other is consistent with nucleophilic catalysis, and of the two kinetically equivalent Mechanisms A and B,

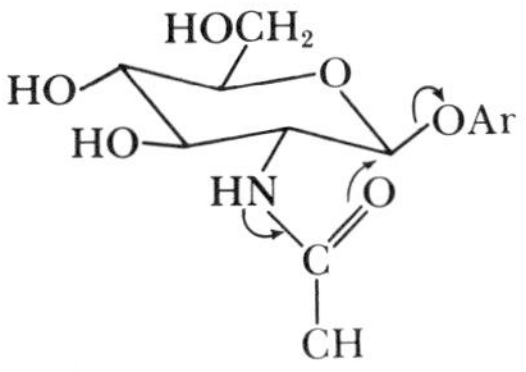

Mechanism A

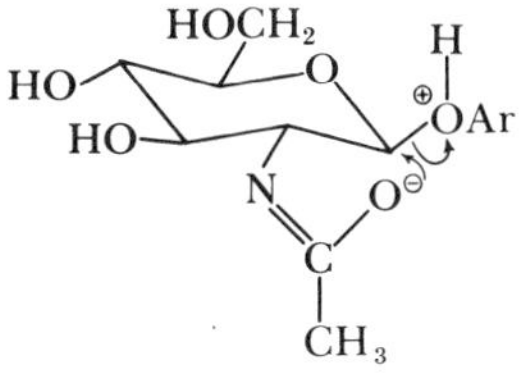

Mechanism B

TABLE 3 INTRAMOLECULAR CATALYSIS OF ACETAL HYDROLYSIS BY THE COOH GROUP

Compound	Catalysis Observed	Ref.	Compound	Catalysis Not Observed	Ref.
1	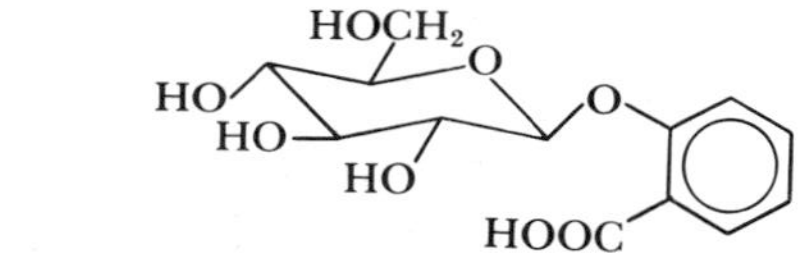	63 74 79	7	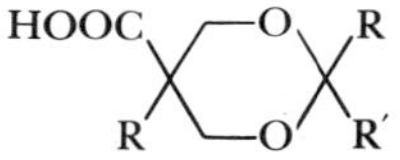 **37**	76
2	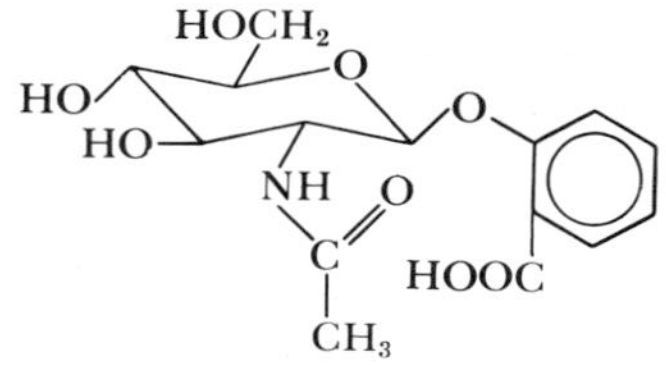	63	8	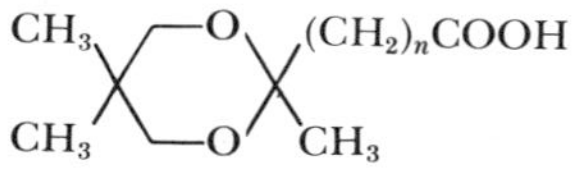 **39**	76
3	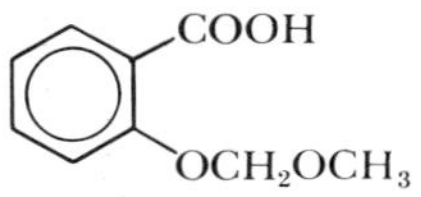 **33**	74 80	9	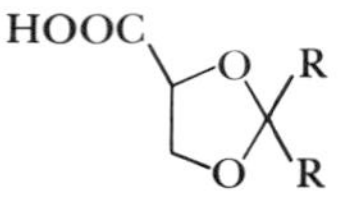	76
4	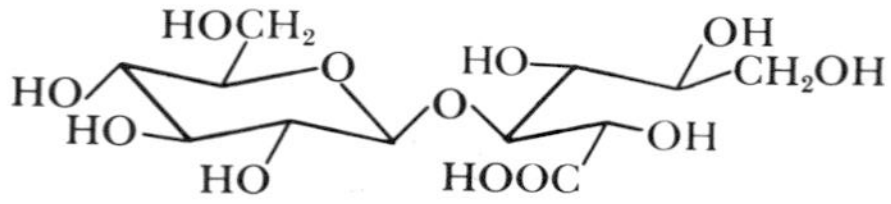	81	10	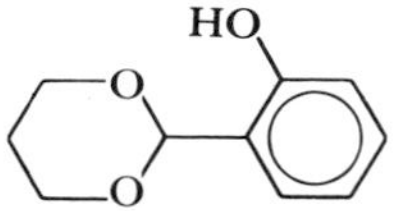	77

77

11

81

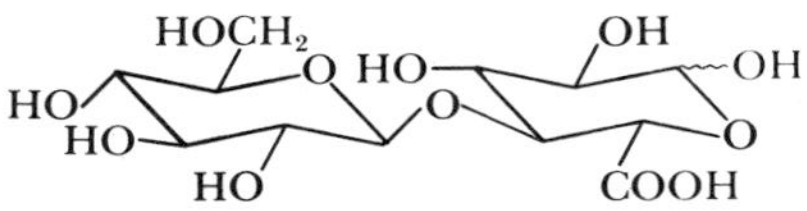

74

12

81

82

13

5

6

NO_2

HO

CH_2COOH

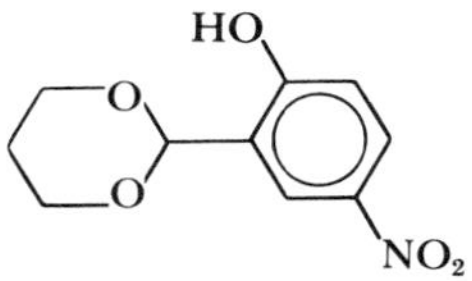

OCH_2OCH_3

COOH

$OC_{10}H_{17}$

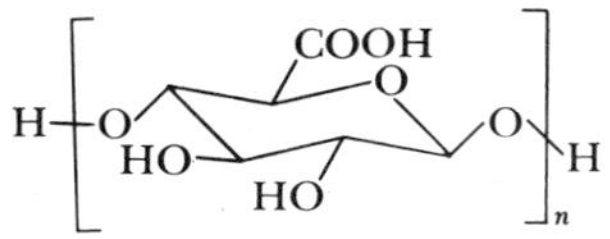

HO

HO

HO

OH

OH

COOH

HOCH$_2$

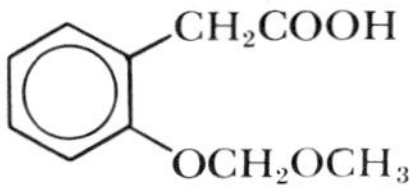

HO

HO

HO

COOH

HO

HO

$\left[\begin{array}{c} H \end{array}\right]_n$

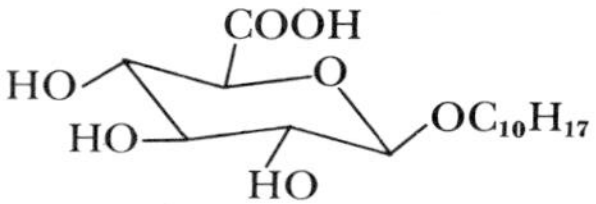

H

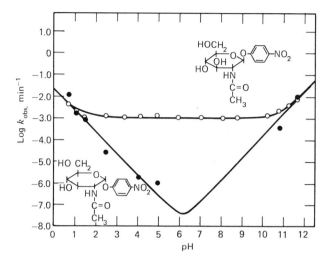

Figure 8   A comparison of pH–rate profiles for the hydrolysis of (●) *p*-nitrophenyl 2-acetamido-2-deoxy-α- and (○)-β-D-glucopyranosides, at 78.2° and ionic strength 0.3 [83].

Mechanism A was preferred, since the specific rate constant required to account for the observed rate of reaction by Mechanism B would be impossibly high [83].

A comparison [84] of the rates of the specific acid catalyzed hydrolysis

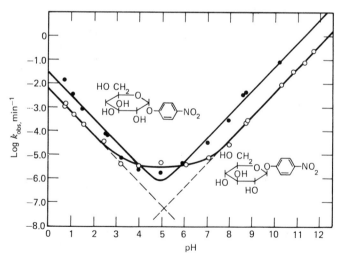

Figure 9   A comparison of pH–rate profiles for the hydrolysis of (●) *p*-nitrophenyl α- and (○) β-D-glucopyranosides, at 78.2° and ionic strength 0.3 [83].

of a series of α- and β-glucoside and 2-acetamido-2-deoxyglucopyranoside derivatives reveals that the methyl glucoside undergoes hydrolysis 50 times as fast as expected. Bruice and Piszkiewicz postulate that the aglycone in this case is small enough to occupy the sterically unfavorable axial position and is thus in the optimum conformation for intramolecular nucleophilic catalysis, in a *trans* diaxial relationship with the nucleophile. This point was considered to be of particular interest by the authors, since in an earlier study[63] they had shown that nucleophilic catalysis in this system depends strongly on the basicity of the leaving group ($\rho^- = 2.6$). Catalysis is thus weaker for poor leaving groups, such as those found in the natural substrates of lysozyme. It appears that a conformational change induced by the enzyme, together with intramolecular general acid catalysis, might restore this loss of catalytic efficiency.

Having confirmed Capon's result[79], that the hydrolysis of 2-carboxyphenyl β-glucopyranoside (compound 1 in Table 3) is subject to intramolecular general acid catalysis (see Fig. 7), Piszkiewicz and Bruice[63] looked for evidence for concerted nucleophilic and general acid catalysis in the hydrolysis of 2-carboxyphenyl 2-acetamido-2-deoxy-β-D-glucopyranoside. The results do not indicate that catalysis is additive. Whereas the β-N-acetylglucosamine nitrophenyl glucosides are hydrolyzed some 300 times as fast as the β-glucoside derivatives, and it is calculated that the catalysis by the neighboring COOH group in 2-carboxyphenyl glucopyranoside is of the order of 6×10^3, 2-carboxyphenyl-N-acetylglucosamine is hydrolyzed only 7 times as fast as 2-carboxyphenyl glucopyranoside. Thus if concerted nucleophilic and general acid catalysis does occur, it is little more effective than simple intramolecular general acid catalysis.

The demonstration of intramolecular general acid catalysis in simple systems is clearly relevant to the mechanism of action of lysozyme. The COOH group of glutamic acid residue 35 is strategically placed in the active site, and the results described for simple systems show how it might act and that large rate enhancements are possible even without the fine control of catalysis available in the enzymic reaction. However, the importance of nucleophilic catalysis is still in doubt. Vernon[85] considers that the geometry of the enzyme–substrate complex does not favor the participation of the carboxylate group of Asp 52 as a nucleophile; and the question of intramolecular participation by the 2-acetamido group has also been reopened. Recent results show that the enzyme does not have an absolute specificity for 2-acetamido-2-deoxyglucosides and that it may cleave the glycosidic bonds of 2-deoxyglucosidic residues even faster than those of substrates with 2-acetamido groups[86]. Also, recent evidence from the laboratories of Rupley and co-workers[87] and Raftery and co-workers[88] supports a mechanism involving a carbonium ion intermediate. Finally, the importance of strain has not been studied in a model system. The possible distortion of the substrate on the enzyme might account for some $3–6$ kcal mole^{-1} [66], and if this were all reflected in the lowering of the activation energy a contribution of $10^2–10^4$ in rate would be possible.

2.4 The Enolization of Ketones

The enolization of ketones in buffered solutions has been shown to be subject to specific acid, specific base, general base, general acid, and combined general acid–general base catalysis[89, 90] of varying degrees of importance. General base catalysis appears to involve slow proton removal from carbon (44), and evidence is available that the observed general acid catalysis in fact represents the general base catalyzed removal of a proton from the protonated substrate (45) [90–93].

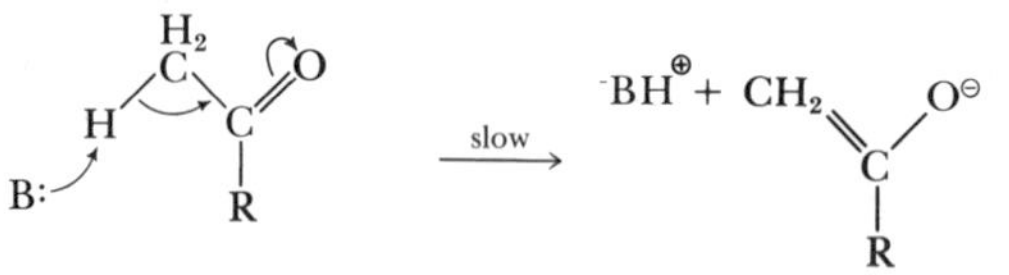

44

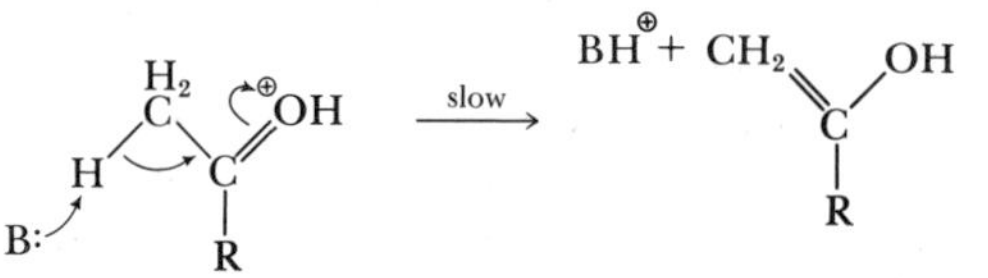

45

Apart from their mechanistic and synthetic interest for the chemist, studies on enolization are relevant to the mechanisms of several enzymic reactions, a particularly well known example being the aldolase-catalyzed enolization of dihydroxyacetone phosphate.

Intramolecular catalysis of enolization has been observed by Bell and Fluendy[94], who studied ketoacids of the form:

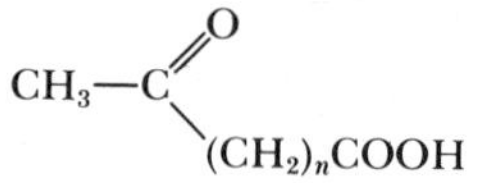

For $n = 2, 3, 4, 5$, and 11 intramolecular general base catalysis by the ionized carboxyl group occurs. Intramolecular general acid catalysis by the undissociated COOH group is observed for $n = 2, 3$, and 4. For

$n = 1$ no catalysis is observed [95, 96]. Catalysis is most efficient for $n = 3$, in which case the cyclic transition state is six membered (including the proton). The dependence of rate upon n was shown to be correlated with Monte Carlo calculations of the probability of forming a cyclic transition state [97]. The effective concentration [23] of the neighboring carboxyl group, calculated from the ratio of the intermolecular and intra-molecular rate constants, is low, of the order of 1 M.

A larger rate enhancement is observed in the enolization of the o-isobutyrylbenzoate anion **46** [98] (see Fig. 10). The ratio of rate con-

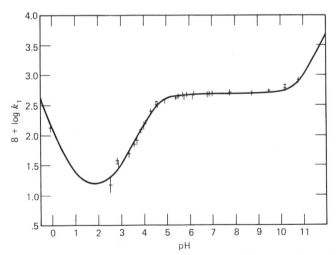

46

stants for this and an equivalent intermolecular reaction in this case is 50 M.

An analysis and criticism of the experimental methods used by earlier workers has been given by Coward and Bruice [99]. Complications arising from lactol–ketone equilibria in such compounds as **46** are discussed. Coward and Bruice also examined the effects of both one and two tertiary amino groups on the equilibration rates of a series of ketones, but found no evidence for concerted general acid–general base catalysis.

Figure 10   pH–rate profile for the iodination of o-isobutyrylbenzoic acid (**46**), in water at 25° and ionic strength 0.5 [98].

2.5   The Hydrolysis of Phosphate Esters

2.5.1   Monoesters.   The best known example of intramolecular catalysis of phosphate ester hydrolysis by the carboxyl group is the rapid hydrolysis of the dianion of salicyl phosphate near pH 5. The accepted mechanism of this reaction rests on a detailed investigation by Bender and Lawlor[100], whose evidence rules out several possibilities. From the remaining possible mechanisms they preferred, essentially by a process of elimination, one involving a preequilibrium proton transfer (Scheme 15). A detailed study of the effects of sub-

47

Scheme 15

stitution on this reaction strongly supports this mechanism[101]. Using the technique employed in the investigation of substituted aspirins[13] (see pp. 7, 8), it was possible to separate the effects of substituents on the carboxyl and leaving groups in salicyl phosphate hydrolysis[101]. The effects on the leaving group are correlated by a Hammett ρ-value of 0.94, which is consistent with several possible mechanisms, but ρ for the effect of substituents on the carboxyl group is significantly negative at -0.48. The group must therefore be less negative in the transition state than in the ground state and is therefore involved as the ionized form. General base catalysis is ruled out because there is no significant deuterium solvent isotope effect[100], and nucleophilic catalysis is not a possibility, since a careful study of the hydrolysis products in $H_2{}^{18}O$ shows there is no incorporation of labeled oxygen into the salicylic acid produced[101], confirming Bender and Lawlor's result[100]. The function of the carboxylate group, as they suggest, is presumably to stabilize the reactive zwitterionic form (**47**) of the substrate.

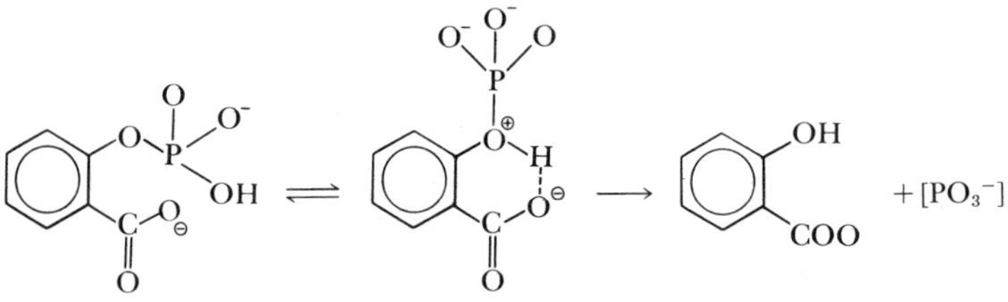

This mechanism fits very well into the general pattern of phosphate monoester hydrolysis. Kirby and Varvoglis[102] consider that the rapid hydrolysis of the monoanions of simple phosphate monoesters involves a preequilibrium proton transfer, followed by the rapid, but rate-determining breakdown of the zwitterionic form (Scheme 16). Bender and Lawlor's mechanism (Scheme 14) for the intramolecular reaction

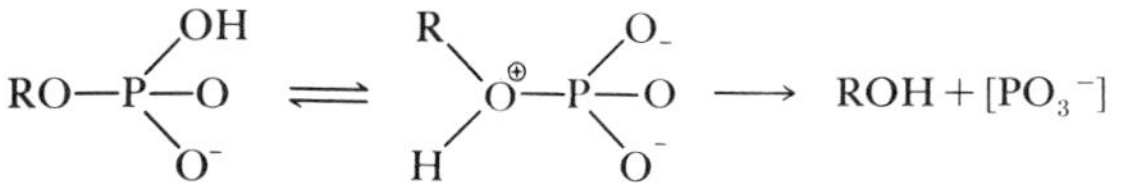

Scheme 16

follows logically from this picture. So also does the otherwise surprising observation by Benkovic and Schray[103] that the hydrolysis of phosphoenol pyruvate (48) is not subject to catalysis by the carboxyl group. The hydrogen bond stabilizing the zwitterionic form in this case would be part of a four-membered ring (49) and thus much less favorable.

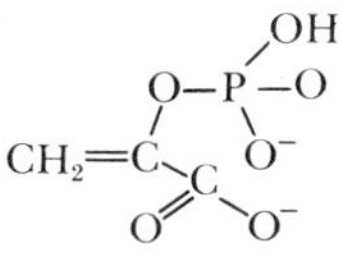

48 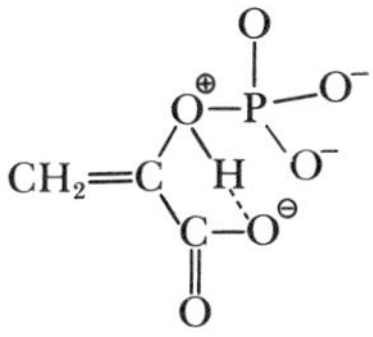   49

2.5.2 Diesters. Arai showed in 1934[104] that phenyl 2-carboxyphenyl phosphate (50a) dianion, though less reactive than salicyl phosphate, is hydrolyzed more rapidly than diphenyl phosphate in a pH-independent reaction. And Clark and Kirby[105] showed that P-methyl phosphoenol pyruvate (51) is hydrolyzed rapidly and selectively to phosphoenol pyruvate under mildly acidic conditions. Both reactions

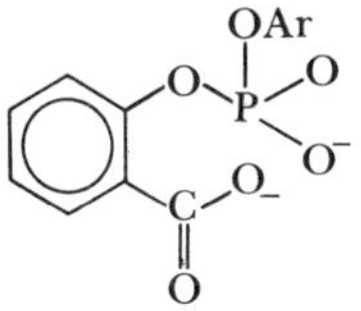

(50a) Ar = (a) phenyl   51
(50b) Ar = (b) 3–nitrophenyl
(50c) Ar = (c) 4–nitrophenyl

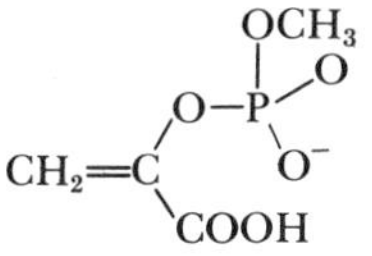

have been investigated more recently in some detail. The hydrolysis of the phosphoenol pyruvate ester is discussed below, together with the similar reaction of the triester.

The hydrolysis of three aryl 2-carboxyphenyl phosphates (50) has been studied by Kirby, Lawlor, and their co-workers[106]. pH–rate profiles are shown in Figure 11 for the hydrolysis at 39° and ionic strength 1.0. Catalysis evidently involves the ionized carboxyl group and is highly efficient. The 3-nitrophenyl ester 50b is hydrolyzed some 10^8 times more rapidly than expected for 3-nitrophenyl phenyl phosphate

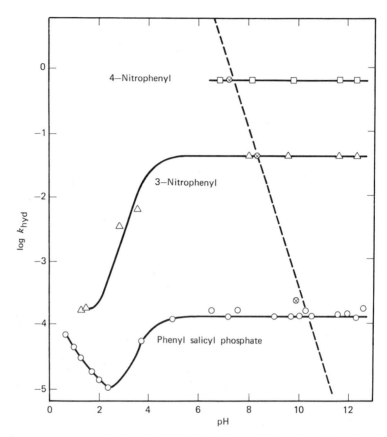

Figure 11   pH–rate profiles for the hydrolysis of three substituted aryl 2-carboxyphenyl phosphates (**50a–50c**). The dashed line represents the linear free energy relationship between the rates of hydrolysis of the dianions and the pK_a's of the leaving groups ArOH[106].

at 39° and pH 4. The mechanism is one of nucleophilic catalysis (Scheme 17). When the phenyl ester is hydrolyzed in the presence of hydroxylamine a hydroxamic acid is produced, as expected if the cyclic acyl phosphate **52** is an intermediate[100]. This intermediate was also detected spectrophotometrically in the hydrolysis of the 3-nitrophenyl ester (**50b**) and was shown to disappear at a rate identical with that measured for the hydrolysis of **52** under the same conditions.

The corresponding intermolecular reaction, nucleophilic attack by carboxylate anions on the phosphorus center of aryl phosphate diester anions, can be detected only with the most reactive esters. The slow reaction of acetate ion with 2,4-dinitrophenyl methyl phosphate anion

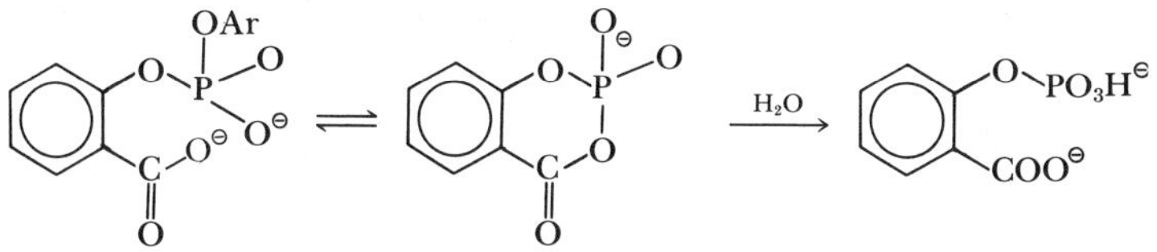

50 ⇌ **52** + $ArO^{\ominus}$

Scheme 17

has been measured at 39°[107] and has been shown to involve the $S_N2(P)$ mechanism. No direct comparison between the inter- and intramolecular reactions is possible because 2-carboxyphenyl 2,4-dinitrophenyl phosphate is prohibitively reactive, while the reaction of acetate with methyl 4-nitrophenyl phosphate anion is too slow to measure. But extrapolation of the linear free energy relationship between the hydrolysis rate and the pK_a of the leaving group for the three aryl 2-carboxyphenyl phosphates **50** (the dashed line in Fig. 11) makes it possible to estimate the rate of hydrolysis of the 2,4-dinitrophenyl ester. The figure obtained $(2.6 \times 10^3 \text{ min}^{-1})$ is enormously larger than the second order rate constant $(7.4 \times 10^{-6} \text{ liter mole}^{-1} \text{ min}^{-1})$ for the reaction of the more basic acetate ion with the anion of 2,4-dinitrophenyl methyl phosphate [107] and would require a notional concentration of some $10^9 M$ catalyst in an intermolecular reaction. This figure is unusually large even for an efficient intramolecular reaction and suggests that the electrostatic repulsion between the reacting anionic centers, which is responsible for an adverse factor of some two orders of magnitude in the rate of the intermolecular reaction, is likely to be less important in the intramolecular reaction. This is an effect which is clearly not of general application, but must be added to the (short) list of factors sensitive to proximity.

2.5.3 Pseudorotation. The important differences between intramolecular displacements of leaving groups which remain attached to the molecule (endocyclic displacements) and those in which the leaving group is a separate species (exocyclic displacements) are powerfully illustrated by the reactions of aspirin and its derivatives described above. Similar considerations must apply to intramolecular displacements at phosphorus, and in these reactions also endocyclic displacement is clearly unfavorable. Neither salicyl phosphate dianion (**53a**) nor the dianion of 2-carboxyphenyl phenyl phosphate (**53b**) is hydrolyzed by the route involving endocyclic displacement (Scheme 18). This is particularly remarkable in the case of the phenyl ester (**53b**), where the

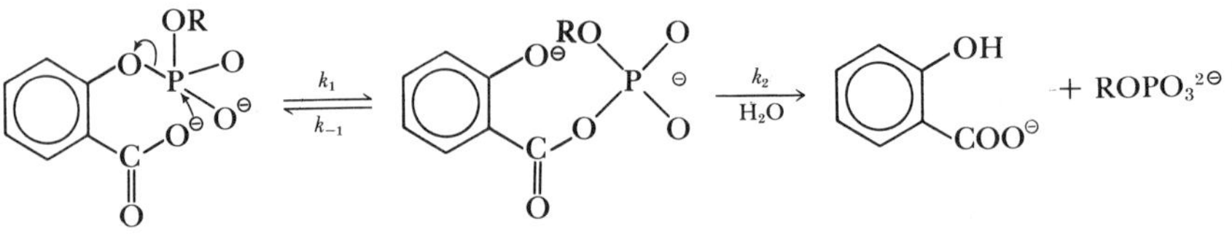

53a R = H
53b R = Ph
                                54
                            Scheme 18

reaction actually involves nucleophilic displacement of a phenoxide anion, which is not intrinsically any better a leaving group than the phenolate ion formed in **54**. The products of the endocyclic and exocyclic displacements, salicylic acid and salicyl phosphate, are both stable at alkaline pH, and the relative amounts formed, and thus the ratio of the rate constants for the two paths, can be estimated with some accuracy. It is found [106] that less than 1% of the theoretical amount of salicylic acid is produced on hydrolysis of **53b**, and part of this can be accounted for by intermolecular displacement by hydroxide ion. Thus exocyclic displacement is favored by a large factor.

These results are readily explained if the acyl phosphate **54** is formed from **53b** in a rapid, but unfavorable, preequilibrium and its hydrolysis is the slow step (k_2) of the mechanism shown in Scheme 17. This would allow the alternative route to dominate even if it were no faster than the endocyclic displacement (k_1). But there is strong evidence that rapid migrations from one nucleophilic center to another (endocyclic displacements) do not occur with phosphate diester anions. Nucleoside methyl and benzyl phosphates, and simpler model compounds, are hydrolyzed in alkali by exocyclic displacement of alkoxide (see p. 75), and no migration is detected in unreacted starting material even when the group displaced is a much poorer leaving group. Apparently there is a geometrical constraint which favors exocyclic displacement over the endocyclic process. This would be simply explained if displacement at the phosphorus center of a phosphate ester anion requires a linear transition state. A possible explanation of why this should be so is embodied in Westheimer's pseudorotation hypothesis [108]. Nucleophilic displacement at phosphorus may involve a pentacovalent intermediate. In the case of 2-carboxyphenyl phenyl phosphate this would have configuration **55**, and phenoxide would be lost from the favored apical position to give the cyclic acyl phosphate **52**. For the salicylate phenolic oxygen to leave it would also have to be in an apical position. It could attain an apical position by a pseudorotation [108], but any pseudorotation of **55** would bring a negatively charged oxygen atom into the other apical position, which is believed to be an energetically

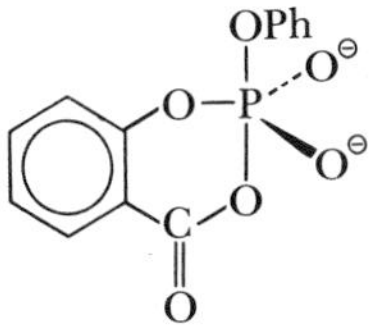

55

highly unfavorable arrangement[108]. Alternatively, since the penta-covalent species bearing two negatively charged atoms is expected to have very little stability[109], its lifetime might be too short for pseudo-rotation to occur. In this case no experimental test is likely to determine whether **55** is a highly reactive intermediate or simply a transition state.

2.5.4 Triesters. Dialkyl and diaryl esters of phosphoenol pyruvate are hydrolyzed rapidly at room temperature and near neutral pH to the monoesters, which are hydrolyzed further in a slower but still efficiently catalyzed reaction, to free phosphoenol pyruvate[105]. This reaction has recently been examined in detail by Benkovic and Schray[103] using the dibenzyl ester (**56**, R = CH$_2$Ph). The pH–rate profiles for hydrolysis

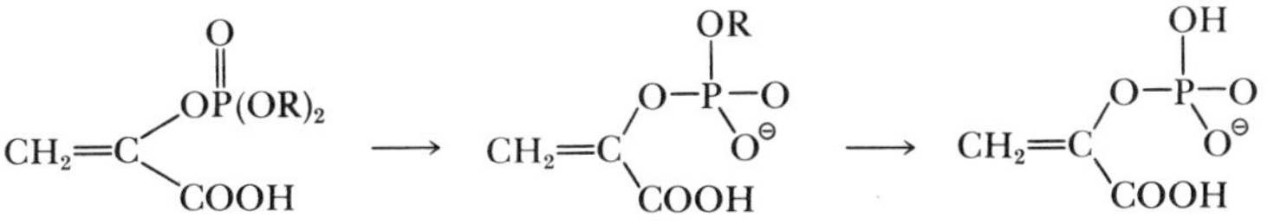

56

$$(15)$$

show that both steps involve the protonated carboxyl group or its kinetic equivalent. The mechanism involves nucleophilic catalysis, since added hydroxylamine traps an acyl-activated intermediate to give the oxime of the hydroxamic acid derived from pyruvic acid. Neither piece of evidence is consistent with the simple nucleophilic mechanism suggested originally by Clark and Kirby[105] for the first step, because the second product in the presence of hydroxylamine is dibenzyl phos-phate. Evidently acyl activation precedes loss of benzyl alcohol in the cyclization reaction, and the simplest explanation is that a pentacovalent intermediate is formed:

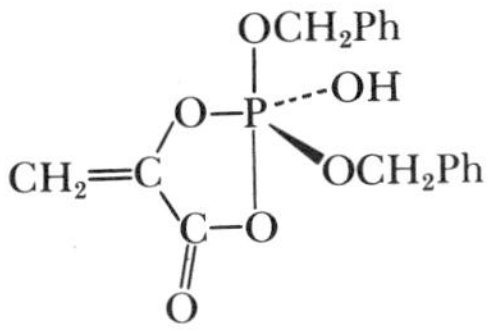

Benkovic and Schray consider that the intermediate trapped by the added hydroxylamine is the acyclic acyl phosphate rather than the pentacovalent intermediate. The trapping reaction shows saturation kinetics, and this fact, combined with the observation that benzyl alcohol is lost from the intermediate predominantly in the hydrolysis reaction, is explained if the acyclic acyl phosphate is formed reversibly, via a pseudorotation, from the pentacovalent intermediate. This would then be a case where exocyclic displacement at phosphorus is favored over endocyclic displacement for the same reasons that the latter type of reaction is unfavorable in intramolecular catalysis of carboxylic ester hydrolysis (Scheme 19).

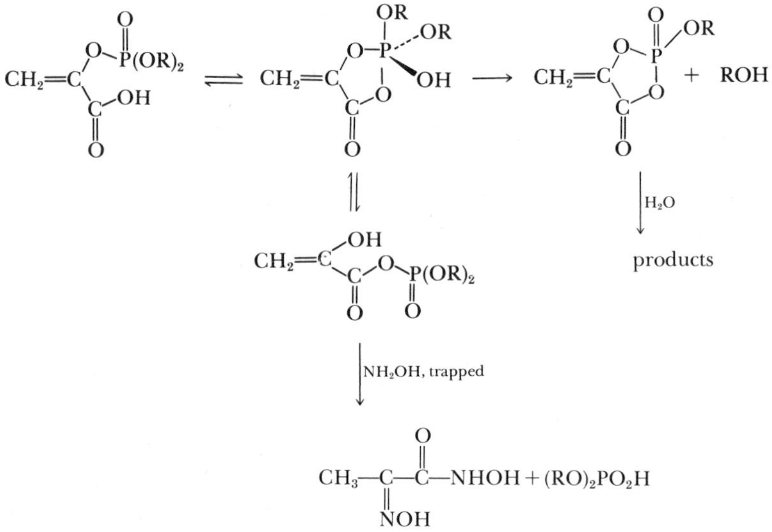

Scheme 19

3 INTRAMOLECULAR CATALYSIS BY THE AMIDE GROUP

The amide group has two possible nucleophilic or basic centers[111–118]. Generally the oxygen atom acts as the nucleophile or base in the neutral amide group, so that delocalization is preserved in the product (eq. 16). In special cases, where delocalization is prevented by a rigid

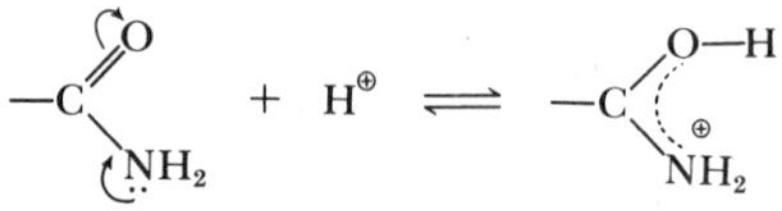

$$\text{(16)}$$

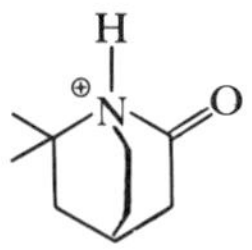

57

molecular geometry, amide character is lost and protonation may occur on nitrogen. An example of this effect is the protonation of the cage compound **57**[119], but the same drastic change of behavior is to be expected of any amide or peptide group when the planar geometry cannot be maintained, as might happen on binding to an enzyme.

The normal amide group is a very weak base, with the pK_a of the conjugate acid in the region of zero, and thus a very weak nucleophile also. Primary and secondary amides are also weakly acidic, with pK_a's in the region 14–15. For example, acetamide has $pK_1 = -0.48$[120] and $pK_2 = 15.1$[121]. Consequently, the amide anion is strongly basic and nucleophilic at nitrogen.

3.1 Catalysis of Ester Hydrolysis

Intramolecular O → N acyl migrations involving the anion of the amide group are common reactions of suitably substituted esters. The newly formed imide group is not, however, a particularly reactive species, and the process may or may not lead to catalysis of acyl transfer reactions or hydrolysis, and it may or may not be reversible. Large rate enhancements are observed in some cases and have led some authors to suggest that the amide groups present in the protein or substrate molecules may be directly involved in the catalytic process in certain enzymic reactions [123].

The neutral amide group is generally too weakly basic towards the carbonyl group to effect catalysis under mild conditions *in vitro*. At slightly elevated temperatures N-p-anisylmaleamic acid rearranges in solution in acetic anhydride to form the isomide[124] (eq. 17).

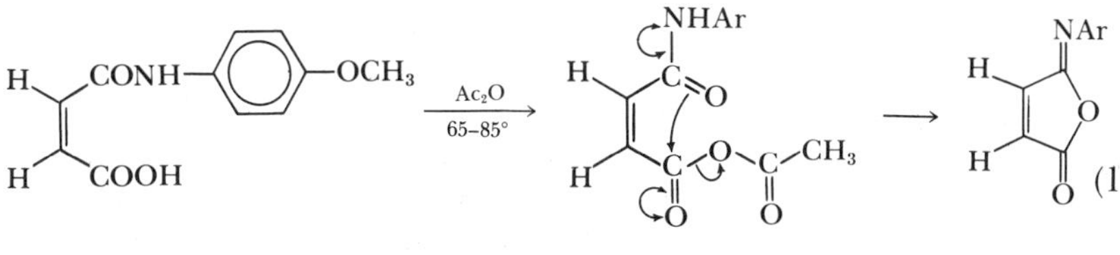

$$
\begin{array}{c}
\text{H}\diagdown\text{CONH-}\langle\bigcirc\rangle\text{-OCH}_3 \\
\text{H}\diagup\text{COOH}
\end{array}
\xrightarrow[65-85°]{\text{Ac}_2\text{O}}
\quad\longrightarrow\quad \text{(17)}
$$

The amide anion is known to be involved in intramolecular reactions in such compounds as the methyl esters of N-carbobenzoxy-L-asparagine and asparagine, which give the same imide intermediate (eq. 18)[125].

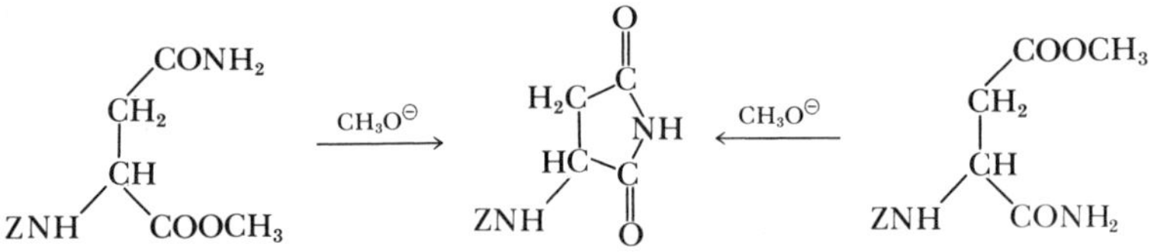

$$
\begin{array}{ccc}
\underset{ZNH}{\overset{\displaystyle CONH_2}{\underset{|}{\overset{|}{CH_2}}}}\!\!\!\!\!\!\!\!\!\underset{\displaystyle COOCH_3}{\overset{|}{CH}} & \xrightarrow{CH_3O^{\ominus}} \quad\text{[imide]}\quad \xleftarrow{CH_3O^{\ominus}} & \underset{ZNH}{\overset{\displaystyle COOCH_3}{\underset{|}{\overset{|}{CH_2}}}}\!\!\!\!\!\!\!\!\!\underset{\displaystyle CONH_2}{\overset{|}{CH}}
\end{array} \tag{18}
$$

This work is reviewed by Bruice and Benkovic[126]. A measure of the efficiency of the amide group as a nucleophilic catalyst under such conditions is available in the work of Behme and Cordes[123] on acetyl salicylamide (eq. 19, **58**). In the pH range 6.2–8.8 the first order rate constant for imide formation increases linearly with hydroxide con-

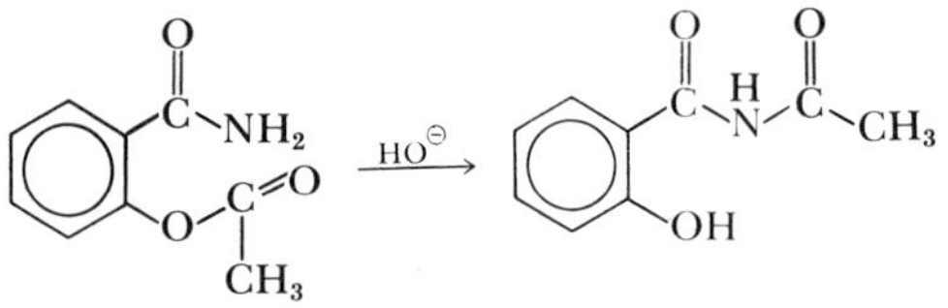

$$\tag{19}$$

58

centration, with a second order rate constant of 1.2×10^6 liter mole^{-1} min^{-1} at 25° and ionic strength 1.0. The reaction is presumed to involve intramolecular nucleophilic attack by the amide anion, that is, to be specific base catalyzed, and is over 60,000 times faster than the intermolecular reaction of $1\,M$ acetamide with p-nitrophenyl acetate under the same conditions.

Slightly lower rates are observed in the corresponding exocyclic displacements[127] (eq. 20). This reaction is markedly insensitive to substitu-

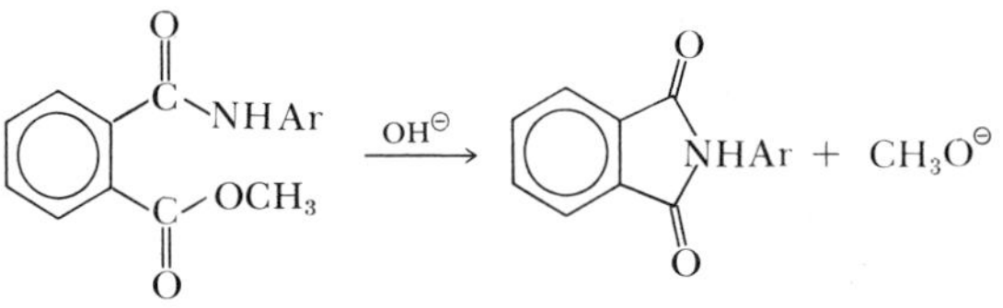

$$\tag{20}$$

tion in the aromatic ring of the anilide group, as the effects of substituents on the pK_a of the group are compensated by their opposing effects on its nucleophilicity.

The *anion* of N-2-carboxyphenyl urea (**59**) cyclizes to the imide in alkali, and the cyclization reaction is first order in both anion and hydroxide[128]. Hegarty and Bruice suggest[128] that the reaction involves intramolecular attack by the ureide anion on the anionic carboxylate group (eq. 21, **60**). This is certainly the simplest explanation of the observed result, but the assignment of mechanism is not unambiguous. With so remarkable a result it would be wise to await fuller details before

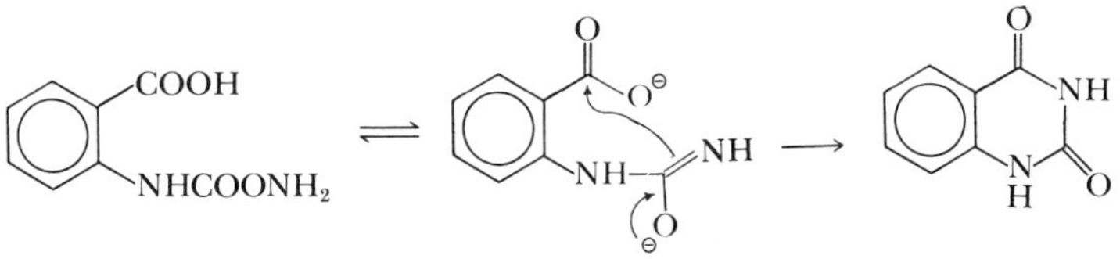

$$\tag{21}$$

59   **60**

ruling out other mechanisms entirely. Other examples of acyl migration from oxygen to amide anion have been described by Topping[129].

Menger and Johnson[130] have demonstrated intramolecular general base catalysis of the hydrolysis of the p-nitrophenyl ester **61** by the anion

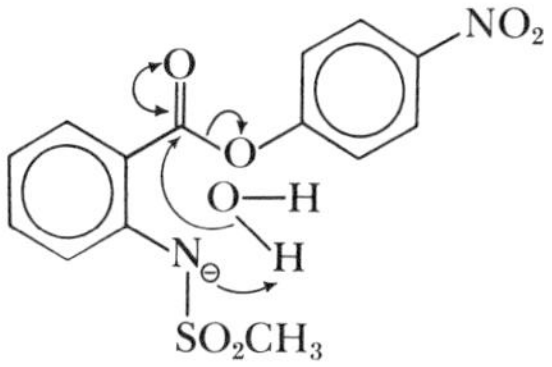

61

of the relatively acidic sulfonamide group. This is a particularly favorable case, because nucleophilic attack by nitrogen is stereochemically precluded. In comparable situations carboxylic amides can be acylated on oxygen. This particular reaction plagues the peptide chemist because it is most favorable for activated esters. For example, oxazolinones are formed from activated esters of hippuric acid in the presence of base, though not from less reactive esters, such as the methyl ester[131–134], and from the p-nitrophenyl esters of N-formyl, acetyl, and cinnamoylglycine, but not from the carbobenzoxy derivative[133]. Both general and specific base catalysis can be observed (eq. 21).

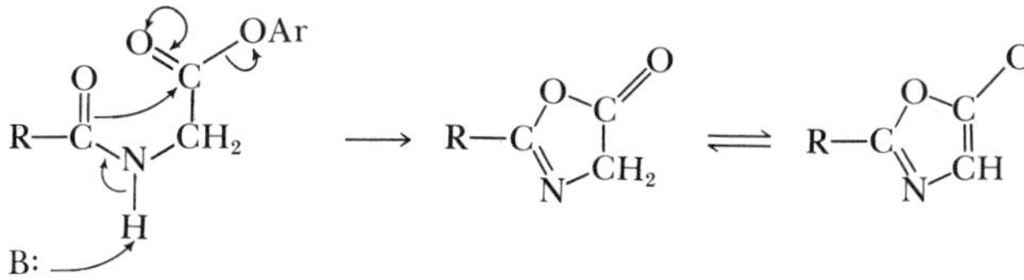

$$\tag{22}$$

4   INTRAMOLECULAR CATALYSIS BY THE AMINO GROUP

Three different mechanisms for intramolecular catalysis of ester hydrolysis are consistent with a kinetic dependence on the free base form of the amino group. These mechanisms are those possible for catalysis by any nucleophilic center and have been described previously for carboxylate catalyzed reactions.

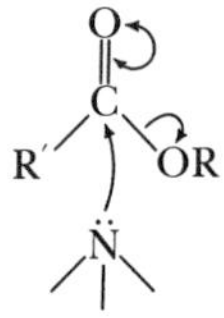

Mechanism A 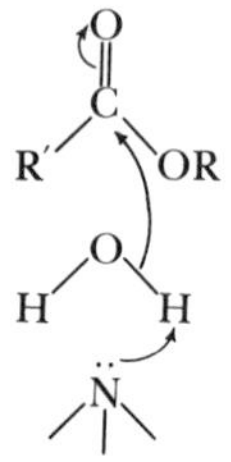   Mechanism B 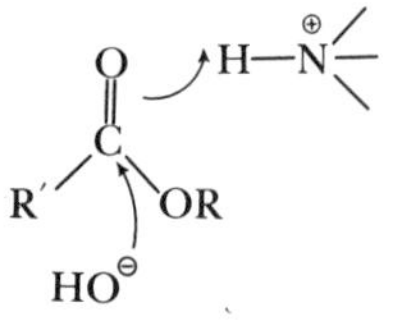   Mechanism C

Intramolecular nucleophilic catalysis (Mechanism A) involves an intermediate and is a unimolecular reaction if this step is rate determining.
It can therefore be readily distinguished, at least in principle, from
general base catalysis (Mechanism B) and specific base–general acid
catalysis (Mechanism C). Mechanisms B and C, on the other hand, cannot
be distinguished by the kinetic properties of the reaction – not except-
ing[135] sensitivity to changes in ionic strength. Although Mechanism B
involves charge formation in the transition state and Mechanism C, as
written, involves its dispersal, the ionic species which react in Mechanism
C are present only in low concentrations. The amine is predominantly
in the neutral form, by definition, and thus the transition state for
Mechanism C also involves the formation of charge compared with the
ground state. This phenomenon was involved in the classical controversy over the mechanism of formation of urea from ammonium cya-
nate, where also salt effects did not in fact distinguish the mechanism
involving NH_3 reacting with HNCO from that in which NH_4^+ reacts
with NCO^-[136]. An illuminating discussion of these matters is available[137].

4.1   Exocyclic Displacement on Esters

Intramolecular nucleophilic catalysis is involved in the hydrolysis of
the phenyl esters of ω-N,N-dimethylamino-butyric (**62**) and valeric (**63**)

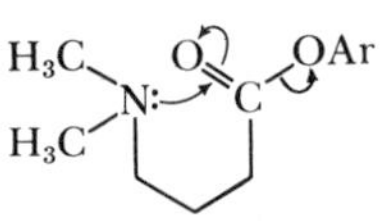

62 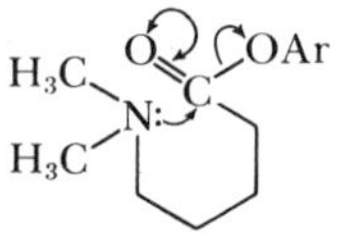   63 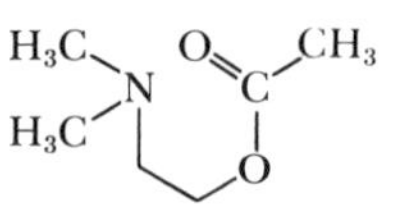   64 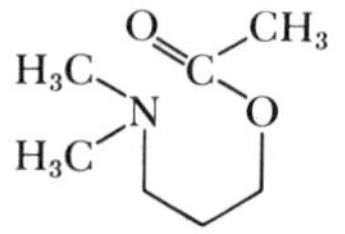   65

acids[138]. These reactions involve exocyclic displacements of good leaving groups, and large rate enhancements are observed, the "effective"
concentration of the amino group being $1\text{–}5 \times 10^3\,M$[138].

Intramolecular general acid catalysis of the attack of hydroxide ion has frequently been suggested to account for the hydrolysis of esters **64** and **65**[135]. These compounds show an enhanced rate of hydrolysis in the free base form, and the rate constant calculated for the Mechanism (C) involving attack of hydroxide ion on the conjugate acid is greater than that found for the alkaline hydrolysis of the corresponding trimethylammonium compounds (**66**). Recent discussions have been critical of the assignment of Mechanism C, and in a later study of these and related reactions involving exocyclic displacement in a similar system (**67**), Aksnes and Froyen[139] favored intramolecular general base catalysis (Mechanism B) for **67** and **68**.

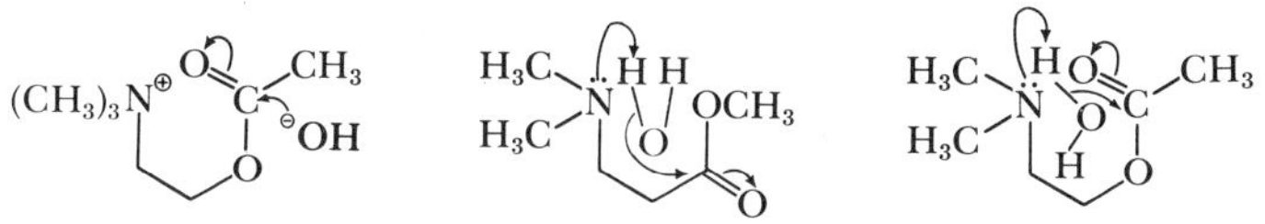

    66   **67**   **68**

The data given by Hansen[140] for the dependence of the rate on the pK_a of the amino group for a series of 2-dialkylaminoethyl acetates (e.g., **64**) are correlated by the Bronsted equation. The Bronsted coefficient, α, calculated for the general acid catalysis, Mechanism C, is 0.17. Alternatively, a Bronsted β of 0.8 is obtained if the mechanism is assumed to be general base catalysis, Mechanism B. There is no close analogy for the α-value, but the value of β is very high for a general base catalyzed reaction, for which β-values in the range 0.3–0.5 are usual, at least for catalysis by carboxylate anions.

4.2   Endocyclic Displacement by the NH$_2$ Group

A number of authors[141–144] have studied the O $\rightarrow$ N acetyl migration reaction of O-acetylethanolamine and the related hydrolysis of 2-methyloxazoline (eq. 23, **69**). The current status of these results is discuss-

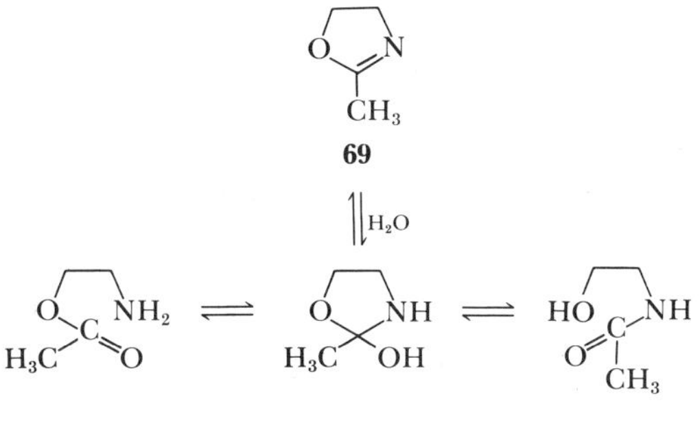

69

$$\Big\Vert \text{H}_2\text{O}$$

(23)

ed by Schmir[145]. The transfer of the acetyl group to nitrogen is only slightly faster than the hydrolysis of the dialkylamino compounds (e.g., **64**). The latter reaction cannot therefore involve nucleophilic catalysis, because tertiary amines are much less reactive than primary or secondary amines towards alkyl esters[144, 145] (by 5–6 orders of magnitude, according to calculations by Fersht and Jencks[146]).

An important recent development in this area is the demonstration by Barnett and Jencks[147] that proton transfer steps can be kinetically significant in acyl migration reactions. The hydration of 2-methyl-Δ^2-thiazoline (**70**) should involve the same tetrahedral intermediate as does the transfer of the acetyl group from S to N in S-acetylmercaptoethylamine (**71**) (Scheme 20). Providing all the proton transfers are fast, the

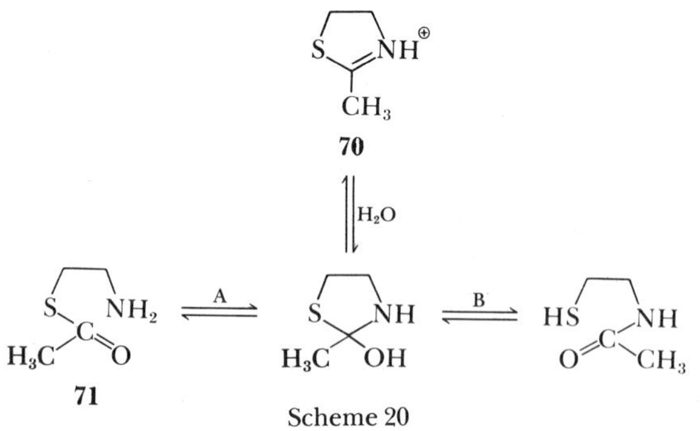

Scheme 20

products of thiazoline hydrolysis will be determined by which step is rate determining in the S → N acetyl transfer reaction. That is, if the tetrahedral intermediate formed by the addition of water to the thiazoline breaks down predominantly to form N-acetylmercaptoethylamine (path B), then it can be deduced that the rate-determining step in the S → N acetyl transfer is the formation of the tetrahedral intermediate (step A). And conversely, if S-acetylmercaptoethylamine is the main initial product of thiazoline hydrolysis, then the rate-determining step in the acetyl transfer reaction must be the breakdown of the tetrahedral intermediate, step B.

In the face of this clear prediction it was disconcerting to find[144] that there is a change of rate-determining step in the S → N acetyl transfer reaction at about pH 2.3, although the product distribution from thiazoline hydrolysis up to this pH is pH independent.* This problem,

---

*That is, the product distribution from thiazoline hydrolysis requires that the relative rates of steps A and B be pH independent; yet their relative rates, as measured in the acetyl transfer reaction, clearly change with pH. These are complex matters, and the baffled reader is advised to sit down quietly with a copy of reference 147.

dubbed by Martin et al.[144] the "thiazoline dilemma," was neatly resolved by Barnett and Jencks[147]. They showed that the rate-determining step of the S → N acetyl transfer reaction above pH 2.3 is a simple
proton transfer, not considered in Scheme 20, which determines the
products of thiazoline hydrolysis in this region of pH. Below pH 2.3
(Scheme 21) k_2 is rate determining; above pH 2.3 k_{HA} is the rate-determining step for acetyl transfer. Both steps are in the thiolester limb of
the reaction (step A of Scheme 20). The product distribution from thiazoline hydrolysis is thus independent of pH below pH 2.3, where k_2 is
rate determining; but the relative rates of steps A and B become dependent on pH in the region where k_{HA} is rate determining, above pH 2.3.
And at pH's above 4.5 step A becomes negligibly slow, so that the sole
product of thiazoline hydrolysis is the N-acetyl compound[147].

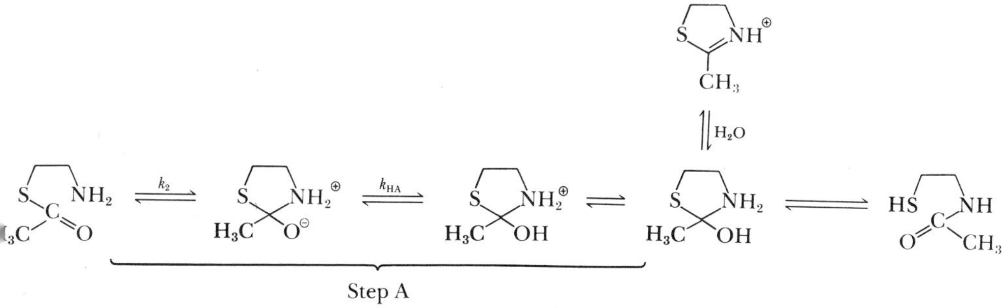

Scheme 21

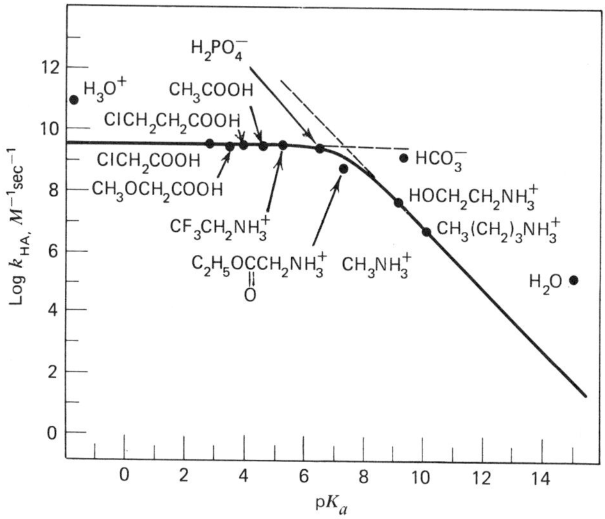

Figure 12   Bronsted plot for general acid catalysis of the S → N acetyl transfer reaction
of S-acetylmercaptoethylamine, at 50° and ionic strength 1.0[147].

The observed catalysis of the acetyl transfer by general acids (represented by k_{HA}) satisfies the criteria for diffusion-controlled proton transfers. The rate constants for proton transfer, from a series of structurally related acids in the thermodynamically favored direction, are virtually independent of acid strength [45]; the Bronsted coefficient $\alpha = 0.03$ [147]. The rate constant for proton transfer from the solvated proton is some 20-fold greater than for the other acids, presumably because of a proton jump mechanism [45]. As the acid strength is decreased and the proton transfer becomes thermodynamically unfavorable, a sharp break in the Bronsted plot appears (Fig. 12), at approximately the point where the pK_a's of the catalytic acid and the substrate are equal, and the slope α changes from 0 to 1.0 [45]. Catalysis by a bifunctional proton donor–acceptor is observed under favorable circumstances [148–150].

4.3 Endocyclic Displacement by the Tertiary Amino Group

The rate of hydrolysis of 8-acetoxyquinoline (72) is independent of pH in the region 4–8, and is 500 times as fast as that of the 7-acetoxy compound at 55° [151]. It was suggested [152], by analogy with the mechanism accepted at the time for aspirin hydrolysis, that the rate enhancement was due to intramolecular nucleophilic catalysis, involving rate-determining formation of an acylated tertiary amine. The aminolysis of esters of 8-hydroxyquinoline is also subject to intramolecular catalysis, and a mechanism involving general base catalysis by the neighboring nitrogen was suggested [153, 154].

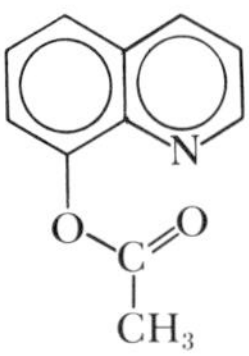

72 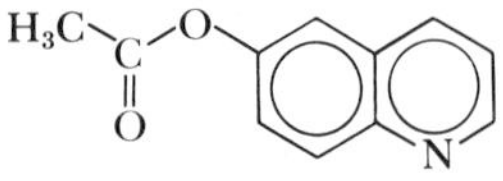   73

A detailed study of the reactions of nucleophiles with 8-acetoxyquinoline and its 6-acetoxy isomer (73) has been made by Felton and Bruice [155, 156]. The kinetic parameters $k_H/k_D = 2.35$ and $\Delta S\ddagger = -29$ eu are consistent with a general base catalysis mechanism for hydrolysis and inconsistent with a nucleophilic catalysis mechanism in which the initial attack is rate determining. These data are consistent, however, with nucleophilic catalysis if a preequilibrium formation of the acylated tertiary amine (74) is followed by a rate-determining step involving intra-

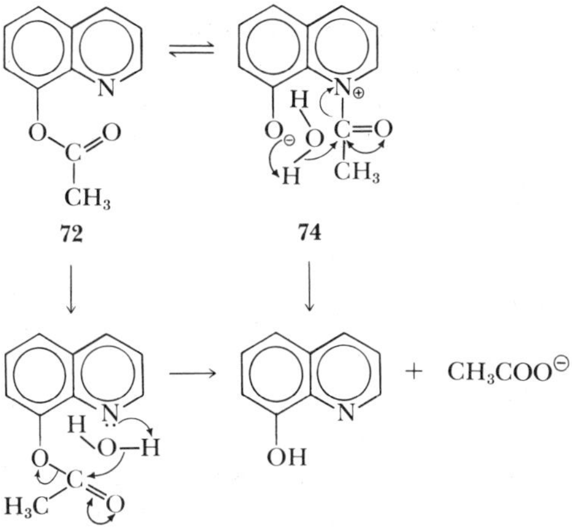

Scheme 22

molecular general base catalysis of its hydrolysis by the phenoxide ion (Scheme 22), as occurs in aspirin derivatives (p. 11). Experimentally the two paths have not been distinguished. The arguments of St. Pierre and Jencks[14] and Fersht[157] support the simple general base catalysis mechanism. If the rate-determining step were the hydrolysis of the intermediate acylated tertiary amine, then the pH–rate profile would be expected to show the enhanced rate for the *cation* rather than the neutral species, since the equilibrium constant K_{H^+} (Scheme 23) is much greater than K and intramolecular general base catalysis by the phenoxide group

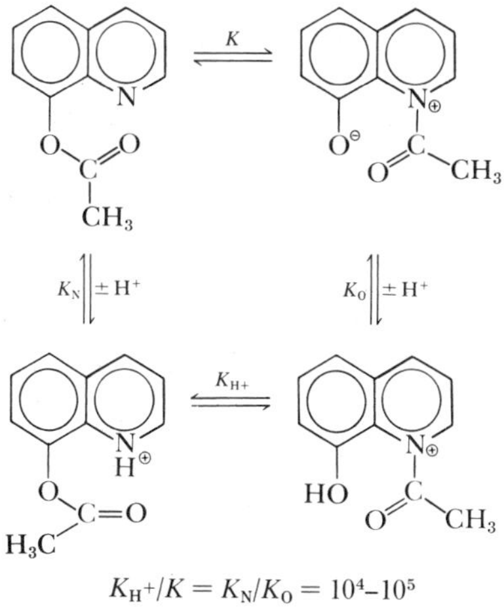

$$K_{H^+}/K = K_N/K_O = 10^4 - 10^5$$

Scheme 23

of **74** is unlikely to lead to a large rate enhancement [21, 22]. For example, for the hydrolysis of the neutral species **72** to be 10 times as fast as that of the cation, the ratio of K to K_{H^+} (Scheme 23) must be of the order of 10^5–10^6 for the nucleophilic mechanism through **74** to be operative. This ratio is unreasonably large.

Felton and Bruice [155] also confirmed that attack by primary and secondary, but not be tertiary, amines is also catalyzed. The rate enhancement is apparent from Figure 13, in which the rate constants for the reactions of amines with 8-acetoxyquinoline are plotted logarithmically against those for the reaction with the 6-acetoxy isomer. The points for nucleophiles without dissociable protons fall on a separate, lower, line, and the difference in ordinate between the two lines is a measure of the magnitude of catalysis. The Bronsted coefficient for the aminolysis of 6-acetoxyquinoline has a value of $\beta = 1.01$, similar to that (1.05) for the similar reaction of phenyl acetate [158]. Rate constants for the aminolysis of 8-acetoxyquinoline are correlated by a β value of 0.70, close to that calculated for general base catalysis of the hydrolysis of methyl formate [159]. As for hydrolysis, there are two kinetically equivalent mechanisms involving general base catalysis:

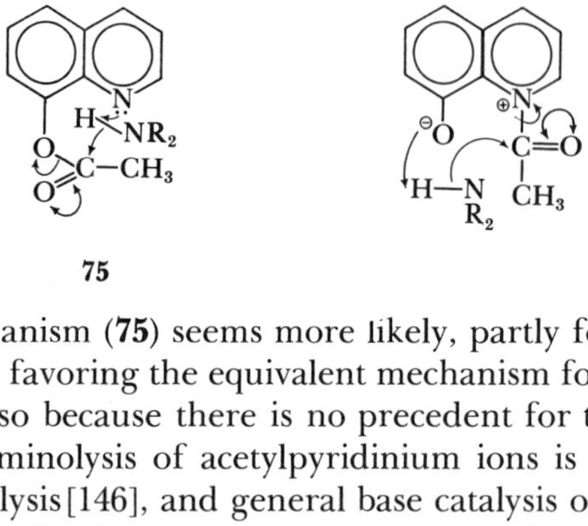

75

The one step mechanism (**75**) seems more likely, partly for the reasons discussed above for favoring the equivalent mechanism for the hydrolysis reaction; and also because there is no precedent for the alternative mechanism. The aminolysis of acetylpyridinium ions is not subject to intermolecular catalysis [146], and general base catalysis of ester aminolysis is most efficient for the poorest leaving groups. It is not clear why intramolecular catalysis of aminolysis should be readily observed in this system, while in the case of aspirin it is observed only in reactions with weakly basic amines, such as semicarbazide [14].

5 INTRAMOLECULAR CATALYSIS BY THE IMIDAZOLE GROUP

The reactions of the imidazole group are of particular interest because the imidazole in the side chain of histidine is clearly implicated in the mechanisms of action of a wide range of enzymes. For example, X-ray

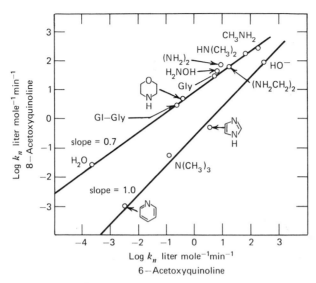

Figure 13   A comparison of the second order rate constants for the reactions of nucleophiles with 8-acetoxyquinoline (**72**) and 6-acetoxyquinoline (**73**)[156].

diffraction studies on the serine proteinases chymotrypsin[56], elastase [160], and subtilisin[161] show that the active sites of all three enzymes contain a serine and a histidine residue.

The hydrolysis of esters by chymotrypsin has been shown to involve an intermediate acylated enzyme[162–167]. The change in the uv spectrum on deacetylation of acetyl chymotrypsin was at one time interpreted as evidence that an acetylimidazole was an intermediate and thus that nucleophilic catalysis by imidazole was involved in the mechanism[168]. But the absorption in the relevant region was later identified as an artifact due to Rayleigh scattering[169], and current opinion is that the imidazole group acts as a general base to catalyze the acylation of serine.

The deacylation of acyl chymotrypsins is understood rather better than the acylation step. Inward and Jencks[170] showed that the reactions of alcohols and amines with furoylchymotrypsin to give esters or amides of furoic acid, had the properties expected for reactions assisted by intramolecular general base catalysis. Other evidence consistent with general base catalysis is the solvent deuterium isotope effect ($k_H/k_D = $ 2–3) for acylation and deacylation. However, the use of the solvent deuterium isotope effect as a criterion for distinguishing between nucleophilic and general base catalysis is not wholly satisfactory even in simple reactions[11, 25] and is more uncertain still in the complex enzymic situation. This is a normal state of affairs in the determination of mechanism and all we can do is gather together the largest possible number of pieces of evidence likely to be relevant. One more such piece of evidence in this instance is the interpretation of the steric parameters for both acylation and deacylation processes as being also consistent with general base catalysis[174–176].

5.1 Catalysis of Ester Hydrolysis

The intermolecular reactions of imidazole with phenyl esters are described by Bruice and Benkovic[177]. Nucleophilic catalysis of hydrolysis is observed for activated esters such as p-nitrophenyl acetate, and as the leaving group becomes poorer, general base catalysis of attack by a second molecule of imidazole can be identified (for example, with phenyl acetate). With alkyl esters general base catalysis of the attack of water becomes dominant.

The first example of intramolecular catalysis of ester hydrolysis by the imidazole group was found using compounds of the type 76[178, 179]. Hydrolysis is accelerated in the pH region where the imidazole group exists as the free base. The acetyl compound can be thought of as an analog of aspirin with the imidazole in place of the carboxylate group, and a mechanism (76–77) involving nucleophilic catalysis was proposed[178], by analogy with the then accepted mechanism of hydrolysis of aspirin. This mechanism (eq. 24) has recently been revised

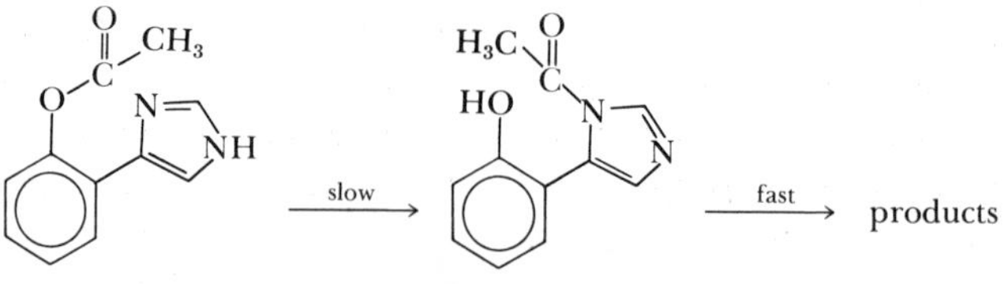

[155, 156] in the light of the evidence that the hydrolysis of aspirin involves intramolecular general base catalysis (see p. 6), and the hydrolysis of 76 is now also thought to involve intramolecular general base catalysis[155, 156]. The evidence on which the reinterpretation was based (the low value of $\Delta S\ddagger$, the deuterium solvent isotope effect, and the effects of an added nucleophile, in this case, hydrazine) is also consistent with the nucleophilic mechanism if the solvolysis of the intermediate acylimidazole (77) is rate determining.

The hydrolysis of the imidazole analog 76 is a more complicated reaction than the hydrolysis of aspirin itself. The nucleophilic mechanism, which involves a simple equilibrium between two anions in the aspirin reaction, in this case leads to two kinetically equivalent forms of the acylimidazole intermediate, **79** and **80** (Scheme 24). Using reasonable

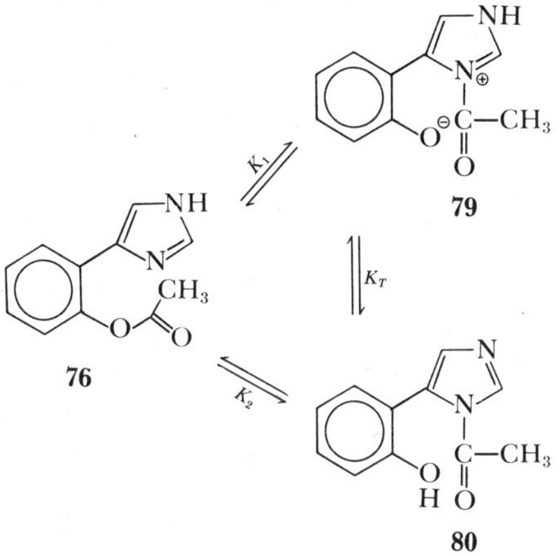

Scheme 24

estimates of the pK_a's involved, we estimate that K_T favors the neutral acetylimidazole over the zwitterionic form (**79**) by a factor of about 10^6. Using the free energy data of Gerstein and Jencks[30] K_1 can be estimated as $10^{-7.5}$ and K_2 as $10^{-1.5}$, whereas for aspirin the single equilibrium constant is calculated to be about 10^{-11}. The increased stability of the zwitterionic species **79** relative to the starting material, compared with that of the mixed anhydride relative to the ester in aspirin hydrolysis (Scheme 2), is a result of the higher basicity of the imidazole group of **76**, relative to that of the carboxyl group of aspirin ($pK_a = 3.3$) and of the greater stability of an acetylimidazole compared with an anhydride (by 1.7 kcal mole^{-1})[30, 31]. Also, the *rate* of hydrolysis of the acetylimidazolium ion is some 20 times greater than that of acetic anhydride [180, 181]. For these reasons the route involving the acetylimidazolium intermediate (**79**) is favored by a factor of 10^4–10^5 compared with the nucleophilic pathway for aspirin hydrolysis.

It is unlikely that the route through the neutral acetylimidazole **80** is important, even though the concentration of **80** may be as high as 1–2% of that of starting material. The observed rate constant for the hydrolysis reaction of **76** (2.1×10^{-3} min^{-1} at 30° in 28% v/v ethanol)[178] is almost as large as that for the hydrolysis of acetylimidazole (5×10^{-3} min^{-1} at 25°)[181]. Also, the ratio of the rate constants for alkaline hydrolysis and spontaneous hydrolysis ($4 \times 10^6 M^{-1}$) for acetyl imidazole[181] is greater than the ratio for the reactions of **76** (1–2×10^5). The water reaction is thus catalyzed, when compared with the hydroxide reaction, so that the mechanism cannot be the rate-determining hydrolysis of **80**.

The mechanism of hydrolysis of the ester **76** is clearly intramolecular general base catalysis of the attack of water on the ester group by the imidazole. The possibility of a contribution from nucleophilic catalysis through **79** cannot be completely ruled out, but is small. The observed rate of hydrolysis is about four times that of aspirin at 25°[14], as expected if both reactions involve intramolecular general base catalysis, because the imidazole group of **76** is a stronger base than the carboxylate group of aspirin.

A mechanism involving intramolecular nucleophilic catalysis with rate-determining intermediate formation has been suggested [179] to account for the observed catalysis by the neutral imidazole group of the hydrolysis of esters 4-(-2'-hydroxyethyl)-imidazole (**81**). In the light of the more recent work discussed above it must be considered that intramolecular general base catalysis is the likely mechanism in this case also, particularly since the leaving group of **81** is much poorer than that of a phenyl ester (Scheme 25). Intramolecular nucleophilic catalysis does occur in the hydrolyses of activated esters of γ-(4-imidazolyl)-butyric

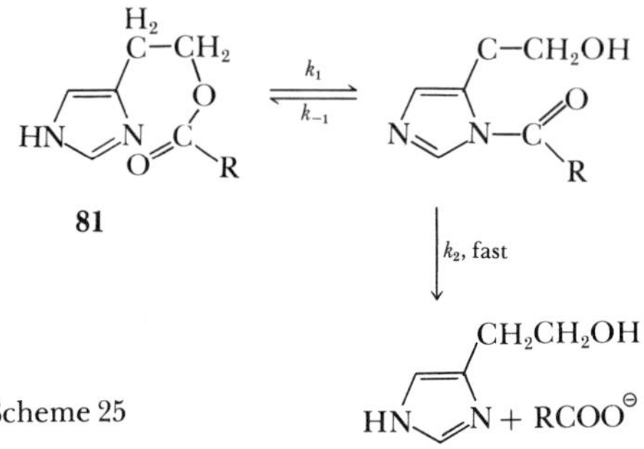

Scheme 25

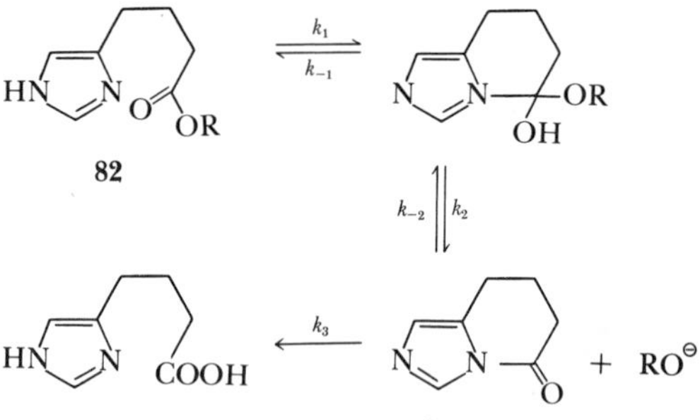

Scheme 26

acid (**82**)[182, 183], which involve exocyclic displacements of good leaving groups. Here the intermediate lactam (**83**) accumulates in solution. Intramolecular catalysis was not detected in the hydrolysis of the methyl ester (**82**, R = CH_3).

5.2   Catalysis of Amide Hydrolysis

Intramolecular catalysis of amide hydrolysis by the imidazole group requires the protonated form of the base[184] and thus is analogous to the same reaction catalyzed by the carboxyl group. Kirk and Cohen [185] have found very effective catalysis in a weakly basic benzimidazole system. Dipeptides **84** are hydrolyzed near pH 0, three or four orders of magnitude faster than the corresponding 2,4-dinitrophenyl derivatives. The most reactive of a series of derivatives is the alanylglycyl compound (**84**, $R_1 = R_2 = CH_3$). This is also one of the least basic derivatives, and thus the conjugate acid form is one of the most strongly acidic ($pK_a = -0.25$), but there is no simple correlation between reactivity and basicity apparent in the series.

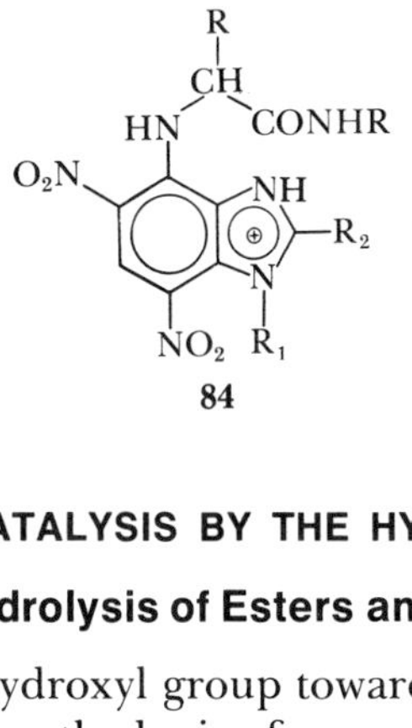

84

6   INTRAMOLECULAR CATALYSIS BY THE HYDROXYL GROUP

6.1   Catalysis of the Hydrolysis of Esters and Amides

The reactivity of the hydroxyl group towards carboxylic acid derivatives can be rationalized on the basis of generalizations similar to those which emerge from the analysis of the corresponding reactions with carboxyl groups. The hydroxyl group too can act as a general acid, albeit a very weak one, in the neutral form, and as a nucleophile or general base when ionized. When it acts as a nucleophile it will not be expected to displace very much more strongly basic leaving groups, such as amide anions. Thus the methyl ester **85** is rapidly hydrolyzed on warming with NaOH in aqueous methanol[186], presumably by specific base catalyzed lactone formation (eq. 25), whereas the amide **86** is quite stable under similar conditions[187].

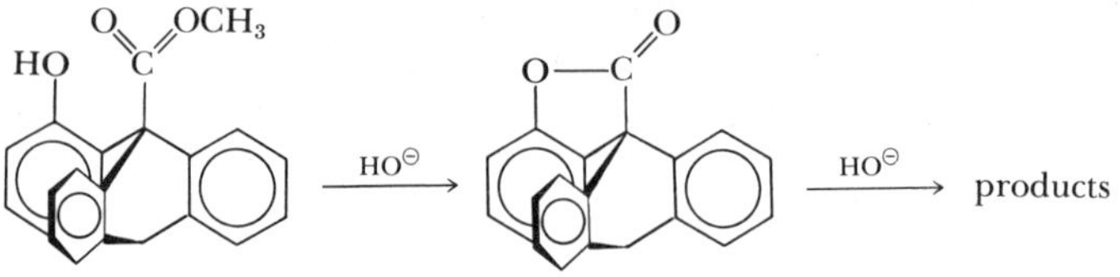

$$\text{85} \xrightarrow{\text{HO}^{\ominus}} \quad \xrightarrow{\text{HO}^{\ominus}} \quad \text{products} \qquad (25)$$

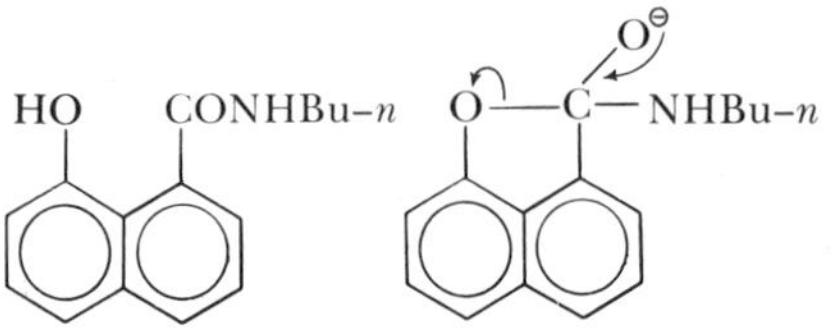

86   87

The simplest explanation for this difference is that a tetrahedral addition intermediate such as **87** will break down essentially exclusively by elimination of the naphthoxide anion, rather than that of the much more strongly basic amide anion. If catalysis by an aryloxide ion is observed, therefore, it is likely to be general base catalysis, as proposed by Bruice and Tanner[20] for the hydrolysis of salicylamide (**88**). The pH–rate profile for hydrolysis of this compound, and for that of its 5-nitro derivative, shows a pH-independent region above the pK_a of the phenolic group, as long as the reaction involving intramolecular catalysis is faster than alkaline hydrolysis. The mechanism of Bruice and Tanner is thought to be correct, although their grounds for rejecting the alternative general acid catalyzed attack of hydroxide ion were not[21] (see p. 8).

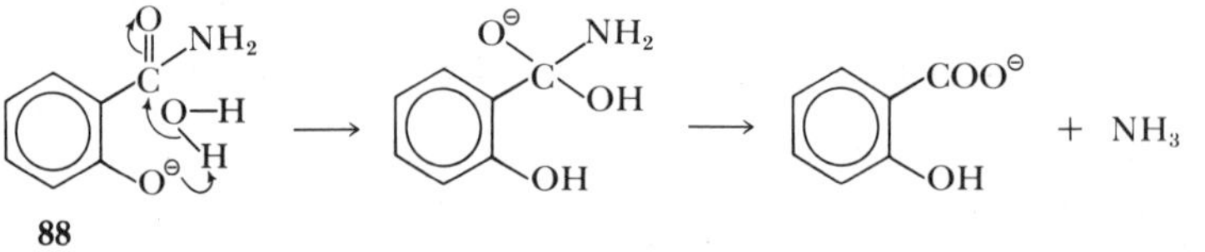

$$\text{88} \longrightarrow \quad \longrightarrow \quad + \text{NH}_3 \qquad (26)$$

It is possible for the more strongly basic alkoxide ions to displace amide anions from neighboring amide groups, at least in an exocyclic displacement reaction. The alkaline hydrolysis of 4-hydroxybutyramide (**89**) is faster by one order of magnitude than that of butyramide itself, and the reaction is thought to involve intramolecular nucleophilic catalysis[188] (eq. 27). Here too the tetrahedral intermediate will break down almost exclusively to starting materials, but the equilibrium constant for its formation is much more favorable than it would be in an

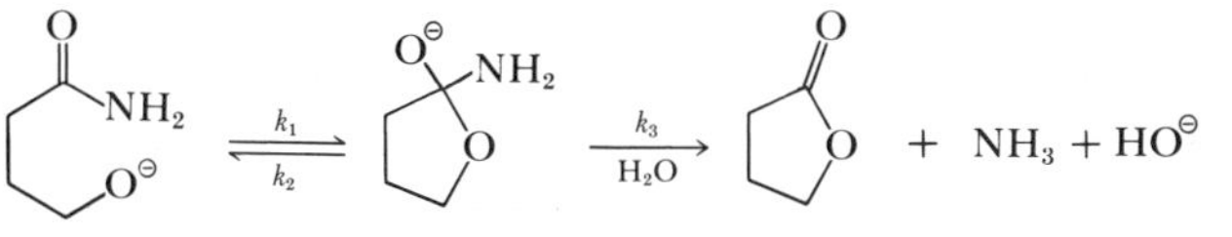

$$\underset{\textbf{89}}{} \xrightleftharpoons[k_2]{k_1} \underset{\textbf{90}}{} \xrightarrow[H_2O]{k_3} + NH_3 + HO^{\ominus} \qquad (27)$$

intermolecular reaction, and there is a slow "leakage" to products by an elimination of amide ion (presumably solvent assisted).

This breakdown step is slow because of the need to eliminate the very strongly basic amide anion, and reactions of this sort are much more favorable if the leaving group is protonated[46]. Thus intramolecular catalysis of the 4-hydroxybutyramide is much more efficient near neutrality, where the rate enhancement is some 800-fold (eqs. 28 and 29).

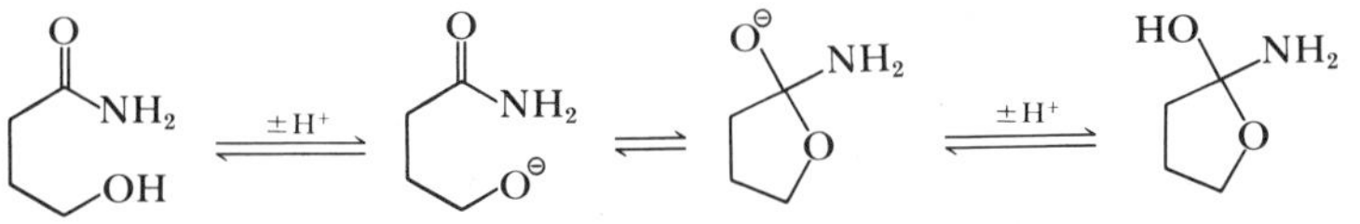

$$\xrightleftharpoons{\pm H^+} \rightleftharpoons \xrightleftharpoons{\pm H^+} \qquad (28)$$

$$\rightleftharpoons \longrightarrow + NH_3 \qquad (29)$$

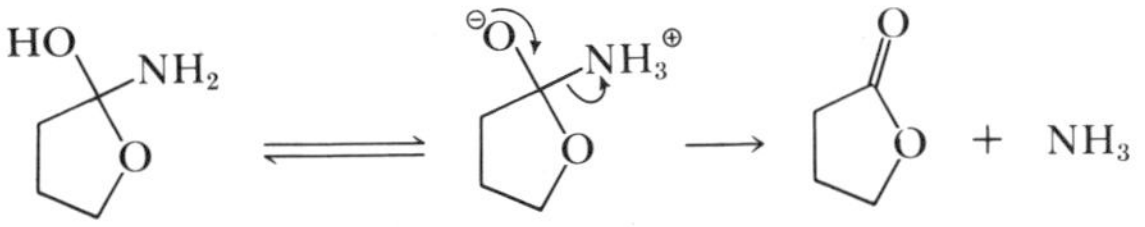

The breakdown of the tetrahedral intermediate might be expected to be subject to general species catalysis also. Cunningham and Schmir [189], working with 4-hydroxybutyranilide, have found very effective catalysis of hydrolysis by phosphate and bi-carbonate buffers. This does not appear to represent simple general acid or base catalysis since imidazole buffers are ineffective. Cunningham and Schmir suggest that

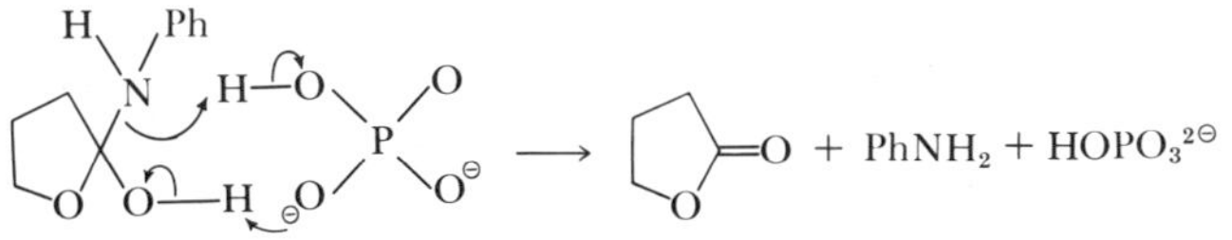

$$ \longrightarrow =O + PhNH_2 + HOPO_3^{2\ominus} \qquad (30)$$

91

this is an instance of bifunctional catalysis (**91**), possible for catalytic species such as phosphate and bicarbonate anions, which carry both a proton and a basic center, but not for a catalyst with only a single catalytic center (eq. 30). At high buffer concentrations the second order rate constant for catalysis by both phosphate and bicarbonate ions approaches a limiting value. This type of observation is itself evidence for the existence of an intermediate in a reaction of this sort and is explained by a change of rate-determining step. The breakdown of the tetrahedral intermediate **91** is rate determining at low catalyst concentrations and is accelerated by catalysis until it is as fast as the formation of the intermediate. At higher concentrations of catalyst this first step becomes rate determining, and if it is not subject to buffer catalysis, the rate becomes independent of buffer concentration. The analysis of the mechanism of this reaction was greatly aided by the authors' previous study[190] of the hydrolysis of 2-phenyliminotetrahydrofuran, which goes by way of the same intermediate[191] (eq. 31).

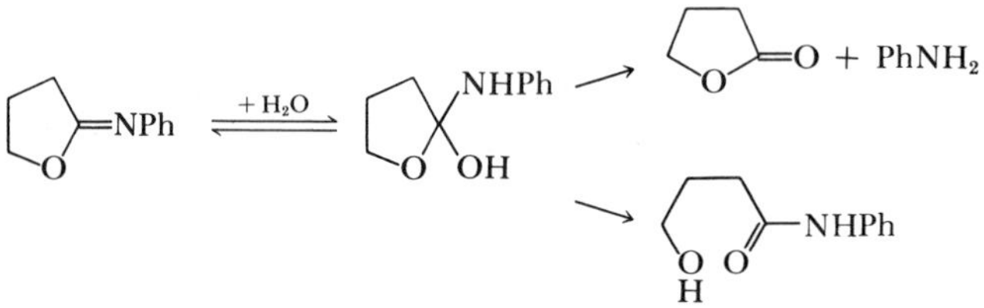

$$\tag{31}$$

In acid the hydrolysis of 4-hydroxybutyramide shows only weak catalysis. Two reaons for this are (*1*) that the neutral hydroxy group is only a weak nucleophile, which may attack a protonated amide, but will not add as such to the neutral group and (*2*) that a protonated carbinolamine (**92**) is much more stable than the zwitterionic species[46].

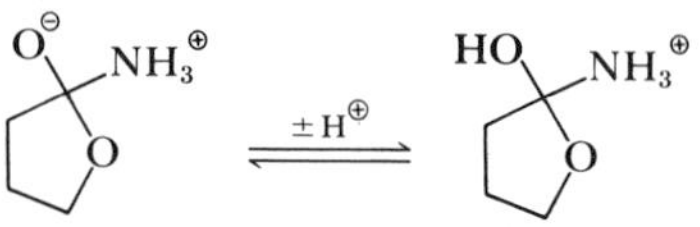

92

Much recent work on the hydroxyl-assisted hydrolysis of esters has involved systems for which intramolecular nucleophilic catalysis is ruled out by the geometry of the molecules involved, as it is, for example, in the case of salicylamide, discussed above. The pioneering study of a system of this sort was that by Bender, Kezdy, and Zerner[22] of the hydrolysis of *p*-nitrophenyl 5-nitrosalicylate. The rapid, pH-independent hydrolysis catalyzed by the ionized phenolic group was interpreted as an example of intramolecular general base catalysis of hydrolysis (eq. 32)[22]. Capon and Ghosh[21] extended this work with an investigation of the hydrolysis of a series of substituted-phenyl salicylates in water at 59.2° and ionic strength 0.05. Their work also supports the general base catalysis mechanism; since the reaction with imidazole is not catalyzed, as it should be if the hydroxyl group acts as a general acid, and since there is a solvent deuterium isotope effect, $k_H/k_D = 1.7$, in the expected region for a reaction of this sort. Also the Hammett

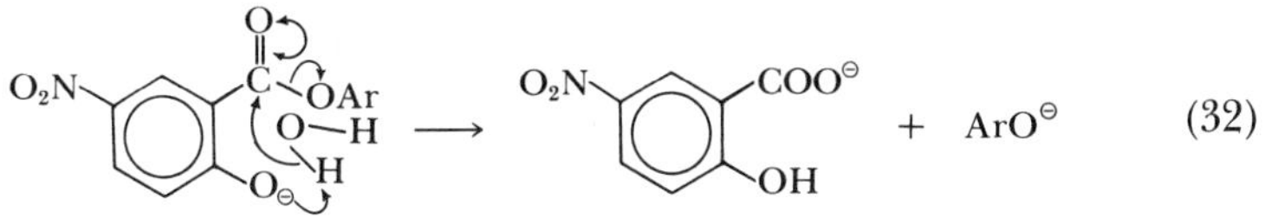

93

p-value of 0.80 is close to that found in analogous reactions. For example, p for the acetate (general base) catalyzed hydrolysis of substituted phenyl acetates is about 1.1 (calculated from the data of Oakenfull, Riley, and Gold[24]) and for the intramolecular general base catalyzed hydrolysis of substituted aspirins it is 0.96[18]. It is noteworthy that these figures lie close to that found by Bruice and Mayahi[191] for the alkaline hydrolysis of substituted phenyl acetates. From the point of view of the carbonyl group under attack, apparently, the general base catalyzed attack of water is closely similar to attack by the solvated hydroxide ion.

Capon and Ghosh[21] also studied the hydrolysis of catechol monobenzoate (**94**) and found that its hydrolysis too is subject to intramolecular catalysis by the ionized phenolic group. Nucleophilic catalysis is ruled out in this case by the symmetry, rather than the geometry, of the starting material, but the hydrolysis of the anion is over a hundred times faster than that of o-methoxyphenyl benzoate near pH 8 and depends on an ionizing group of $pK_a = 8.7$, identical to that measured for the hydroxyl group of **94**. Nucleophilic attack by imidazole is not catalyzed, however, and this fact, together with the observed solvent deuterium isotope effect, is consistent with a mechanism (eq. 33) involving general base catalysis of hydrolysis, similar to that described for the hydrolysis of phenyl salicylate. Although the ionized hydroxyl group of

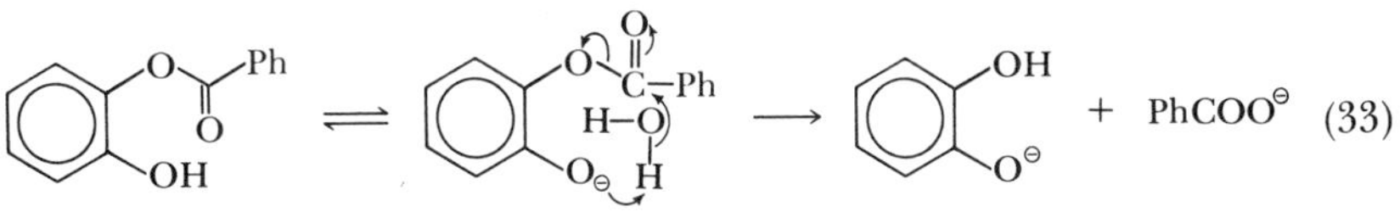

94

salicylate esters does not catalyze their reactions with nucleophiles such as imidazole, it might be expected that attack by neutral nucleophiles which have a dissociable proton on the nucleophilic center would be catalyzed. Two groups have studied the aminolysis of a salicylate ester in an aprotic solvent and found significant contributions from intramolecular catalysis by the ionized hydroxyl group under these (particularly favorable) conditions.

The n-butylaminolysis of methyl salicylate in dry dioxane is second order in n-butylamine and is characterized by values of $\Delta H\ddagger = 8.2$ kcal mole^{-1} and $\Delta S\ddagger = -48.5$ eu [192]. The reaction with methyl p-hydroxy-benzoate could not be detected after 190 hr at 70° and is clearly a great deal slower. These data are consistent with a termolecular reaction involving intramolecular catalysis by the hydroxyl group, and more recent experiments by Menger and Smith [193] have helped to clarify the mechanism involved. These workers showed that the n-butyl-aminolysis of phenyl salicylate in acetonitrile is kinetically a very similar reaction and is 132 times as fast as the corresponding reaction of the methoxy compound (95). Added n-butylammonium chloride slows the reaction with phenyl salicylate (although it accelerates the aminolysis of the methoxy compound), which makes it unlikely that the reaction involves the anion of the ion pair (96), since the addition of the ammonium chloride suppresses the ionization of the phenolic group. The

tertiary amine triethylenediamine which has almost the same basicity in acetonitrile as does n-butylamine, also catalyzes the aminolysis, with a rate constant very close to that for n-butylamine. Clearly the molecule of tertiary amine replaces one molecule of n-butylamine.

Menger and Smith [193] propose a mechanism (Mechanism A) in which the general base catalyzed attack of n-butylamine is assisted by general acid catalysis by the neighboring hydroxyl group. This is

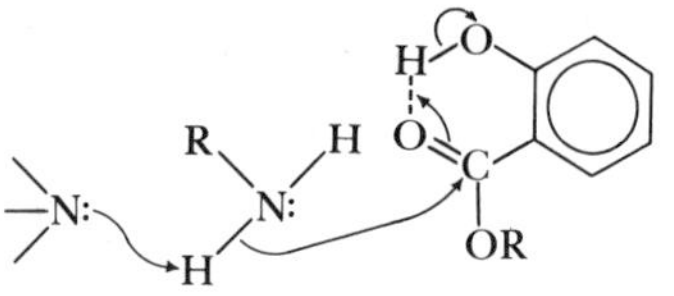

Mechanism A

certainly the simplest explanation of the facts and, in particular, of the observation that catalysis by the n-butylammonium ion is not observed for the hydroxy compound. It is presumed that the hydroxy group fulfills the requirement satisfied by this cation in the reaction of the methoxy compound. The explanation does, however, require that the reaction of the methoxy compound involve an unprecedented four-molecule collision.

When the hydroxyl group involved in catalysis is in the alkyl rather than the acyl portion of the ester, nucleophilic attack leads not to hydrolysis, but to the migration of the acyl group. This is a well known problem in the synthesis of partially acylated polyhydroxy compounds. For example, a recent attempted synthesis of the pyridoxol benzoate **97** gave the rearranged product **98** when the protecting isopropylidene group was removed[194] (eq. 34). This type of acyl migration reaction

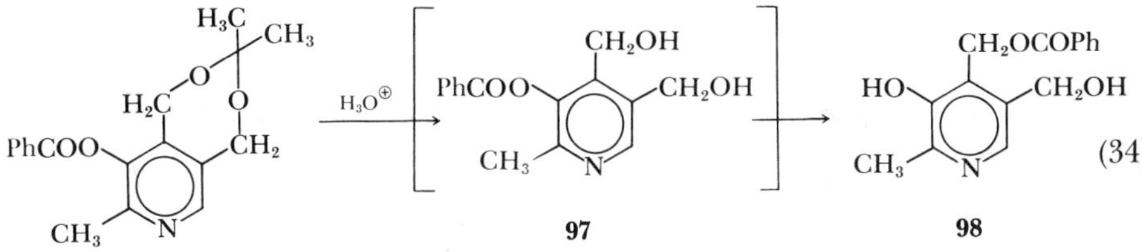

(34)

97   **98**

has received a good deal of attention in recent years because it leads to rapid equilibration of 2' and 3'-aminoacyl ribonucleotides and thus complicates the determination of structure of aminoacyl-s-RNA's. Thus McLaughlin and Ingram[195], working with a number of 2' and 3'-aminoacyl adenosine derivatives, found that the acyl migration reaction becomes too fast to measure at 15° at pH's above 7 and gives an equilibrium mixture with the 3'-isomer predominating by a factor of 3:1. The reaction was taken to be specific base catalyzed, and a second order rate constant for hydroxide catalysis of 3.6×10^6 liter mole^{-1} min^{-1} at 15° was measured for the migration to the 3'-position in 2'-O-N-acetyl-valyl adenosine[195]. Similar results were obtained by Reese and his co-workers[196] using formyl and acetyl adenosines and uridines. In all these cases migration is faster than hydrolysis by a very large factor, of the order of 10^5–10^6 near pH 7 (Scheme 27).

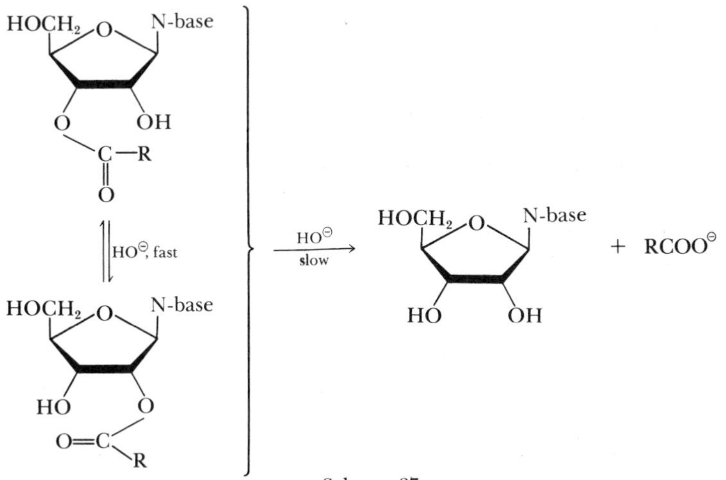

Scheme 27

In such cases any observed catalysis of hydrolysis must involve the hydroxyl group as a general acid or base. Rate enhancements are usually small in aliphatic compounds, though factors as large as 10^3 have been observed by Kupchan and co-workers[197] for the methanolysis of a diaxial 1,3-dihydroxy steroid. The reaction is subject to general base catalysis by the triethylamine buffer used, and the mechanism proposed (99) is supported by the observation that the suitably placed tertiary amino group of a structurally related alkaloid can replace the molecule of base, as in 100. The rapid hydrolysis of the acetate group of

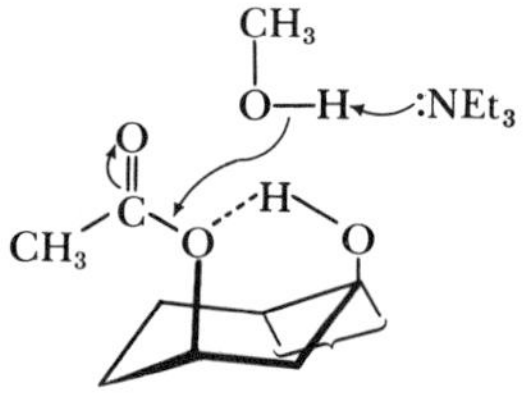

99 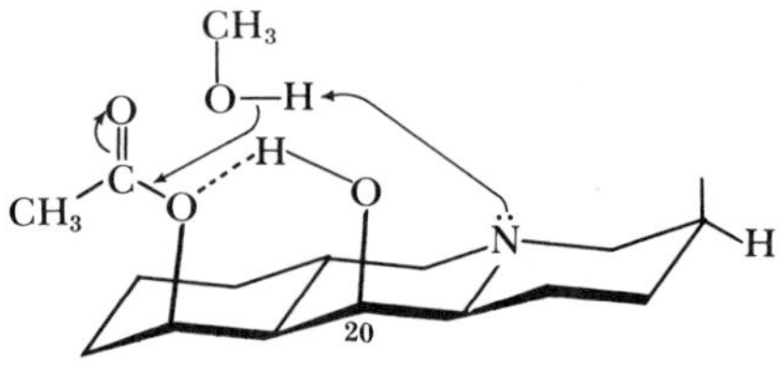   100

100 depends on the combined presence of the hydroxyl and amino groups, and systems of this sort, where the stereochemistry is rigidly defined, make it possible to design and study highly complex catalytic reactions. In this particular case an alternative mechanism that cannot be completely ruled out involves the rapid migration of the acetyl group to the 20-position; the migration would be expected to be a rapid reaction, and the 20-acetate has available potentially more effective pathways for hydrolysis. But in a nonrigid system the problem would be far more complex, and the advantages of a well defined stereochemistry are no less convincing.

6.2   The Hydrolysis of Phosphate Esters

6.2.1   Phosphate Monoesters. For reasons which are partly under-stood, displacements at the phosphorus center of phosphate esters in which the transition state bears a negative charge require a linear geometry. This restriction has been discussed above in connection with intramolecular catalysis by the ionized carboxyl group (see p. 48), but applies equally to base catalyzed reactions involving the hydroxyl group. Thus simple migrations of the phosphate group to neighboring hydroxyl functions is not observed under neutral or basic conditions. For example, no rearrangement accompanies the alkaline hydrolysis of ribonucleic acids[198]. In strongly acidic media, however, migrations are observed [199]. This reaction has been carefully studied using the 1-phosphate ester of propan-1,2-diol[200] and both isomers of glycerophosphoric acid (**101** and **102**)[200, 201]. These reactions follow first order kinetics and are specific acid catalyzed, being faster in D_2O than in H_2O. The most important route involves the cyclic diester (**103**) as an intermediate [200, 201]. The cyclization is readily followed by measuring the rate of

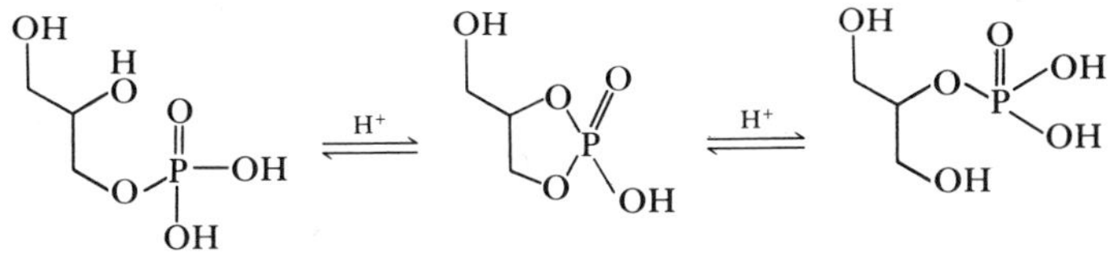

$$\text{(35)}$$

101   **103**   **102**

^{18}O exchange into glycerophosphoric acid when the equilibrium is set up in enriched water. The total migration reaction can be followed by periodate oxidation, which is specific for 1-glycerophosphoric acid [200, 201] or by an nmr method[201] and is several orders of magnitude faster than hydrolysis. Kugel and Halmann[200] found that three of the phosphate oxygens of propan-1,2-diol-1-dihydrogen phosphate are exchanged rapidly with the solvent at 100°, but Fordham and Wang's analysis of their[201] rate data for oxygen-18 exchange into the equilibrium mixture of 1- and 2-glycerophosphoric acids requires a path for migration without oxygen exchange. This is interpreted by the authors in terms of a concerted displacement of the ester oxygen by the attacking hydroxyl group (**104 ⇌ 102**), although conditions would appear to be favorable for the formation of a pentacovalent intermediate (**105**, Scheme 28).

Another polyhydroxyalkyl phosphate to receive considerable recent attention is glucose-6-phosphate (**106**)[202–204]. Two groups have

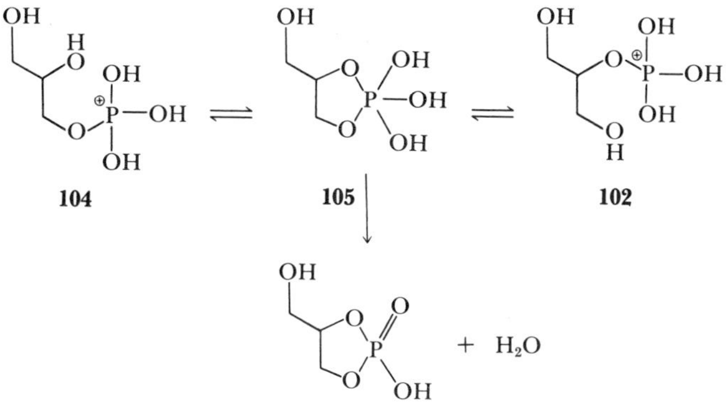

104   **105**   **102**

Scheme 28

studied the hydrolysis of this ester at 100° and found that it is hydrolyzed at a rate normal for a monoalkyl phosphate in the acid and monoanion regions. A faster oxygen-exchange reaction is observed in ^{18}O-enriched water in strong acid, and Degani and Halmann[202] consider that this is a consequence of a migration from the 6- to the 4-position of glucose by way of a six-membered cyclic diester (**107**), similar to the more rapid migrations observed for glycerophosphoric acid in this region (Scheme 29). Unusually for a phosphate ester, the dianion is hydrolyzed five

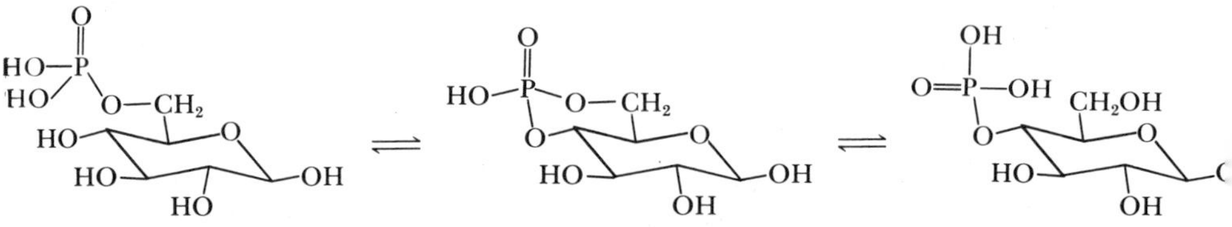

Scheme 29

times as rapidly as the monoanion, and Bunton and Chaimovich[203] suggest a mechanism involving general acid catalysis by the glycosidic hydroxyl (**108**), which is the most strongly acidic hydroxyl group remaining ($pK_a = 10.8$ in glucose). This mechanism requires the molecule

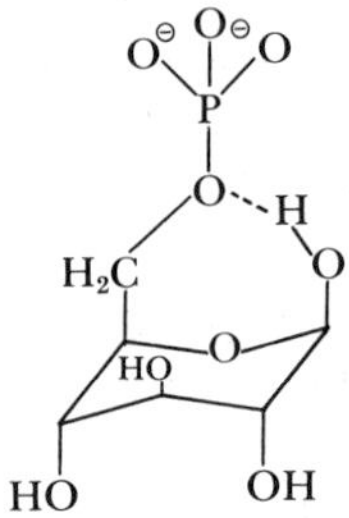

108

to adopt the unfavorable all-axial conformation, but finds a close analogy in the hydrolysis of salicyl phosphate, which is catalyzed by a similar mechanism, and more effectively by more weakly acidic carboxyl groups (see p. 44).

　　Glucose-6-phosphate is hydrolyzed faster still in strongly alkaline solution, and Bunton and Chaimovich[203] suggested a mechanism involving intramolecular catalysis by the hydroxyl group here too. But a careful investigation of the products of hydrolysis, by Degani and Halmann[204] showed that under these conditions the disappearance of glucose-6-phosphate (assayed enzymically, using glucose-6-phosphate dehydrogenase) is faster than the appearance of inorganic phosphate and thus that an intermediate must be involved. Product analysis showed that the slow step of the reaction must be the reversible, base catalyzed rearrangement of glucose-6-phosphate to fructose-6-phosphate. This is converted partly to 6-phosphoglucometasaccharinic acid, which is stable, but mainly cleaved to dihydroxyacetone and glyceraldehyde-3-phosphate. This ester, as is common with other phosphates β to a carbonyl group, releases inorganic phosphate in a rapid, base-catalyzed elimination process[204].

6.2.2 Phosphate Diesters. The most important intramolecular reaction of phosphate esters is that in which the free 2'-hydroxyl group of a ribonucleoside-3'-phosphate diester (**109**) or of a ribonucleic acid molecule carries out the base- or ribonuclease-catalyzed displacement of the second esterifying group to form the ribonucleoside-2',3'-cyclic phosphate (**110**) (eq. 36). Under these conditions no migration of the

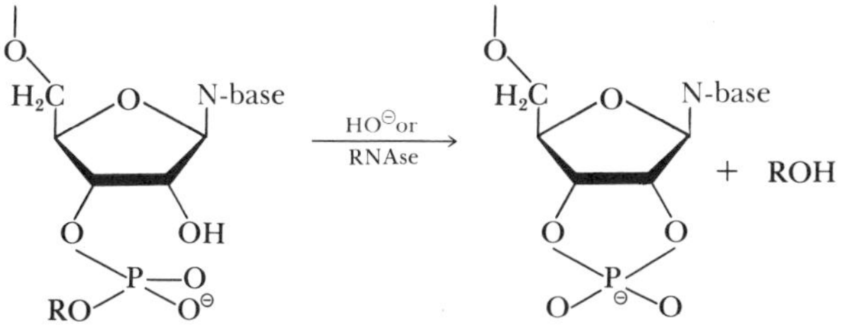

$$\text{(36)}$$

phosphate ester group to the 2'-position is observed[198], and the same result is obtained for the model compound (**111**)[205]. Endocyclic

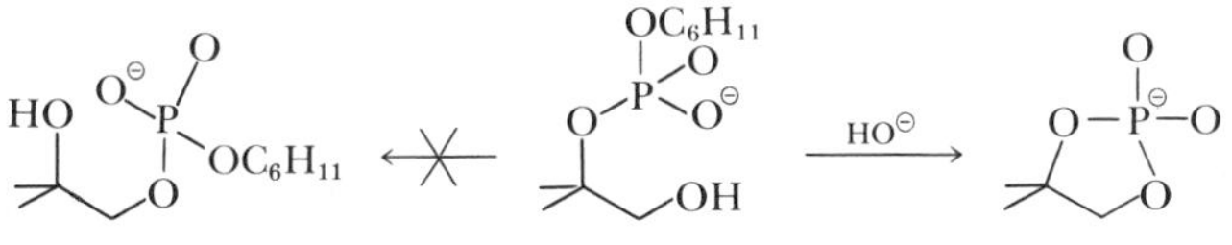

$$\text{(37)}$$

displacement (migration) should be particularly favored in this model, because cyclohexyloxide is a very poor leaving group, much more strongly basic than the glycol oxygen, and this experiment constitutes a stringent test of the stereospecificity of displacement in this system.

Although migration is not observed as a consequence of having such a poor leaving group on phosphorus, a second alternative intramolecular displacement does occur in the alkaline hydrolysis of compounds of the type **111**[205]. Cyclohexyl *erythro*-3-hydroxy-2-butyl phosphate (**112**) on

alkaline hydrolysis at 100° gives cyclohexyl phosphate as the major
product, and no ^{18}O from the enriched solvent appears in this product.
The reaction must therefore involve carbon–oxygen cleavage. The
stereochemistry of the product (meso-butane-2,3-diol, **113**) rules out an
S_N2 displacement by hydroxide, so the hydrolysis involves intramolecular
nucleophilic attack by the neighboring hydroxyl group to form the
epoxide **114**[205]. Brown and Usher also studied the effect of changing

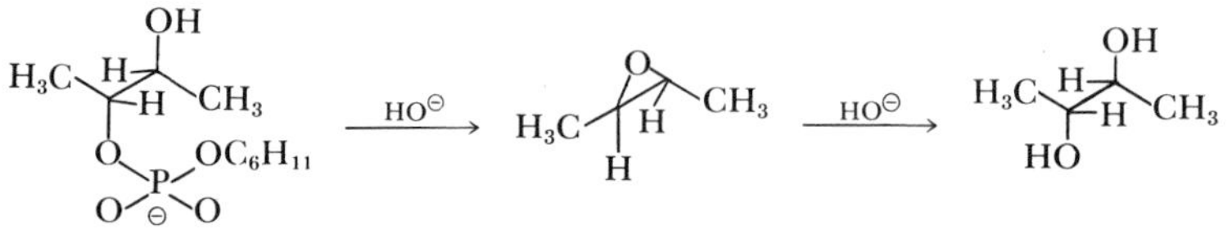

(38)

        112   **114**   **113**

the leaving group on the rate of the exocyclic displacement at phosphorus[206]. They found, as expected, that the rate of the reaction is
faster for the esters of more strongly acidic alcohols and phenols ROH
(eq. 39). The logarithms of the second order rate constants for the

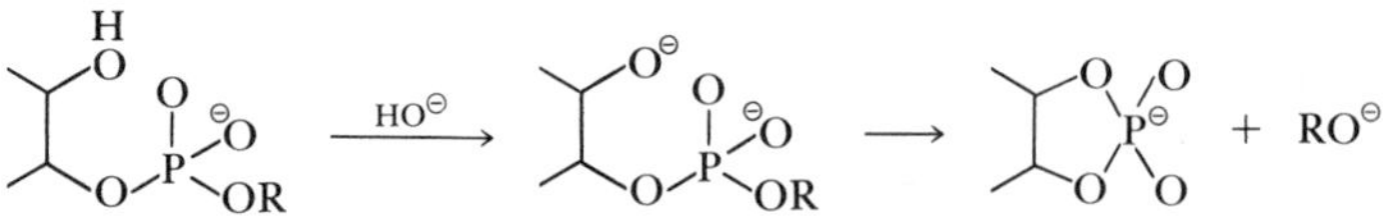

(39)

hydroxide-catalyzed reaction at 80° give a straight line when plotted
against the pK_a of the conjugate acid of the leaving group (ROH)
given by equation 40[206]:

$$\log k_2 = (4.43 - 0.56\, pK_a)\ \text{liter mole}^{-1}\, \text{sec}^{-1} \qquad (40)$$

The sensitivity to the leaving group, as measured by the slope of this
line, is smaller than that observed for the reactions of methyl aryl
phosphate anions with tertiary amines[207], as would be expected if the
transition state is reached early in the bond formation process, and is
consistent with the specific base catalyzed mechanism of equation 39.

More recently Usher and his co-workers[208] have turned their
attention to the cis-tetrahydrofuran-3,4-diol phosphate system (**115**), in
which the stereochemistry is fixed and closely similar to that of a nucleoside phosphate. They find that the elimination of phenol from the
phenyl ester of this phosphate (**115**) is subject to several types of catalysis.
As well as the expected specific acid and hydroxide ion catalysis, general
species catalysis by morpholine and imidazole buffers is also observed.
Data cited for catalysis by morpholine show clearly that the catalytic
species is the conjugate acid (eq. 41). The authors consider that their data

$$+ \text{ PhOH} \qquad (41)$$

115

require an intermediate, which might be a pentacovalent species, on the path to the cyclic ester. A full interpretation of this very interesting reaction must wait until more complete results are available.

REFERENCES

1. T. C. Bruice and S. J. Benkovic, *Bioorganic Mechanisms*, Vol. 1, Benjamin, New York, 1966.
2. D. E. Koshland, *J. Cellular Comp. Physiol.*, **47**, Suppl. 1, 245 (1959).
3. D. E. Koshland, *J. Theoret. Biol.*, **2**, 75 (1962).
4. W. P. Jencks, *Catalysis in Chemistry and Enzymology*, McGraw-Hill, New York, 1969.
5. R. Huisgen and H. Ott, *Tetrahedron*, **6**, 253 (1959).
6. A. J. Parker, *Quart. Rev.*, **1962**, 163.
7. E. Tommila and M.-L. Murto, *Acta Chem. Scand.* **17**, 1947 (1963).
8. E. Tommila, *Suomen Kemistilehti*, **37B**, 117 (1964).
9. W. P. Jencks, *Catalysis in Chemistry and Enzymology*, McGraw-Hill, New York, 1969, p. 20.
10. E. H. Cordes, in *Progress in Physical Organic Chemistry*, Vol. 4, A. Streitwieser, Jr., and R. W. Taft, Eds., Interscience, New York, 1967, p. 1.
11. S. L. Johnson, in *Progress in Physical Organic Chemistry*, Vol. 5, A. Streitwieser, Jr., and R. W. Taft, Eds., Interscience, New York, 1967, p. 237.
12. L. J. Edwards, *Trans. Faraday Soc.*, **46**, 723 (1950); **48**, 696 (1952).
13. A. R. Fersht and A. J. Kirby, *J. Amer. Chem. Soc.*, **89**, 4857 (1967).
14. T. St. Pierre and W. P. Jencks, *J. Amer. Chem. Soc.*, **90**, 3817 (1968).
15. A. R. Fersht and A. J. Kirby, *J. Amer. Chem. Soc.*, **90**, 5826 (1968).
16. M. L. Bender, F, Chloupek, and M. C. Neveu, *J. Amer. Chem. Soc.*, **80**, 5384 (1958).
17. E. R. Garrett, *J. Amer. Chem. Soc.*, **79**, 3401, 5206 (1957); **80**, 4044 (1958); **82**, 711 (1960); *J. Org. Chem.*, **26**, 3660 (1961).
18. A. R. Fersht and A. J. Kirby, *J. Amer. Chem. Soc.*, **89**, 4853 (1967).
19. H. H. Jaffé, *Chem. Rev.*, **53**, 191 (1953); *Science*, **118**, 246 (1953); *J. Amer. Chem. Soc.*, **76**, 4261 (1954).
20. T. C. Bruice and D. W. Tanner, *J. Org. Chem.*, **30**, 1668 (1965).
21. B. Capon and B. C. Ghosh, *J. Chem. Soc.*, *Sect. B.*, **1966**, 472.
22. M. L. Bender, F. J. Kezdy, and B. Zerner, *J. Amer. Chem. Soc.*, **85**, 3017 (1963).
23. W. P. Jencks, *Catalysis in Chemistry and Enzymology*, McGraw-Hill, New York, 1969, p. 11.
24. V. Gold, D. G. Oakenfull, and T. Riley, *J. Chem. Soc.*, *Sect. B*, **1968**, 515.
25. A. R. Fersht and A. J. Kirby, *J. Amer. Chem. Soc.*, **90**, 5818 (1968).
26. T. C. Bruice and U. K. Pandit, *J. Amer. Chem. Soc.*, **82**, 5858 (1960).
27. J. W. Thanassi and T. C. Bruice, *J. Amer. Chem. Soc.*, **88**, 747 (1966).

28. E. Gaetjens and H. Morawetz, *J. Amer. Chem. Soc.*, **80**, 2591 (1958); T. C. Bruice and W. C. Bradbury, *ibid.*, **90**, 3808 (1968).
29. W. P. Jencks and M. Gilchrist, *J. Amer. Chem. Soc.*, **86**, 4651 (1964).
30. W. P. Jencks and J. Gerstein, *J. Amer. Chem. Soc.*, **86**, 4655 (1964).
31. W. P. Jencks, F. Barley, R. Barnett, and M. Gilchrist, *J. Amer. Chem. Soc.*, **88**, 4464 (1966).
32. W. P. Jencks and M. Gilchrist, *J. Amer. Chem. Soc.*, **90**, 2622 (1968).
33. A. R. Fersht and W. P. Jencks, *J. Amer. Chem. Soc.*, **91**, 2125 (1969).
34. D. S. Kemp and T. D. Thibault, *J. Amer. Chem. Soc.*, **90**, 7154 (1968).
35. D. S. Kemp and R. B. Woodward, *Tetrahedron*, **21**, 3014 (1965).
36. R. Paulsen, J. H. Pitman, and T. Higuchi, *J. Org. Chem.*, **34**, 2097 (1969).
37. D. C. Williams and J. R. Whitaker, *Biochemistry*, **7**, 2562 (1968).
38. A. R. Fersht and A. J. Kirby, *J. Amer. Chem. Soc.*, **90**, 5833 (1968).
39. H. Morawetz and I. Oreskes, *J. Amer. Chem. Soc.*, **80**, 2591 (1958).
40. T. C. Bruice and S. J. Benkovic, *Bioorganic Mechanisms*, Vol. 1, Benjamin, New York, 1966, p. 173.
41. M. L. Bender, *J. Amer. Chem. Soc.*, **79**, 1258 (1957).
42. M. L. Bender, Y. Chow, and F. Chloupek, *J. Amer. Chem. Soc.*, **80**, 5380 (1958).
43. B. Zerner and M. L. Bender, *J. Amer. Chem. Soc.*, **83**, 2267 (1961).
44. J. Brown, S. C. K. Su, and J. Shafer, *J. Amer. Chem. Soc.*, **88**, 4468 (1966).
45. M. Eigen, *Angew. Chem. Intern. Ed. Engl.*, **3**, 1 (1964).
46. W. P. Jencks, in *Progress in Physical Organic Chemistry*, Vol. 2, Interscience, New York, 1964, p. 63.
47. R. M. Beesley, C. K. Ingold, and J. F. Thorpe, *J. Chem. Soc.*, **107**, 1080 (1915); C. K. Ingold, *ibid.*, **119**, 305 (1921).
48. E. L. Eliel, *Stereochemistry of Carbon Compounds*, McGraw-Hill, New York, 1962, p. 196 ff.
49. T. Higuchi, L. Eberson and A. K. Herd, *J. Amer. Chem. Soc.*, **88**, 3805 (1966).
50. T. C. Bruice and W. C. Bradbury, *J. Amer. Chem. Soc.*, **87**, 4838, 4846, 4851 (1965).
51. T. C. Bruice and W. C. Bradbury, *J. Amer. Chem. Soc.*, **90**, 3808 (1968).
52. P. J. G. Butler, J. I. Harris, B. S. Hartley, and R. Leberman, *Biochem. J.*, **103**, 78P (1967).
53. H. B. F. Dixon and R. N. Perham, *Biochem. J.*, **109**, 312 (1968).
54. A. J. Kirby and P. W. Lancaster, unpublished work.
55. T. C. Bruice and S. J. Benkovic, *Bioorganic Mechanisms*, Vol. 1, Benjamin, New York, 1966, pp. 212–258.
56. B. W. Matthews, P. B. Sigler, R. Henderson, and D. M. Blow, *Nature*, **214**, 652 (1967).
57. C. C. F. Blake, D. F. Koenig, G. A. Mair, A. C. T. North, D. C. Phillips, and U. R. Sarma, *Nature*, **206**, 757 (1965).
58. C. C. F. Blake, G. A. Mair, A. C. T. North, D. C. Phillips, and U. R. Sarma, *Proc. Roy. Soc. (London)*, **167B**, 365 (1967).
59. D. C. Phillips, *Proc. Natl. Acad. Sci. U.S.*, **57**, 484 (1967).
60. L. N. Johnson and D. C. Phillips, *Nature*, **206**, 761 (1965).
61. D. C. Phillips, *Sci. Am.*, **215**, 78 (1966).
62. C. C. F. Blake, L. N. Johnson, G. A. Mair, A. C. T. North, D. C. Phillips, and U. R. Sarma, *Proc. Roy. Soc. (London)*, **167B**, 378 (1967).
63. D. Piszkiewicz and T. C. Bruice, *J. Amer. Chem. Soc.*, **90**, 2156 (1968).
64. J. A. Rupley and V. Gates, *Proc. Natl. Acad. Sci. U.S.*, **57**, 496 (1967).
65. B. Capon, *Chem. Rev.*, **69**, 407 (1969).
66. D. M. Chipman and N. Sharon, *Science*, **165**, 454 (1969).
67. P. Jollés, *Angew. Chem. Intern. Ed. Engl.*, **8**, 227 (1969).
68. T. H. Fife and L. K. Jao, *J. Amer. Chem. Soc.*, **90**, 4081 (1968).

69. E. Anderson and B. Capon, *Chem. Commun.*, **1969**, 390.
70. R. H. De Wolfe, K. M. Ivanetich, and N. F. Perry, *J. Org. Chem.*, **34**, 848 (1969).
71. E. Anderson and B. Capon, *J. Chem. Soc., Sect. B*, **1969**, 1033.
72. E. Anderson and T. H. Fife, *J. Amer. Chem. Soc.*, **91**, 7163 (1969).
73. N. C. Deno, in *Progress in Physical Organic Chemistry*, Vol. 2, S. G. Cohen, A. Streitwieser, Jr., and R. W. Taft, Eds., Interscience, New York, 1964, p. 129.
74. B. Capon, M. C. Smith, E. Anderson, R. H. Dalm, and G. H. Sankey, *J. Chem. Soc., Sect. B*, **1969**, 1038.
75. M. L. Bender, F. J. Kezdy, and B. Zerner, *J. Amer. Chem. Soc.*, **85**, 3017 (1963).
76. T. C. Bruice and D. Piszkiewicz, *J. Amer. Chem. Soc.*, **89**, 3568 (1967).
77. M. L. Bender and M. S. Silver, *J. Amer. Chem. Soc.*, **85**, 3006 (1963).
78. R. W. Taft, in *Steric Effects in Organic Chemistry*, M. S. Newman, Ed., Wiley, New York, 1956, p. 556.
79. B. Capon, *Tetrahedron Letters*, **1963**, 911.
80. B. Capon and M. C. Smith, *Chem. Commun*, **1965**, 523.
81. O. Smidsrod, A. Haug, and B. Larsen, *Acta Chem. Scand.*, **20**, 1026 (1966).
82. B. Capon and B. C. Ghosh, *Chem. Commun.*, **1965**, 586.
83. D. Piszkiewicz and T. C. Bruice, *J. Amer. Chem. Soc.*, **89**, 6237 (1967).
84. D. Piszkiewicz and T. C. Bruice, *J. Amer. Chem. Soc.*, **90**, 5844 (1968).
85. C. A. Vernon, *Proc. Roy. Soc.* (*London*), **167B**, 389 (1967).
86. M. A. Raftery and T. Rand-Meir, *Biochemistry*, **7**, 3281 (1968).
87. J. A. Rupley, V. Gates, and R. Bilbery, *J. Amer. Chem. Soc.*, **90**, 5633 (1968).
88. F. W. Dahlquist, T. Rand-Meir, and M. A. Raftery, *Proc. Natl. Acad. Sci. U.S.*, **61**, 1194 (1968).
89. T. C. Bruice and S. J. Benkovic, *Bioorganic Mechanisms*, Vol. 1, Benjamin, New York, 1966, p. 334 ff.
90. G. E. Lienhard and T.-C. Wang, *J. Amer. Chem. Soc.*, **91**, 1146 (1969).
91. C. G. Swain, E. C. Stivers, J. F. Reuwer, and L. J. Schaad, *J. Amer. Chem. Soc.*, **80**, 5885 (1958).
92. C. G. Swain, A. J. de Milo, and J. P. Cordner, **80**, 5983 (1958).
93. C. G. Swain and A. S. Rosenberg, *J. Amer. Chem. Soc.*, **83**, 2154 (1961).
94. R. P. Bell and M. A. D. Fluendy, *Trans. Faraday Soc.*, **59**, 1623 (1962).
95. W. J. Albery, R. P. Bell, and A. L. Powell, *Trans. Faraday Soc.*, **61**, 1194 (1965).
96. R. P. Bell and H. F. F. Ridgewell, *Proc. Roy. Soc.* (*London*), **298A**, 178 (1967).
97. M. A. D. Fluendy, *Trans. Faraday Soc.*, **59**, 1681 (1963).
98. E. T. Harper and M. L. Bender, *J. Amer. Chem. Soc.*, **87**, 5625 (1965).
99. J. K. Coward and T. C. Bruice, *J. Amer. Chem. Soc.*, **91**, 5339 (1969).
100. M. L. Bender and J. M. Lawlor, *J. Amer. Chem. Soc.*, **85**, 3010 (1963).
101. R. H. Bromilow and A. J. Kirby, unpublished work.
102. A. J. Kirby and A. G. Varvoglis, *J. Amer. Chem. Soc.*, **89**, 415 (1967).
103. S. J. Benkovic and K. J. Schray, *J. Amer. Chem. Soc.*, **91**, 563 (1969); also to be published.
104. J. Arai, *J. Biochem.* (*Tokyo*), **20**, 474 (1934).
105. V. M. Clark and A. J. Kirby, *J. Amer. Chem. Soc.*, **85**, 3705 (1963).
106. S. A. Khan, A. J. Kirby, M. Wakselman, J. M. Lawlor, and D. P. Horning, *J. Chem. Soc., Sect. B*, **1970**, 1182.
107. A. J. Kirby and M. Younas, *J. Chem. Soc., Sect. B*, **1970**, 1165.
108. F. H. Westheimer, *Acts Chem. Res.*, **1**, 70 (1968).
109. A. J. Kirby and S. G. Warren, *The Organic Chemistry of Phosphorus*, Elsevier, Amsterdam, 1967, pp. 8–10 and 276 ff.
110. S. J. Benkovic and K. J. Schray, *J. Amer. Chem. Soc.*, **91**, 5653 (1969).

111. E. M. Arnett, in *Progress in Physical Organic Chemistry*, Vol. 1, S. G. Cohen, A. Streitwieser, Jr., and R. W. Taft, Eds., Interscience, New York, 1963, p. 223.
112. R. J. Gillespie and T. Birchall, *Can. J. Chem.*, **41**, 148 (1963).
113. R. E. Lyle, *Chem. Eng. News*, **44**, 72 (1966).
114. G. A. Olah and M. Calin, *J. Amer. Chem. Soc.*, **90**, 401 (1968).
115. R. Stewart and L. J. Muenster, *Can. J. Chem.*, **39**, 401 (1961).
116. J. L. Sudmeier and K. E. Schwartz, *Chem. Commun.*, **1968**, 1646.
117. V. C. Armstrong, D. W. Farlow, and R. B. Moodie, *J. Chem. Soc., Sect. B*, **1968**, 1099.
118. R. B. Moodie, *Chem. Commun.*, **1968**, 1366.
119. H. Pracejus, *Chem. Ber.*, **92**, 988 (1959).
120. N. Hall, *J. Amer. Chem. Soc.*, **52**, 5115 (1930).
121. G. Branch and J. Clayton, *J. Amer. Chem. Soc.*, **50**, 1680 (1928).
122. T. C. Bruice and S. J. Benkovic, *Bioorganic Mechanisms*, Vol. 1, Benjamin, New York, p. 187–201.
123. M. T. Behme and E. H. Cordes, *J. Org. Chem.*, **29**, 1255 (1954).
124. C. K. Sauers, *J. Org. Chem.*, **34**, 2275 (1969).
125. E. Sondheimer and R. W. Holley, *J. Amer. Chem. Soc.*, **76**, 2467 (1954); **79**, 3767 (1957).
126. T. C. Bruice and S. J. Benkovic, *Bioorganic Mechanisms*, Vol. 1, Benjamin, New York, 1966, pp. 189–190.
127. J. A. Shafer and H. Morawetz, *J. Org. Chem.*, **28**, 1899 (1963).
128. A. F. Hegarty and T. C. Bruice, *J. Amer. Chem. Soc.*, **91**, 4924 (1969).
129. R. M. Topping, *Chem. Commun.*, **1966**, 698, R. M. Topping and D. E. Tutt, *J. Chem. Soc., Sect. B*, **1967**, 1346; **1969**, 104.
130. F. M. Menger and C. J. Johnson, *Tetrahedron*, **23**, 19 (1967).
131. J. de Jersey, A. A. Kortt, and B. Zerner, *Biochem. Biophys. Res. Commun.*, **23**, 745 (1966).
132. J. de Jersey, M. T. C. Runnegar, and B. Zerner, *Biochem. Biophys, Res. Commun.*, **25**, 383 (1966).
133. J. de Jersey, P. Willadsen, and B. Zerner, *Biochemistry*, **8**, 1959 (1969).
134. R. W. Hay and P. J. Morris, *Chem. Commun.*, **1967**, 663.
135. T. C. Bruice and S. J. Benkovic, *Bioorganic Mechanisms*, Vol. 1, Benjamin, New York, 1966, p. 134.
136. I. Weil and J. C. Morris, *J. Amer. Chem. Soc.*, **71**, 1664 (1949).
137. W. P. Jencks, *Catalysis in Chemistry and Enzymology*, McGraw-Hill, New York, 1969, p. 182 ff.
138. T. C. Bruice and S. J. Benkovic, *J. Amer. Chem. Soc.*, **85**, 1 (1963).
139. G. Aksnes and P. Froyen, *Acta Chem. Scand.*, **20**, 1451 (1966).
140. B. Hansen, *Acta Chem. Scand.*, **16**, 1927 (1962).
141. G. R. Porter, H. N. Rydon, and J. A. Schofield, *J. Chem. Soc.*, **1960**, 2686.
142. B. Hansen, *Acta Chem. Scand.*, **17**, 1307 (1963).
143. R. B. Martin and A. Parcell, *J. Amer. Chem. Soc.*, **83**, 4835 (1961).
144. R. B. Martin, R. I. Hedrick, and A. Parcell, *J. Org. Chem.*, **29**, 3197 (1964).
145. G. L. Schmir, *J. Amer. Chem. Soc.*, **90**, 3478 (1968).
146. A. R. Fersht and W. P. Jencks, *J. Amer. Chem. Soc.*, **92**, 5442 (1970).
147. R. Barnett and W. P. Jencks, *J. Amer. Chem. Soc.*, **91**, 2538 (1961).
148. E. Grunwald, C. F. Jumper, and S. Meiboom, *J. Amer. Chem. Soc.*, **85**, 522 (1963).
149. E. Grunwald and S. Meiboom, *J. Amer. Chem. Soc.*, **85**, 2047 (1963).
150. Z. Luz and S. Meiboom, *J. Amer. Chem. Soc.*, **85**, 3923 (1963).
151. D. Elliott, L. C. Howick, B. G. Hudson, and W. K. Noyce, *Talanta*, **9**, 723 (1962).
152. R. H. Barca and H. Freiser, *J. Amer. Chem. Soc.*, **88**, 3744 (1966).
153. H. D. Jakubke and A. Voigt, *Chem. Ber.*, **99**, 2419 (1966).

154. J. H. Jones and G. T. Young, *Chem. Commun.*, **1967**, 35.
155. S. M. Felton and T. C. Bruice, *Chem. Commun.*, **1968**, 907.
156. S. M. Felton and T. C. Bruice, *J. Amer. Chem. Soc.*, **91**, 6721 (1969).
157. A. R. Fersht, Ph.D. Thesis, Cambridge, 1968.
158. T. C. Bruice, A. Donzel, R. W. Huffman, and A. R. Butler, *J. Amer. Chem. Soc.*, **89**, 2106 (1967).
159. G. M. Blackburn and W. P. Jencks, *J. Amer. Chem. Soc.*, **90**, 2638 (1968).
160. D. Shotton and H. C. Watson, *Phil. Trans. Roy. Soc. London, Ser. B*, **257**, 111 (1970).
161. C. S. Wright, R. A. Alden, and J. Kraut, *Nature*, **221**, 235 (1969).
162. B. S. Hartley and B. A. Kilby, *Biochem. J.*, **50**, 672 (1952).
163. B. S. Hartley and B. A. Kilby, *Biochem. J.*, **56**, 288 (1954).
164. H. Gutfreund and J. M. Sturtevant, *Biochem. J.*, **63**, 656 (1956).
165. H. Gutfreund and J. M. Sturtevant, *Proc. Natl. Acad. Sci. U.S.*, **42**, 719 (1956).
166. F. J. Kezdy and M. L. Bender, *Biochemistry*, **1**, 1097 (1962).
167. R. M. Epand and I. B. Wilson, *J. Biol. Chem.*, **238**, 1718 (1963).
168. G. H. Dixon and H. Neurath, *J. Biol. Chem.*, **225**, 1049 (1957).
169. J. F. Wootton and G. P. Hess, *J. Amer. Chem. Soc.*, **83**, 4234 (1961).
170. P. W. Inward and W. P. Jencks, *J. Biol. Chem.*, **240**, 1986 (1965).
171. M. L. Bender and F. J. Kezdy, *J. Amer. Chem. Soc.*, **86**, 3704 (1964).
172. M. L. Bender and G. A. Hamilton, *J. Amer. Chem. Soc.*, **84**, 2570 (1962).
173. W. P. Jencks, *Ann. Rev. Biochem.*, **32**, 639 (1963).
174. T. H. Fife and J. B. Milstien, *Biochemistry*, **6**, 2901 (1967).
175. J. B. Milstien and T. H. Fife, *J. Amer. Chem. Soc.*, **90**, 2164 (1968).
176. J. B. Milstien and T. H. Fife, *Biochemistry*, **8**, 623 (1969).
177. T. C. Bruice and S. J. Benkovic, *Bioorganic Mechanisms*, Vol. 1, Benjamin, New York, 1966, p. 46.
178. G. L. Schmir and T. C. Bruice, *J. Amer. Chem. Soc.*, **80**, 1173 (1958).
179. U. K. Pandit and T. C. Bruice, *J. Amer. Chem. Soc.*, **82**, 3386 (1960).
180. J. F. Kirsch and W. P. Jencks, *J. Amer. Chem. Soc.*, **86**, 833, 837 (1964).
181. W. P. Jencks and J. Carriuolo, *J. Biol. Chem.*, **234**, 1272 (1959).
182. T. C. Bruice and J. M. Sturtevant, *Biochim. Biophys. Acta*, **30**, 208 (1958).
183. T. C. Bruice, *J. Amer. Chem. Soc.*, **81**, 5444 (1959).
184. T. C. Bruice and J. M. Sturtevant, *J. Amer. Chem. Soc.*, **81**, 2860 (1959).
185. K. L. Kirk and L. A. Cohen, *J. Org. Chem.*, **34**, 390 (1969).
186. P. D. Bartlett and F. D. Greene, *J. Amer. Chem. Soc.*, **76**, 1088 (1954).
187. F. M. Menger and H. T. Brock, *Tetrahedron*, **24**, 3454 (1968).
188. T. C. Bruice and F.-H. Marquardt, *J. Amer. Chem. Soc.*, **84**, 365 (1962).
189. B. A. Cunningham and G. L. Schmir, *J. Amer. Chem. Soc.*, **89**, 917 (1967).
190. B. A. Cunningham and G. L. Schmir, *J. Amer. Chem. Soc.*, **88**, 555 (1966).
191. T. C. Bruice and M. F. Mayahi, *J. Amer. Chem. Soc.*, **82**, 3067 (1960).
192. R. L. Snell, W. K. Kwok and Y. Kim, *J. Amer. Chem. Soc.*, **89**, 6728 (1967).
193. F. M. Menger and J. H. Smith, *J. Amer. Chem. Soc.*, **91**, 5346 (1969).
194. W. Korytnyk and B. Paul, *J. Org. Chem.*, **32**, 3791 (1967).
195. C. S. McLaughlin and V. M. Ingram, *Biochemistry*, **4**, 1448 (1965).
196. B. E. Griffin, M. Jarman, C. B. Reese, J. E. Sulston, and D. R. Trentham, *Biochemistry*, **5**, 5638 (1966).
197. S. M. Kupchan, S. P. Eriksen, and M. Friedman, *J. Amer. Chem. Soc.*, **88**, 343 (1966).
198. D. M. Brown, D. I. Magrath, A. H. Neilson, and A. R. Todd, *Nature*, **177**, 1124 (1956).
199. M. C. Bailly, *Bull. Soc. Chim. France*, **9**, 314, 340, 365 (1942).
200. L. Kugel and M. Halmann, *J. Amer. Chem. Soc.*, **88**, 3566 (1966).

201. W. D. Fordham and J. H. Wang, *J. Amer. Chem. Soc.*, **89**, 4197 (1967).
202. Ch. Degani and M. Halmann, *J. Amer. Chem. Soc.*, **88**, 4075 (1966).
203. C. A. Bunton and H. Chaimovich, *J. Amer. Chem. Soc.*, **88**, 4082 (1966).
204. Ch. Degani and M. Halmann, *J. Amer. Chem. Soc.*, **90**, 1313 (1968).
205. D. M. Brown and D. A. Usher, *J. Chem. Soc.*, **1965**, 6547.
206. D. M. Brown and D. A. Usher, *J. Chem. Soc.*, **1965**, 6558.
207. A. J. Kirby and M. Younas, *J. Chem. Soc., Sect. B*, **1970**, 1165.
208. D. G. Oakenfull, D. I. Richardson, and D. A. Usher, *J. Amer. Chem. Soc.*, **89**, 5491 (1967).

THE PROTON IN BIOLOGICAL REDOX REACTIONS

GORDON A. HAMILTON

Department of Chemistry, The Pennsylvania State University
University Park, Pennsylvania

1 INTRODUCTION

Although a large percentage of the reactions occurring in biological systems involve oxidation or reduction of one or more reactants, the mechanisms of these reactions are generally not understood to the same extent as the mechanisms of most other enzymic reactions. Probably the main reason for this is that chemical reactions analogous to the enzymic redox reactions have usually not been characterized and studied as thoroughly as chemical reactions analogous to other enzymic reactions. During the past few decades physical organic chemists have elucidated in detail the mechanisms of isomerizations, displacements, eliminations, condensations, etc. Although the sequence of steps occurring in nonredox enzymic reactions may be complicated, it is surprising how many of these reactions can be described as some combination of the above well known chemical steps. There are a few exceptions to this, for example, the group of reactions involving the coenzymic form of vitamin B_{12}.

A particularly important type of reaction which appears to occur in the great majority of nonredox enzymic reactions is proton transfer which is aided by general acid and base catalysis. In fact, this type of catalysis is probably responsible for a large amount of the rate increase observed with enzymic reactions when compared to nonenzymic ones. Because redox reactions on the surface appear to be different from other types of reactions, the consensus seems to have arisen that they have different types of mechanisms. It will be the general thesis of this article that the mechanisms of many enzymic redox reactions are closely related to other enzymic reactions and, in particular, that proton transfers and acid and base catalysis probably play as large a role in enzymic redox reactions as in other enzymic reactions.

No attempt will be made in this article to survey possible mechanisms for all enzymic redox reactions. Rather, individual reactions which appear representative and are of present interest to the author will be chosen for discussion. One group of biological redox reactions whose mechanisms will not be considered in detail are those involving oxidation and reduction of metal ion complexes by electron transfer. Many of the reactions discussed involve the oxidation or reduction of relatively stable organic molecules, and in most of these cases there is little evidence for free radical intermediates. It should not be concluded from this that free radicals are not involved in some enzymic redox reactions. There is considerable evidence for such intermediates in a number of enzymic redox reactions. As the present author has stated previously[1], free radical intermediates are reasonable when complex conjugated molecules are involved and the radicals can be stabilized by resonance. However, there are many enzymic redox reactions for which it is unreasonable to suggest free radical intermediates because they would be thermodynamically very unstable, and it is not obvious how the enzyme could localize sufficient energy to form them. In such cases most of the evidence points to an ionic mechanism. It is the purpose of this article to focus attention on possible ionic mechanisms by which these reactions might proceed. In a recent article[1] some such mechanisms for a number of redox reactions catalyzed by metalloenzymes were considered. In the present article related concepts are developed and applied to other enzymic redox reactions (not involving metal ions) in addition to the metalloenzyme catalyzed reactions.

2 CLASSIFICATION OF ORGANIC REDOX REACTIONS

Before beginning a discussion of individual reactions, or groups of reactions, it seems appropriate to define what is meant by redox reactions involving organic molecules and to classify various types of organic redox reactions[1, 2]. According to the classical definition, an oxidation reaction is one in which electrons are removed from an atom or molecule, and a reduction reaction is one in which electrons are added to the atom or molecule. For most inorganic reactions this is a satisfactory definition. For example, ferrous ion is oxidized to ferric ion by removing one electron, and ferric ion is reduced to ferrous ion by adding an electron. In such cases we say that the atom changes its valence when it is oxidized or reduced. The same definition is less satisfactory for most redox reactions of organic compounds. Organic compounds frequently remain uncharged when they have undergone oxidation or reduction. Also, most stable organic compounds have all their electrons paired so that when an oxidation or reduction occurs it is a two-electron, or some multiple of a two-electron, redox reaction. The main elements in organic compounds, namely, carbon, hydrogen, and oxygen, do not change their valence in the same sense as inorganic metal ions and generally maintain the same number of bonds to neighboring elements in the compound even though the compound may be oxidized or reduced. A few elements (for example, nitrogen and sulfur) commonly found in organic compounds do change valence in some redox reactions.

It is clear, therefore, that a definition for organic redox reactions different from that for inorganic reactions is required. Yet a simple definition for the organic reactions does not appear to have been developed. Traditionally, it appears that an organic compound was considered to be oxidized if, during a reaction, a known inorganic oxidizing agent was consumed. This is probably the main reason that the oxidation–reduction terminology, as applied to organic compounds, arose. However, now, many reactions which do not involve inorganic reagents are also considered redox reactions.

Because inorganic oxidizing agents perform a variety of reactions on organic compounds, several different types of organic reactions are considered to be oxidation reactions. Attempts to arrive at an all-inclusive definition for such reactions lead only to complicated statements which are not conceptually helpful. However, all organic oxidation reactions can be included in one or more of four main categories. Therefore, a suitable definition for an organic oxidation reaction is that it can be included in these categories. Organic reductions, being just the reverse of oxidations, are thus also defined. The categories with some representative examples of reactions are given below.

1. Dehydrogenation
  a. Simple dehydrogenation

$$CH_3-CH_2OH \longrightarrow CH_3-CHO + (2H) \tag{1}$$

$$HOOCCH_2CH_2COOH \longrightarrow HOOCCH=CHCOOH + (2H) \tag{2}$$

  b. Dehydrogenation with bond cleavage

$$R_2C-CR_2 \longrightarrow R_2C=O + O=CR_2 + (2H) \tag{3}$$
$$\underset{HO\quad OH}{|\quad\ |}$$

  c. Dehydrogenation with bond formation

$$RSH + HSR \longrightarrow RSSR + (2H) \tag{4}$$

$$\text{``Oxidative coupling'' reactions (ref. 3)} \tag{5}$$

2. Replacement of hydrogen (or sometimes a metal or electropositive element) by a more electronegative element

$$R_3CH \xrightarrow{\text{Br}_2} R_3CBr + HBr \tag{6}$$

$$R_3CH \xrightarrow{[O]} R_3COH \tag{7}$$

$$\text{Phenylalanine} \xrightarrow{[O]} \text{tyrosine} \tag{8}$$

$$R_3CMgBr \xrightarrow{\text{Br}_2} R_3CBr + MgBr_2 \tag{9}$$

3. Addition of electronegative elements with cleavage of a carbon–carbon bond (or bonds)

$$CH_2{=}CH_2 \xrightarrow{\ Br_2\ } BrCH_2CH_2Br \tag{10}$$

$$CH_2{=}CH_2 \xrightarrow{\ [O]\ } CH_2\!\!\diagdown\!\!\underset{O}{}\!\!\diagup\!\!CH_2 \tag{11}$$

$$Catechol \xrightarrow{\ O_2\ } cis,cis\text{-muconic acid} \tag{12}$$

$$CH_3{-}\underset{\underset{O}{\|}}{C}{-}COOH \xrightarrow{\ [O]\ } CH_3{-}COOH + CO_2 \tag{13}$$

4. Addition of electronegative elements to nitrogen, sulfur, etc.

$$R_3N \xrightarrow{\ [O]\ } R_3N^+{-}O^- \tag{14}$$

$$R_2S \xrightarrow{\ [O]\ } R_2SO \xrightarrow{\ [O]\ } R_2SO_2 \tag{15}$$

Most of the oxidation reactions given as examples above are two-electron oxidations, but a couple (for example, the conversions of R_2S to R_2SO_2, and catechol to cis,cis-muconic acid) are four-electron oxidations. Some enzymic reactions which involve a multiple of a two-electron oxidation or reduction must be included in two of the above categories. For example, the reaction catalyzed by lactic acid oxygenase (eq. 16)

$$CH_3{-}\underset{\underset{OH}{|}}{CH}{-}COOH + O_2 \longrightarrow CH_3{-}COOH + CO_2 + H_2O \tag{16}$$

involves both a dehydrogenation (category 1) and the addition of an electronegative element with cleavage of a carbon–carbon bond (category 3). Also, the reaction catalyzed by p-hydroxyphenylpyruvic acid oxidase (eq. 17) must be included in categories 2 and 3. Thus, although all

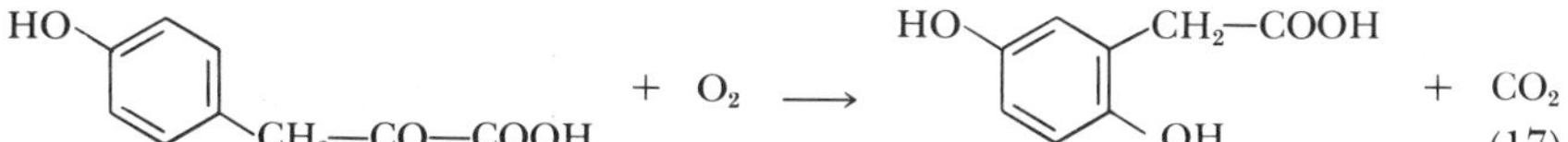

$$HO{-}\langle\text{ring}\rangle{-}CH_2{-}CO{-}COOH + O_2 \longrightarrow HO{-}\langle\text{ring}\rangle\binom{CH_2{-}COOH}{OH} + CO_2 \tag{17}$$

enzymic redox reactions of organic compounds can be included in the above categories, it is not always possible to include each individual reaction in only one category.

3 GENERAL COMMENTS ON MECHANISMS

One should not conclude from the above classification of redox reactions that the mechanisms of the reactions in a given category are related or that mechanisms of reactions in different categories are unrelated. The divisions are not based on any mechanistic distinctions, but rather on stoichiometry. The main purpose of the classification is to clarify what organic redox reactions are. It is, in fact, most likely that several reactions in different categories are mechanistically very similar. For example, it seems probable that alkane hydroxylases and phenylalanine hydroxylase (category 2), epoxidases (category 3), and enzymes catalyzing the formation of amine oxides from amines (category 4) all proceed via an "oxene" mechanism (see Sects. 14 and 15), although the characteristics of the "oxene" may be slightly different in each case.

In considering possible mechanisms for biological redox reactions several early investigators felt that one general mechanism would hold for all such reactions, but they did not agree on what the general mechanism was[1]. In 1954, Westheimer[4] focused attention on the fact that many redox reactions involving organic compounds proceed by ionic mechanisms which are well known in organic chemistry. A classic example is the conversion of an aldehyde and hydroxylamine to a carboxylic acid and ammonia by the steps shown in equation 18. Each

$$RCHO + NH_2OH \xrightarrow{-H_2O} RCH{=}NOH \xrightarrow{-H_2O} R{-}CN$$

$$\xrightarrow{2H_2O} RCOOH + NH_3 \qquad (18)$$

step in this sequence is known to occur by an ionic mechanism. Yet, it is clear that, in the overall transformation, the aldehyde is oxidized to the acid and hydroxylamine is reduced to ammonia. However, it is not immediately obvious in which step the oxidation–reduction has occurred. Another example is the conversion of an alcohol to an aldehyde by nitric acid followed by base (eq. 19). Again each step is

$$RCH_2OH + HNO_3 \xrightarrow{-H_2O} RCH_2{-}ONO_2 \xrightarrow{OH^-} R{-}CHO + NO_2^- \qquad (19)$$

ionic in nature, but clearly an oxidation–reduction reaction has occurred. Since there are many such examples in organic chemistry, it seems reasonable to expect similar mechanisms to be involved in biological redox reactions as well.

4 DEHYDROGENATIONS – GENERAL CONSIDERATIONS

Mechanistically, dehydrogenation–hydrogenation reactions are probably the simplest of redox reactions involving organic compounds. For this reason general mechanisms for these reactions will be considered first. As will be seen, many of the principles developed for such reactions are also applicable to other types of redox reactions. When considering mechanisms for transferring hydrogen from one molecule to another, one should bear in mind that the hydrogen can be transferred as a proton with no electrons, as a hydrogen atom with one electron, or as a hydride ion with two electrons. Each of these species is believed to be involved in various organic reactions. Since dehydrogenation of organic compounds is a two-electron process (i.e., two hydrogen nuclei and two electrons are removed), there are a total of six general mechanisms by which dehydrogenation can be accomplished[1]. Because some of these mechanisms will be referred to frequently in this article, they have been given symbols; in deriving these A has been used to symbolize a hydrogen atom, Hy a hydride ion, and P a proton. The six general mechanisms are given below.

1. The AA mechanism: both hydrogens transferred as hydrogen atoms ($H^{\cdot} + H^{\cdot}$).

2. The PHy mechanism: one hydrogen transferred as a proton and the other as a hydride ion ($H^+ + H^-$).

3. The PP mechanism: both hydrogens transferred as protons, but to maintain electrical neutrality two electrons must be transferred by some other mechanism ($H^+ + H^+ + 2e$).

4. The AP mechanism: one hydrogen transferred as a hydrogen atom, the other as a proton, and one electron by another path ($H^{\cdot} + H^+ + 1e$).

5. The AHy mechanism: one hydrogen transferred as a hydrogen atom, the other as a hydride ion, and one electron by another path $H^{\cdot} + H^- - 1e$).

6. The HyHy mechanism: both hydrogens transferred as hydride ions, and two electrons by another path ($H^- + H^- - 2e$).

The AA mechanism is relatively common in organic chemistry. For example, free radical chain reactions, in which oxidized and reduced products are formed, usually proceed by the AA mechanism. The reduction of various carbonyl compounds by metal hydrides presumably occurs by the PHy mechanism. In both of these, two electrons are transferred with the two hydrogens, and thus there is no requirement for an additional pathway for transferring electrons. However, in each of the other mechanisms, electrons must be transferred in one direction or another in order to maintain electrical neutrality. Thus, for these mechanisms to be possible, there must be some mechanism available for transferring electrons. Pathways by which this might occur will now be elaborated more fully. The discussion will focus especially on the PP mechanism because this mechanism appears to occur frequently in biochemical systems. However, comments on electron transfer, made concerning the PP mechanism, will hold for the AP, AHy, and HyHy mechanisms as well.

The PP mechanism is particularly interesting because both hydrogens are removed or added as protons. Thus, such a reaction can be related to many other acid- and base-catalyzed reactions which take place in aqueous solution. Consider, for example, the tautomerization of acetone to the enol (eq. 20). The reaction is catalyzed by acids and bases, and the

$$
\text{CH}_3\text{—}\overset{\overset{\text{O}}{\|}}{\text{C}}\text{—CH}_2\text{—H} \rightleftharpoons \text{CH}_3\text{—}\overset{\overset{\text{OH}}{|}}{\text{C}}\text{=CH}_2 \tag{20}
$$

mechanism involves the transfer of electrons through the molecule as shown. The point which should be emphasized is that essentially a hydrogen atom has been transferred from a carbon to the oxygen, and thus one can consider this an internal oxidation–reduction reaction; the carbon has been oxidized and the oxygen reduced. This may seem like a far-fetched redox reaction, but if taken one step further it becomes clear that an oxidation and reduction has occurred. For example, in the overall reaction shown in equation 21, a carbonyl group is reduced to an alcohol and at the same time an alcohol is oxidized to a carbonyl group.

$$
\underset{\overset{|}{\text{H}}}{\overset{\text{HO}}{\text{R—C}}}\text{—}\overset{\overset{\text{O}}{\|}}{\text{C}}\text{—R'} \rightleftharpoons \overset{\text{H}}{\underset{}{\text{R—C}}}\overset{\overset{\text{O}}{\diagdown}}{=}\overset{\overset{\text{OH}}{|}}{\text{C}}\text{—R'} \rightleftharpoons \overset{\overset{\text{O}}{\|}}{\text{R—C}}\text{—}\underset{\overset{|}{\text{H}}}{\overset{\overset{\text{OH}}{|}}{\text{C}}}\text{—R'} \tag{21}
$$

This isomerization occurs readily in the absence of enzymes with either acid or base catalysis. However, there are also a large number of related enzymic reactions (for example, isomerizations involving aldose and ketose interconversions). It is quite clear that both the enzymic [5] and nonenzymic reactions proceed by ionic mechanisms; the internal hydrogenation–dehydrogenation reaction occurs by two successive proton and electron transfers as shown. This is one intramolecular example of the PP mechanism.

A related example is that shown in equation 22. A molecule such as the azocatechol (1) would be expected to be in equilibrium with the

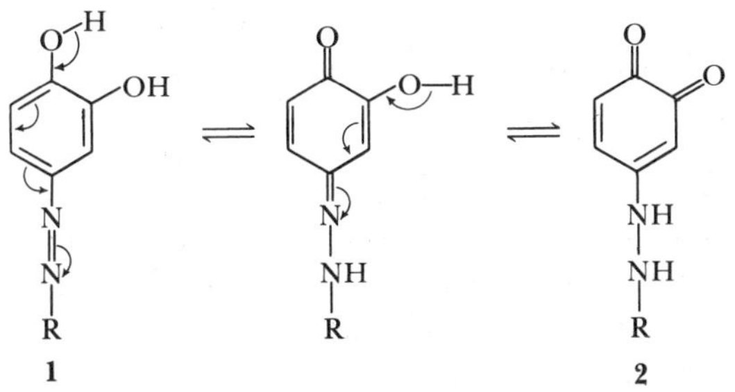

$$(22)$$

hydrazocatechol quinone (**2**). It is clear that this involves an oxidation–reduction; the catechol part of the molecule has been oxidized to catechol quinone and the azo part has been reduced to the hydrazo oxidation level. In hydroxylic solvents the equilibrium shown in equation 22 should be approached extremely rapidly because proton transfers (acid and/or base catalyzed) to or from oxygen or nitrogen are very rapid under such conditions[6] and the electrons can be transmitted readily through the pi system of the molecule. This is another intramolecular example of the PP mechanism. However, most redox reactions are intermolecular, involving two or more molecules. Can the same mechanism apply to the equilibrium shown in equation 23? In hydroxylic solvents the protons can still be transferred readily, but since

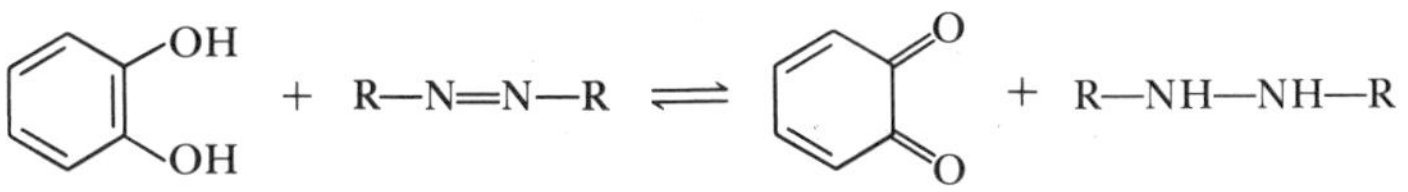

$$(23)$$

the molecules are not connected by a pi system, the mechanism for transferring electrons, shown in equation 22, is not available to this intermolecular case.

One could imagine the equilibrium shown in equation 23 being established by the AA or PHy mechanism. Such possibilities are shown in equations 24 and 25, respectively. In equation 24 both hydrogens are

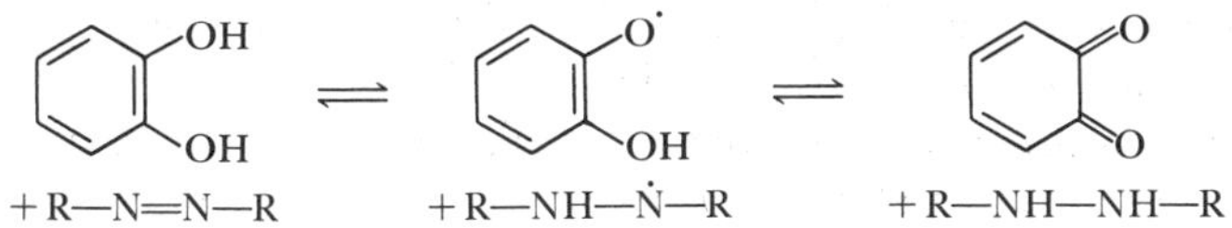

$$(24)$$

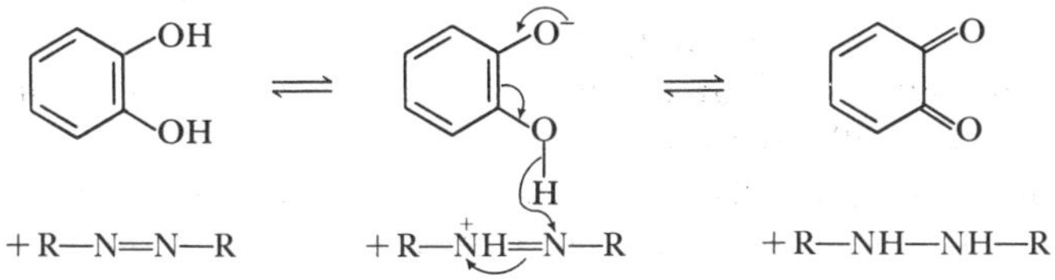

$$(25)$$

transferred as hydrogen atoms (as required by the AA mechanism), and in equation 25 one hydrogen is transferred as a proton and the other as a hydride ion (as required by the PHy mechanism). These are not completely unreasonable mechanisms for this reaction, but one would not expect them to occur as readily as a mechanism similar to that of equation 22 if it were possible. Hydrogen atom and hydride ion transfers can occur, but they are relatively slow compared with proton transfers between oxygen and nitrogen acids and bases (See Sect. 16). The latter usually occur at a rate which is diffusion controlled [6].

There appear to be three general ways by which the equilibrium shown in equation 23 could be established by the PP mechanism — both hydrogens being transferred as protons. One possibility which will be referred to as the PPR mechanism (R symbolizing radical) is outlined in equation 26. Free radicals, very similar to those required by the AA mechanism

$$
\begin{array}{c}
\underset{\text{+ R—N=N—R}}{\text{(catechol, OH, OH)}} \rightleftharpoons \underset{\text{+ R—\overset{+}{N}H=N—R}}{\text{(O}^-\text{, OH)}} \rightleftharpoons \underset{\text{+ R—NH—\overset{\cdot}{N}—R}}{\text{(O}^{\cdot}\text{, OH)}} \\
\\
\underset{\text{+ R—NH—NH—R}}{\text{(O, O)}} \rightleftharpoons \underset{\text{+ R—NH—\overset{+\cdot}{N}H—R}}{\text{(O}^{\cdot}\text{, O}^-\text{)}}
\end{array}
\tag{26}
$$

(eq. 24), would be necessary intermediates. The difference between the mechanisms shown in equations 24 and 26 is that in 24 the hydrogens are transferred as hydrogen atoms, whereas in 26 both hydrogens are transferred as protons, and then electrons are transferred in other steps. The sequence of proton and electron transfers could be different from that shown in equation 26 without changing the basic mechanism. In order for either mechanism (eq. 24 or 26) to be possible, the intermediate radicals must be relatively stable. For the example shown, the semiquinone radicals are fairly stable, but the nitrogen radicals would probably require some resonance stabilization by the R groups (for example, if R was an aryl group) to be able to form under physiological conditions. In cases where organic free radicals are formed in biological redox reactions it appears that they usually form by the PPR mechanism, but the AA mechanism cannot be eliminated in some cases. Some examples will be considered later.

Another way by which the equilibrium, shown in equation 23, could be established by transferring both hydrogens as protons is shown in equation 27. This will be referred to as the PPC mechanism (C symbolizing covalent compound). In this mechanism free radical intermediates are not involved, but an intermediate compound must be formed. The

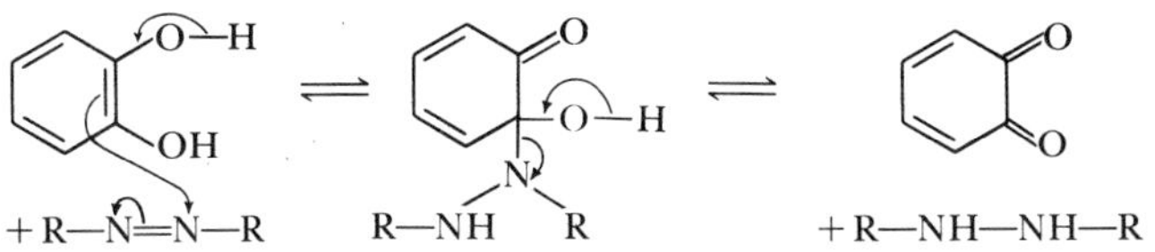

$$(27)$$

formation and subsequent decomposition of this intermediate provide a pathway for electron transfer from one reactant to the other, as shown in equation 27.

In the third mechanism for establishing the equilibrium shown in equation 23 by proton transfers, a metal ion is necessary as a catalyst. This will be referred to as the PPM mechanism (M symbolizing metal ion). The present author has considered this mechanism in detail elsewhere[1]. With a metal ion (for example, M^{2+}) present, the mechanism shown in equation 28 is possible. Metal ions (especially transition metal

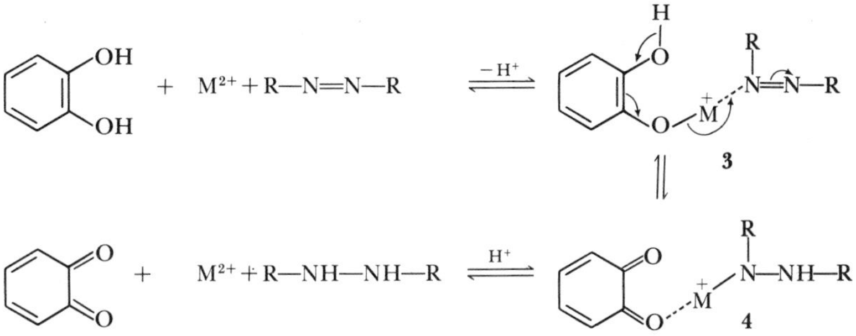

$$(28)$$

ions) readily form complexes similar to **3** in hydroxylic solvents. In the conversion of **3** to **4**, the catechol is oxidized and the azo compound reduced by the simple transfer of a proton from the oxygen to the nitrogen through the solvent and the transfer of electrons through the complex as shown. There is a considerable body of evidence[1] that such electron transfers through transition metal complexes are very rapid reactions. By forming pi-bonds with both ligands, the metal ion effectively extends the conjugation so that one pi-orbital overlaps both ligands. As in the case of the PPC mechanism, this mechanism for dehydrogenation also does not require organic free radicals as intermediates.

The above examples have been used for illustration purposes, especially to clarify the various possibilities for the PP mechanism. As indicated earlier, this particular mechanism seems to occur very frequently in biological hydrogenation–dehydrogenation reactions, and some actual examples will now be considered. It can be concluded from these considerations that, for the PP mechanism to be possible, one of the following must occur:

1. Free radicals must be intermediates (the PPR mechanism).
2. A covalent compound formed from the oxidant and reductant must be an intermediate (the PPC mechanism).
3. A metal ion must be present as a catalyst (the PPM mechanism).

If free radicals are unlikely, and the experimental evidence suggests they are not present, then a metal ion is required or an intermediate covalent compound is indicated. If, in such a case, metal ions are known to be absent, it is most likely that a covalent compound is an intermediate. Most of the actual examples to be discussed are those in which free radicals are not indicated as intermediates, and, thus, presumably they proceed by metal ion catalysis or through the intermediacy of a covalent compound.

5 PYRIDOXAL PHOSPHATE MEDIATED HYDROGENATION–DEHYDROGENATIONS

Two different reactions which are catalyzed by enzymes requiring pyridoxal phosphate as a coenzyme can be classified as hydrogenation–dehydrogenation reactions. These are the transaminase and amine oxidase catalyzed reactions. The transaminase reaction does not require a metal ion, but the amine oxidase reaction does. Usually, transamination is not thought of as a redox reaction, but it is such a reaction. Since it is clearly an example of the PPC mechanism in a biological reaction, it will be discussed here mainly for illustration purposes; the general mechanism has been known for some time [7, 8].

The transamination reaction is shown in equation 29. In order to convert **5** to **7**, a dehydrogenation must occur; **7** can be thought of as

$$R\text{—}\underset{\underset{\textbf{5}}{\overset{|}{NH_2}}}{CH}\text{—}COOH + R'\text{—}\underset{\underset{\textbf{6}}{\overset{\|}{O}}}{C}\text{—}COOH \rightleftharpoons R\text{—}\underset{\underset{\textbf{7}}{\overset{\|}{O}}}{C}\text{—}COOH + R'\text{—}\underset{\underset{\textbf{8}}{\overset{|}{NH_2}}}{CH}\text{—}COOH \quad (29)$$

arriving from the imine (**9**) by hydrolysis (eq. 30), and certainly **9** is the dehydrogenation product of **5**. For the same reasons, the conversion of **6** to **8** requires a hydrogenation. Of course, the overall process is complicated by the transfer of the nitrogen from one molecule to the other,

$$R—\underset{\underset{NH}{\|}}{C}—COOH + H_2O \rightleftharpoons R—\underset{\underset{O}{\|}}{C}—COOH + NH_3 \tag{30}$$

$$\qquad\quad\mathbf{9}\qquad\qquad\qquad\qquad\mathbf{7}$$

but that does not change the fact that a hydrogenation and a dehydrogenation have occurred.

It is now quite clear that an intermediate in the transaminase-catalyzed reactions is pyridoxamine phosphate (**11**). Thus, **5** reacts with the enzyme-bound pyridoxal phosphate (**10**) to give **7** and pyridoxamine

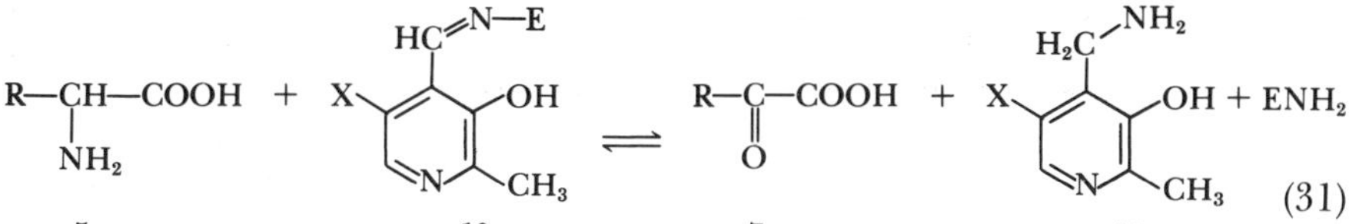

$$\mathbf{5}\qquad\qquad\qquad\mathbf{10}\qquad\qquad\mathbf{7}\qquad\qquad\mathbf{11} \tag{31}$$

phosphate (eq. 31), and **6** subsequently reacts with the pyridoxamine phosphate to give **8** and regenerate the enzyme-bound coenzyme (this last reaction is just the reverse of eq. 31 with a different α-ketoacid). For the same reasons as indicated in the previous paragraph, the reaction shown in equation 31 is also a hydrogenation–dehydrogenation reaction. Therefore, the overall transamination shown in equation 29 is a sequence of two dehydrogenation–hydrogenations. However, the mechanism of each of these is the same because one is just the reverse of the other with a different R group.

The generally accepted mechanism for the enzymic reaction shown in equation 31 is outlined in equation 32[7, 8]. The resting enzyme apparently has the pyridoxal phosphate covalently attached by a Schiff base linkage as indicated in **10**[9]. The reaction of the amino acid with **10** to give **12** is certainly an ionic reaction; in the overall process a proton is transferred from the amino group of **5** to the Schiff base nitrogen, and the amino group carries out a nucleophilic attack on the Schiff base carbon. Although it is not indicated in equation 31, this step could be catalyzed by suitably placed acid or base groups on the enzyme. The conversion of **12** to **13** is also an ionic reaction; again a proton is transferred from one nitrogen to the other, and this might also be catalyzed by groups on the enzyme. Thus, in the formation of **13** from the reactants all the hydrogens originally on the amino group of **5** are transferred as protons. There is a considerable body of evidence that the hydrogen originally on the α-carbon of the amino acid is also transferred as a proton[10–12]. Thus, acid and base groups on the enzyme

$$(32)$$

presumably catalyze the conversion of **13** to **14** and **14** to **15** as shown in equation 32. There is some doubt that **14** is an obligatory intermediate, and possibly the functions of two or more of the enzymic acid and base groups indicated in equation 31 are carried out by a single group. However, there can be little doubt that the basic character of the conversion of **13** to **15** is as indicated; namely, the hydrogen is transferred from one carbon to the other as a proton. The subsequent reactions of **15** with H_2O to give **16**, and of **16** to give **11** and **7** are well known ionic reactions. As before, all hydrogen transfers are accomplished by transferring protons, and one or both of these steps may be acid and/or base catalyzed by the enzyme.

Therefore, there is compelling evidence that the dehydrogenation–hydrogenation reaction catalyzed by transaminases occurs completely by proton shifts through the solvent and by electron shifts through a covalent compound formed between the oxidant and reductant. Consequently, this is a very definite enzymic example of the PPC mechanism discussed in the previous section. The enzyme presumably increases the rate of this redox reaction by having suitably placed acidic and basic groups which can donate or extract a proton at the appropriate stages of the reaction. This is closely related to the mechanism by which other enzymes increase the rates of other ionic reactions, such as those catalyzed by hydrolases.

The amine oxidases which require pyridoxal phosphate as a coenzyme catalyze reactions of the stoichiometry shown in equation 33. Several of these enzymes with differing specificities, and from a number of sources, have been characterized[13–20]. In each case the presence of

$$RCH_2NH_2 + O_2 + H_2O \longrightarrow R\!\!-\!\!CHO + NH_3 + H_2O_2 \qquad (33)$$

Cu(II) is required for catalytic activity. Since the products, RCHO and NH_3, can be thought of as arising by hydrolysis of $RCH\!\!=\!\!NH$ and since O_2 is hydrogenated to H_2O_2, the reaction is a dehydrogenation–hydrogenation reaction. Among pyridoxal phosphate-containing enzymes, the amine oxidases are unique because they require a transition metal ion as well as the pyridoxal coenzyme for activity. Since these are also the only pyridoxal phosphate enzymes which use O_2 as a reactant, one suspects that the copper is probably important in facilitating the transfer of hydrogen to O_2. It is known from electron spin resonance measurements that the copper on the amine oxidases is in the Cu(II) state and apparently does not change its valence during the enzymic catalysis[18, 21–23].

A model system for these enzymic reactions has been studied extensively in the author's laboratory[24, 25] and by Hill and Mann[26]. It is observed that many amino acids and amino acid derivatives are oxidized readily by O_2 at room temperature and pH 9 to α-keto derivatives and ammonia when catalytic amounts of pyridoxal and some metal ions, especially Mn(III) or Co(II), are present. Schirch[27] has observed that with some substrates Cu(II) is an effective catalyst in the model system, and in the author's laboratory Gillis[28] has recently found that Cu(II) is a good catalyst for the amino acid oxidation at pH's > 11. At pH 9, however, Cu(II) is essentially inactive in this system. From a detailed kinetic and product study, especially of the Mn(III)-catalyzed model reaction, the following were concluded: (1) the metal ion does not change its valence during the catalysis, (2) the reaction does not proceed by a free radical chain mechanism, (3) pyridoxamine is not an intermediate in the oxidation, and (4) the mechanism is related to mechanisms proposed for other pyridoxal-catalyzed reactions[7, 8]. A mechanism for the model reaction consistent with all the experimental information was proposed [24, 25]. Since the model and enzymic reactions seem similar, a related mechanism for the enzymic catalysis was suggested[1]. This is shown in equation 34.

As with other pyridoxal phosphate enzymes, the pyridoxal phosphate is probably attached to the enzyme by a Schiff base linkage. For mechanistic reasons, it is suggested that the phenolic hydroxyl group of the pyridoxal phosphate is bound to the Cu(II) which is attached to the enzyme. The reaction of the amine with the enzyme to give **17** presumably occurs by an ionic mechanism as previously indicated for the formation of **13** (eq. 32). There is direct evidence that the Schiff base (**17**) of the substrate with the enzyme-bound pyridoxal phosphate is formed [29]. The conversion of **17** to **18** is similar to the conversion of **13** to **14** in the transaminase-catalyzed reaction (eq. 32). Again, suitably placed acid and base groups on the enzyme would be expected to increase the rate of this step, as well as that of the earlier step in this sequence. It is suggested that **18**, or an intermediate like **18**, can complex with O_2 to give **19**, which by a shift of a proton through the solvent and electrons through the pi-system of the complex as shown would give **20**. The subsequent steps to give H_2O_2, RCHO, and NH_3 and regenerate the amine oxidase would be expected to occur rapidly. Each of these steps is ionic and requires one or more proton transfers. The sequence of some of the steps shown in equation 34 could be changed without altering the basic mechanism significantly.

In this mechanism the amine is essentially oxidized when **18** is formed; if the Schiff base part of **18** hydrolyzed, the aldehyde and a quinoid tautomer of enzyme-bound pyridoxamine phosphate would be formed. The formation of **18** occurs by a mechanism exactly the same as in the

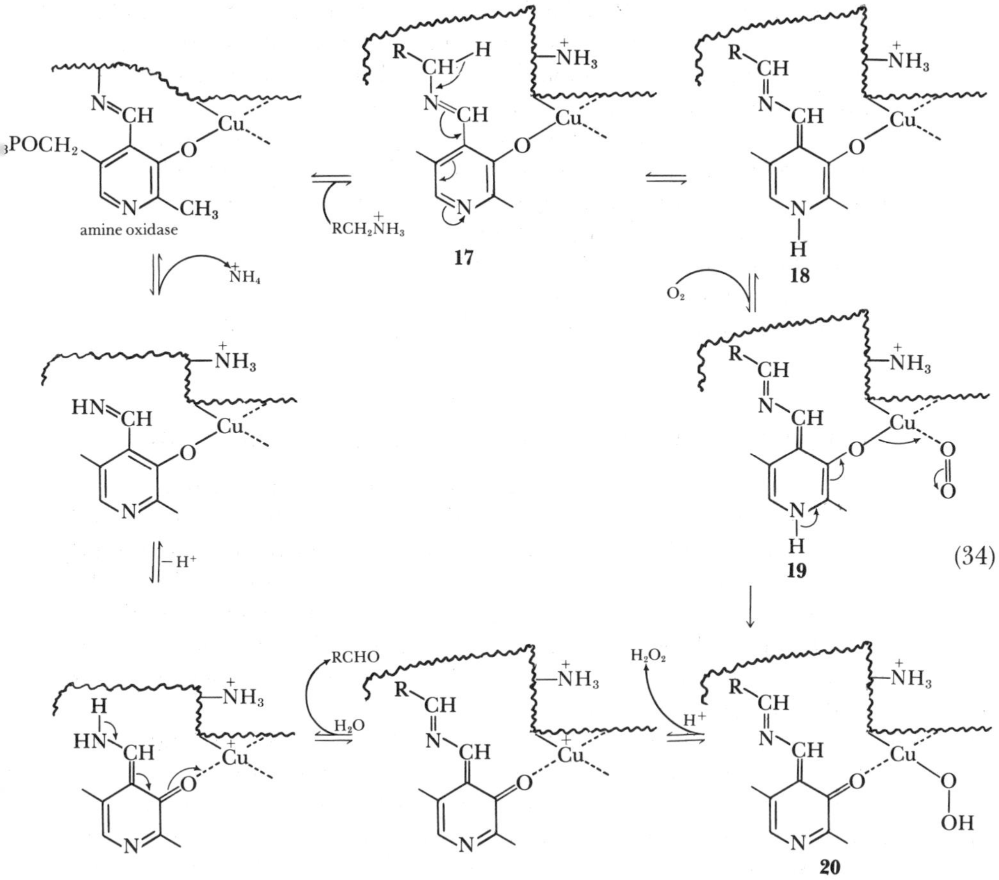

(34)

transaminase-catalyzed reaction, i.e., by the PPC mechanism. It is not suggested in equation 34 that a proton gets on the carbon alpha to the pyridine ring to give a pyridoxamine derivative similar to 15 (eq. 32); it is known in the model system that pyridoxamine is not an intermediate. Rather it is suggested that the quinoid tautomer of the pyridoxamine derivative (18) is oxidized directly in the reaction with O_2. It is reasonable that a compound like 15 would not be an intermediate because there would not necessarily be any benefit to adding a proton to the carbon in one step and removing it in the next. Proton transfers to and from carbon are very slow relative to proton transfers to or from oxygen and nitrogen bases and acids (see Sect. 16). Thus, the subsequent oxidation step can occur more rapidly with the quinoid tautomer (18) than it can with a pyridoxamine derivative related to 15.

In the conversion of **19** to **20** the quinoid tautomer is oxidized and the complexed oxygen is reduced. Since this occurs merely by the shift of a proton through the solvent and electrons through the pi-system of the complex, this is an example of the PPM mechanism. By the time **20** is formed the redox part of the reaction has been completed; the subsequent steps merely give the products and the enzyme by well known ionic reactions. Thus, the overall hydrogen transfer in the amine oxidase catalyzed dehydrogenation–hydrogenation appears to occur by a combination of the PPC and PPM mechanisms. Of course, these are very similar because they both involve proton and electron transfers, but the method for transferring electrons is somewhat different in each case.

6 ROLE OF METAL IONS IN THE PPM MECHANISM INVOLVING O_2

One may inquire why the transfer of hydrogen to molecular oxygen in these amine oxidase reactions requires a metal ion as a catalyst and proceeds by the PPM mechanism. Why does O_2 not form a compound with the reducing agent and oxidize it by the PPC mechanism? The reason for this is that O_2 in the ground state is a triplet molecule — it has two unpaired electrons. If O_2 reacted directly with some organic compound in the absence of a metal ion, then a triplet molecule would be the initial product. Essentially all organic molecules are singlets (they have all their electrons paired) and the reaction of a singlet with a triplet to give a singlet molecule is a spin-forbidden process. A triplet is a diradical and would not be expected to form readily unless the radicals could be stabilized by resonance in a highly conjugated system. Of course, if a triplet molecule was the initial product it could go eventually to a more stable singlet by spin inversion. However, the time required for spin inversion is long (1 to 10^{-9} sec) relative to the time in which a chemical reaction occurs (approximately 10^{-13} sec) when the reactants have sufficient energy to react. Therefore, the spin inversion cannot accompany the initial reaction of O_2 with the organic compound; the greater stability of a singlet product cannot be obtained until after spin inversion, and thus O_2 cannot react readily with most organic molecules at physiological temperatures in the absence of a transition metal ion. This is probably the main reason that O_2 can exist in the atmosphere in the presence of organic compounds, for example, people. In terms of thermodynamics the reaction of O_2 with organic compounds is very exothermic. However, at ordinary temperatures there is no mechanism for the reaction because O_2 in the ground state is a triplet molecule.

Oxygen can be converted from its ground triplet state to a singlet state if energy is supplied, usually in the form of light in the presence of a photosensitizer. Such singlet state O_2 has a reactivity which is quite different from that of triplet O_2[30]. For example, singlet O_2 reacts very rapidly (in a fraction of a second) with alkenes at room temperature, and it is clear that this is not a free radical reaction. However, it is unlikely that singlet O_2 is involved in too many biological reactions because the first singlet state is 22 kcal/mole higher in energy than ground state triplet. It is not apparent how an enzyme could supply electronic energy of that magnitude. Furthermore, few enzymic reactions involving O_2 show any evidence for singlet oxygen being involved.

It is possible to obtain ionic reactions of O_2 without requiring the formation of singlet O_2 if the O_2 is complexed to a transition metal ion which itself has unpaired electrons[1]. In such complexes the d-orbitals of the metal ion overlap with the p-orbitals on oxygen, and consequently one cannot speak of the number of unpaired electrons on O_2, but only of the number on the complex as a whole. Therefore, an ionic reaction such as the conversion of **19** to **20** is possible if the initially formed compound, **20**, has the same number of unpaired electrons as the reactant, **19**. The O_2 is not reacting as singlet O_2, but, because it is complexed to a transition metal ion, it can react by an ionic mechanism as shown. For the above reasons, therefore, the hydrogenation of oxygen in equation 34 is probably forced to occur by the PPM mechanism rather than the PPC mechanism.

7 FLAVIN-MEDIATED HYDROGENATION–DEHYDROGENATIONS

One or another of the flavin coenzymes (FMN or FAD) participates in an enormous number of enzymic redox reactions[31–34]. In some such reactions metal ions are also required, while in others they are not necessary. Free radicals are definitely intermediates in some cases, while in others, the evidence strongly indicates that free radicals are not involved. Given this type of evidence it seems quite clear that there is no one general mechanism for flavin reactions. In fact, the central position which flavoenzymes occupy in many biological pathways may be because of this versatility of mechanism. A metabolic pathway, for example, faced with the necessity of using a reductant which can only react by an ionic (or 2-electron) mechanism and requiring the subsequent steps to occur in one-electron stages would need a coenzyme which can be reduced by an ionic mechanism and then be reoxidized by one-electron steps. Such appears to be the role frequently played by the flavin coenzymes.

There can be no doubt that the flavin coenzymes are capable of being oxidized and reduced in one-electron stages[35–37]. Thus, the semi-quinone form, FH (see eq. 35), has been observed by various spectral methods. In equation 35 and in the following illustrations, R refers to the side chain in either FMN or FAD; there appears to be little effect of this side chain on the chemistry or biochemistry of the flavin part of the molecule. Depending on the conditions, FH is present in varying

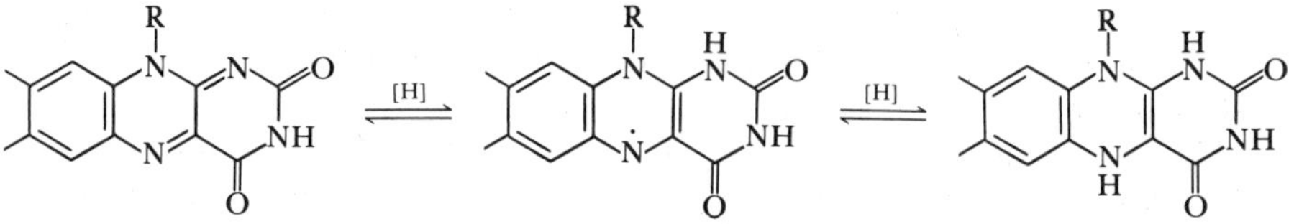

$$\text{(35)}$$

amounts at equilibrium when F and FH_2 are mixed in solution or are present on an enzyme. Presumably the semiquinone form is relatively stable because the odd electron can be delocalized throughout the conjugated system of the molecule. In equation 35 only the neutral forms of the various oxidation states of the flavin are shown, but depending on the pH of the solution, ionized forms are also possible. It is not clear what the mechanism of the reaction of F with FH_2 to give FH is. It may involve a hydrogen atom transfer, but more likely would proceed in two steps: a proton transfer followed by an electron transfer or vice versa.

In flavoenzymes, which also contain transition metal ions that are readily oxidized and reduced by one electron, there is considerable evidence that the reaction of the oxidized form of the metal [e.g., Fe(III)] with FH or FH_2 to give F or FH, respectively, [as well as Fe(II)], is a very facile reaction. Most probably such reversible reactions occur by electron transfer to or from the metal ion and subsequent or prior proton transfer to the medium. Electron transfer of this type certainly would be expected to occur very readily if the metal ion is complexed to the flavin [1]. In some model systems the evidence indicates that, when a metal ion is complexed to flavin, the electron transfer occurs so readily that a completely delocalized system is formed, i.e., the extra electron is in a molecular orbital which overlaps both the flavin pi-system and the metal ion orbitals[36, 37]. Such one-electron transfers from a reduced form of flavin (FH or FH_2) to transition metal ions appear to be very important in a number of flavoenzyme reactions, for example, in the feeding of electrons into the oxidative phosphorylation pathway.

Although the flavin semiquinone (FH) is a well defined and relatively stable species which is certainly involved in some flavin reactions, there is little evidence that it is an intermediate in some other flavoenzyme reactions [31–33, 38]. In particular, many dehydrogenations of the type shown in equation 36 appear to occur without the semiquinone being an intermediate. In Table 1 are listed some typical examples of substrates

TABLE 1 SOME CLASSES OF SUBSTRATES DEHYDROGENATED BY FLAVOENZYMES

AH_2	A	Enzyme Examples
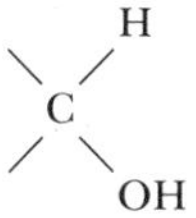	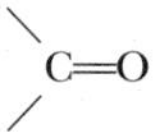	Glucose oxidase, lactate dehydrogenase, choline dehydrogenase
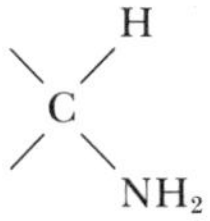	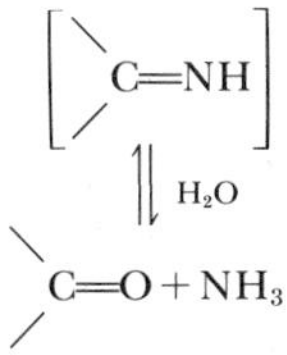	Amino acid oxidases, amine oxidases
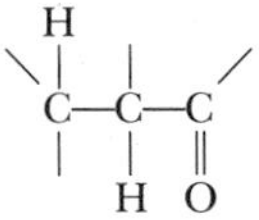	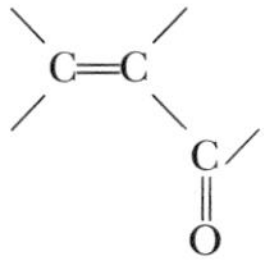	Acyl-CoA dehydrogenases, succinate dehydrogenase, ketone-α,β-dehydrogenases
$NADH + H^+$	NAD^+	NADH dehydrogenases
		Lipoamide dehydrogenase, thioredoxin reductase, glutathione reductase

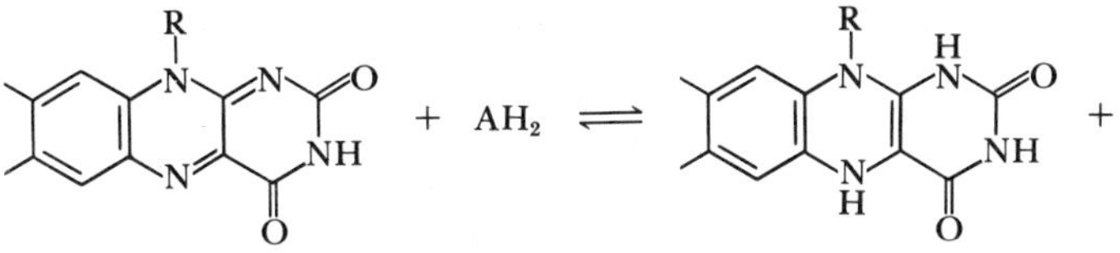

which are dehydrogenated by flavoenzymes. In the enzymic reactions, FH_2 is then reoxidized to F by some oxidizing agent. Sometimes O_2 is the oxidant, either reacting directly with FH_2 to give F and H_2O_2 or indirectly (for example, the oxidative phosphorylation pathway) to give F and H_2O. Other reagents, such as oxidized transition metal complexes, react with the FH_2 in other cases. Probably many of these reoxidation reactions occur by the electron transfer mechanism discussed above, with the semiquinone form of flavin being an intermediate. However, for the reactions given in equation 36 and Table 1, the semiquinone form is apparently not an intermediate, and the reactions must therefore occur by another pathway.

It is reasonable that the reactions given in Table 1 would not occur by a radical mechanism. Although the flavin semiquinone is a stable species, the radicals formed from most of the different AH_2's would not be stable. F and FH have comparable stabilities, but the difference between the energy of AH_2 and AH is enormous, so the reaction given by equation 37 could not occur under ordinary conditions. Thus, a mechanism not involving AH as an intermediate seems indicated.

$$F + AH_2 \longrightarrow FH + AH \qquad (37)$$

Some model reactions related to the enzymic reactions shown in Table 1 have been reported. Thus, it has been known for some time that NADH will react with F under physiological conditions to give NAD^+ and FH_2[39–42]. Fox and Tollin[42] have recently studied this reaction in some detail and came to the conclusion that FH is not an intermediate in the model reaction; apparently F is reduced by NADH to FH_2 by some ionic mechanism. Gascoigne and Radda[43] observed that dithiols are oxidized to disulfides by F, and the evidence[43, 44] indicates that FH is not an intermediate. Very recently Weatherby and Carr[45] reported that dihydrophthalate derivatives are dehydrogenated by F in the presence of base. This can be considered a model reaction for succinate dehydrogenase and acyl-CoA dehydrogenases. Again, no evidence for radical intermediates was presented. In the author's laboratory, Brown[46] has observed that, in the presence of base catalyst, certain amino acid derivatives react rapidly with F in organic solvents to give reduced forms of flavin and, presumably, α-imino acid derivatives (α-ketoacids are identified on workup in water). Under the same conditions, Brown has found that an α-hydroxy acid derivative will also reduce F. Insufficient work on these models for amino acid oxidases and alcohol dehydrogenases has been completed to allow a definite distinction between a radical and nonradical mechanism, but the evidence favors a nonradical mechanism.

If radicals are not intermediates in these model or enzymic dehydrogenations, then presumably the reactions must involve a hydride ion transfer or proceed by the PPC or PPM mechanism. For reasons which will be considered later in more detail (see Sect. 16), this author believes that mechanisms for biological redox reactions involving hydride ion transfers are very rare, if they occur at all. All the reactions of Table 1 can be rationalized readily by the PPC or PPM mechanism, and these will now be considered. For the PPM mechanism to be operative a metal ion must be present as a catalyst. It is known that some of the flavo-enzymes mentioned in Table 1 do not have metal ions and thus the PPM mechanism is not possible for those cases. Also, it is probable that, in those flavoenzymes (listed in Table 1) where metal ions are required, the metal ion is involved in the reoxidation of FH_2 to F rather than in the reaction of AH_2 with F to give A and FH_2. Therefore, possibly all the substrate dehydrogenations listed in Table 1 occur by the PPC mechanism.

For the PPC mechanism to be operative a covalent compound must be formed between F and AH_2, so that some mechanism is available for transmitting electrons from one molecule to the other. For an ionic reaction there are only two general mechanisms by which such a compound could be formed: by an electrophilic attack of AH_2 on F or by a nucleophilic attack of AH_2 on F. For many of the AH_2's listed in Table 1 there is no reasonable way in which the AH_2 could act as an electrophile. Thus, the covalent compound is presumably formed by a nucleophilic attack of AH_2 on F. Therefore, there must be an electrophilic site on F which can be attacked readily by the nucleophilic AH_2. An inspection of the flavin molecule (F) indicates that the most reactive electrophilic site on the molecule is at position 4a. Other positions which are possible

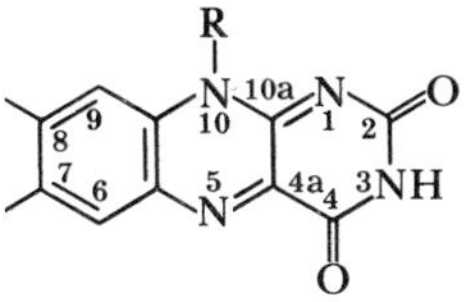

F

electrophilic sites are positions 2, 4, and 10a, but these should be much less susceptible to nucleophilic attack. The carbon at 10a is an amidine carbon and those at 4 and 2 are amide or urea-type carbons. Such electro-philic sites should be attacked only with difficulty by nucleophiles. How-ever, position 4a is like a Schiff base carbon; in fact, its reactivity towards nucleophiles is increased somewhat by the inductive effects of the adjacent amide and amidine groups. Of course, F is stabilized by aromatic resonance, and this decreases the reactivity of any of its positions towards nucleophiles. However, the important point is that position 4a should be by far the most reactive position on F towards nucleophiles.

Given the above considerations it then becomes easy to present mechanisms for all the reactions of Table 1. In each case the mechanism is the PPC mechanism in which both hydrogens are transferred as protons. For the dehydrogenation of alcohols by the various flavo-enzymes the mechanism would be as shown in equation 38. The addition

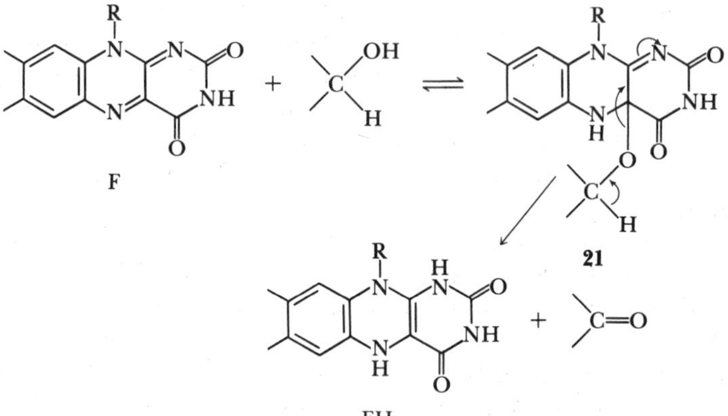

$$(38)$$

of the alcohol to F to give **21** is an ionic reaction and should occur readily; it is similar to the addition of alcohols or water to a Schiff base. The proton transfer could be aided by general acid and/or general base catalysis by the enzyme or the reaction might be a rapid equilibrium process. In the conversion of **21** to FH_2 and the carbonyl compound it is suggested that again the hydrogen is transferred as a proton, and the electrons are transferred through the covalent intermediate as shown. As in the previous step, this reaction could be speeded up by having a suitably placed general acid group on the enzyme available for protonat-ing on nitrogen 1 of the flavin ring and having a general base group for removing the proton from the carbon.

A mechanism essentially the same as that given in equation 38 can be written for the dehydrogenation of amines or amino acids by flavo-enzymes. The only difference is that NR replaces O in the mechanism. There is evidence for such a two-step mechanism in the reaction catalyzed by D-amino acid oxidase. Hellerman and his co-workers[47] observed that, when a series of substituted phenylglycines was oxidized, the rate-determining step for compounds with electron-donating sub-stituents was different from that for compounds with electron-with-drawing substituents. Their data are consistent with the first step of the mechanism of equation 38 being rate determining for compounds with electron-withdrawing substituents and the second step being rate determining for the other compounds. One would expect a small effect of substituents on the rate of the first step and a large effect on the rate of the second step. This is what was observed. Furthermore, the effect of substituents on the second step is in the direction expected for the mechanism shown; electron-withdrawing substituents would be expected to aid proton loss as observed.

Additional evidence for the proposed mechanism for amino acid oxidases comes from the work of Porter and Bright[48] with L-amino acid oxidase. They observed that V_{max} showed both a solvent deuterium isotope effect and an isotope effect in the cleavage of the α-carbon–hydrogen bond. This is expected if the second step of the proposed mechanism is rate determining and protonation of the flavin must accompany the loss of a proton from the α-carbon.

In equation 39 is shown a mechanism for the third category of flavoenzyme dehydrogenations given in Table 1. This follows from the

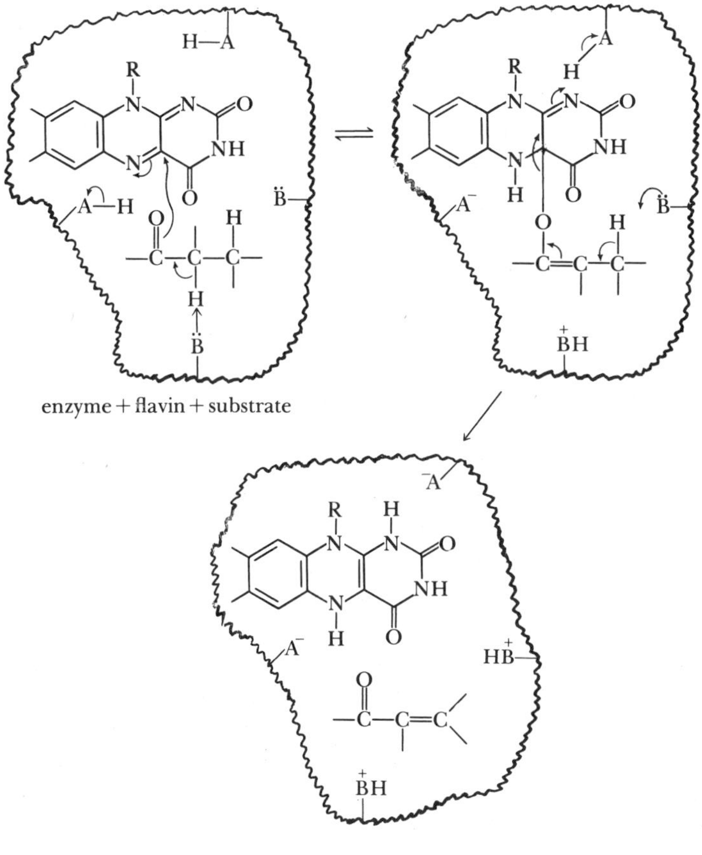

enzyme + flavin + substrate

(39)

same general principles outlined above. As illustrated, suitably placed acid and base groups on the enzymes would aid such a reaction, the acid groups for protonating on the flavin nitrogens and the base groups for removing the protons from the carbons of the substrate. Few investigators have looked for acid and base groups in these and other enzymic redox reactions discussed in this article. An effort should be made to locate and identify such groups.

The stereochemistry of hydrogen removal for the reactions shown in equation 39 is *trans*[49–51, and references therein]. This is consistent with the proposed mechanism. Furthermore, a kinetic deuterium isotope effect of differing magnitudes has been found for the removal of each of the hydrogens from the substrate, and one hydrogen of succinate exchanges more rapidly than the others with the medium[50, 51]. The proposed two-step mechanism can explain the results obtained.

A mechanism for NADH dehydrogenations by flavoenzymes (and by flavin in the model system) is shown in equation 40. As can be seen, the proposed mechanism is of the same type as that suggested for the other

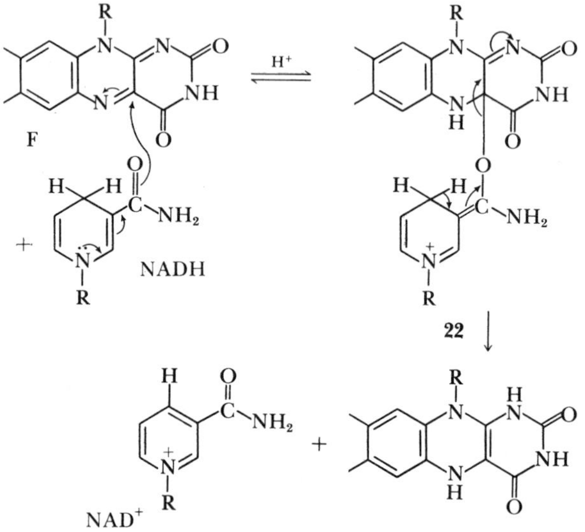

(40)

dehydrogenations; the hydrogen is lost from the 4-position of the dihydropyridine ring as a proton. The intermediate (**22**) should be formed fairly readily, although in its formation the resonance stabilization of the carboxamide group of NADH is lost. However, **22** has the considerable resonance stabilization of a vinylogous amidinium ion. This should compensate partially for the loss of the amide resonance.

Another possibility (not illustrated) for the mechanism of the NADH reaction with flavin would involve the nitrogen of the dihydropyridine ring of NADH attacking the 4a-position of flavin to give a quaternary ammonium compound. Subsequent proton transfer and carbon–nitrogen cleavage would give NAD^+ and FH_2. This is similar to the mechanism suggested for amine oxidases; in the case of tertiary amines the intermediate covalent compound would also be a quaternary ammonium compound.

A mechanism for dithiol dehydrogenations by flavoenzymes (and by flavins in the model system) is shown in equation 41. The first step in this sequence is very similar to that suggested for the alcohol and amine dehydrogenations and should occur very readily. Recent work in the

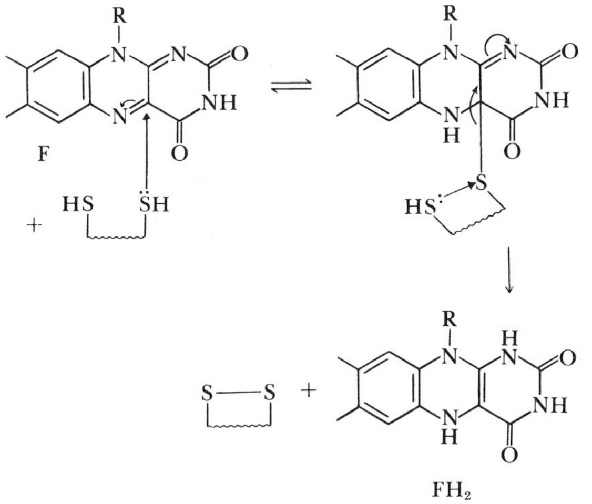

$$(41)$$

FH_2

author's laboratory[52] on a related system, which will be discussed more thoroughly in section 9, indicates that the second step should also occur readily. Nucleophilic substitutions of sulfur on sulfur are very common, and FH_2 should be a good leaving group if the nitrogen is protonated prior to, or concurrently with, the nucleophilic displacement. Again, an enzyme could be expected to increase the rate of such a reaction by general acid catalysis.

One may inquire whether there is any indication from the known voluminous chemistry of flavins that compounds related to the suggested intermediates form readily. Hemmerich et al.[53] observed that in aqueous solution the 5-N-benzyl derivative (**23**) of flavin exists as **24**. Therefore, nucleophiles certainly add readily to position 4a if the 5-nitrogen has a positive charge. This nitrogen could obtain a positive

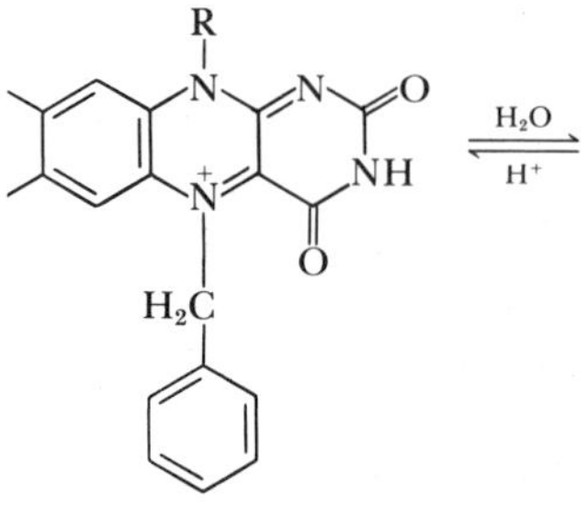

(42)

23   **24**

charge in the enzymic reaction by protonation by a suitably placed acid group on the enzyme. In aqueous solution evidence for addition compounds of water (or alcohols, amines, or thiols) and flavins is presumably not obtained because at equilibrium the free form of flavin is the most stable. However, this does not eliminate the possibility that addition compounds are intermediates in the dehydrogenation reactions.

In aqueous base it is known that the flavin ring system is destroyed and **25** is obtained [54]. This does not mean that position 10a of flavin is the more reactive toward nucleophiles. Rather, **25** is formed because its formation from flavin is an irreversible reaction. If the base attacked position 4a to give **26**, and even if this opened up to **27**, no overall change would be observed, because one would expect both steps to be readily reversible under the reaction conditions. Thus, hydroxide could react with F many times to give **26** before it ultimately reacts at position 10a to give the irreversible product **25**.

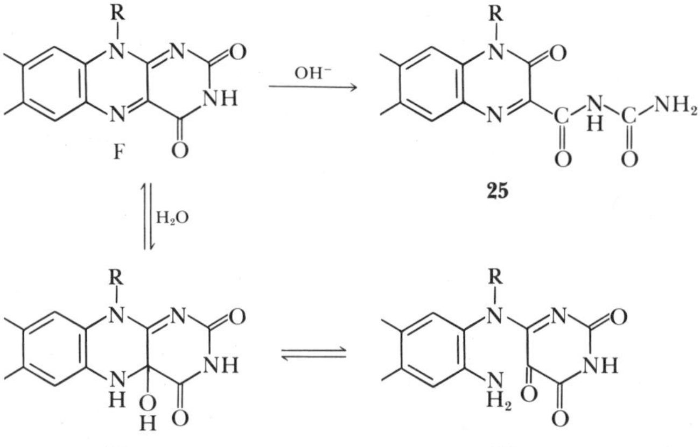

F   **25**

(43)

26   **27**

It should be noted that several others[31-34] have suggested that a covalent compound is formed between flavin and the compound being reduced in some model and enzymic dehydrogenations. Furthermore, many have suggested that some of the reactions given in Table 1 occur by a pathway which does not involve free radical intermediates. However, in general, these suggestions have not been as specific as outlined herein, no general mechanism was offered for all the reactions shown in Table 1, and the possibility that all these dehydrogenations occur by proton transfer (i.e., the PPC mechanism) was not recognized. This last point is particularly important because now the enzymic reactions can be seen to be closely related to nonredox enzymic reactions which proceed by acid and base catalysis.

As indicated earlier, in many enzymic reactions the reduced flavin (FH_2) reacts directly with O_2 to give F and H_2O_2. One would not expect this reaction to occur by the PPC mechanism (i.e., involving only proton transfers and no free radical intermediates) because of the fact that O_2 is a triplet, while FH_2 is a singlet, and the direct reaction of the two to give a covalent singlet compound in one step is not possible (see Sect. 6). Recent evidence[55] indicates that, in fact free radicals are formed when O_2 reacts with FH_2; presumably, these are the semiquinone FH, and the OOH radical.

If metal ions are involved in any of the dehydrogenations listed in Table 1, it is conceivable that the reaction may occur by the PPM mechanism. This is shown in general terms in equation 44; M^{2+} refers to the metal ion catalyst and HXYH refers to AH_2 where X is the end of the AH_2 which can complex with a metal ion (i.e., O, NH, S, etc.). By this mechanism also, both H's are transferred as protons and no free radical

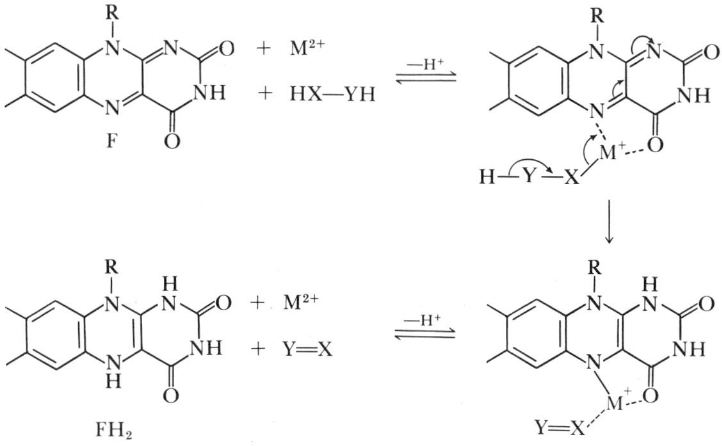

(44)

intermediates are formed. However, as indicated earlier, it is unlikely that metal ions are involved in very many of the dehydrogenations listed in Table 1.

8 HYDROGENATION–DEHYDROGENATIONS INVOLVING NICOTINAMIDES

A considerable body of literature exists on possible mechanisms for enzymic reactions involving NAD$^+$–NADH and NADP$^+$–NADPH [56–60]. Consequently, they will be considered only briefly here. Although enzymes are frequently specific for only one of NAD$^+$ or NADP$^+$, the nicotinamide ring has virtually the same reactivity in both coenzymes. In the following discussion NAD$^+$ will be used as the example of the nicotinamides. The reactions which have received the most study are the reversible dehydrogenation of alcohols to carbonyl compounds (eq. 45) and the dehydrogenation of amines to imines (and eventually hydrolysis to carbonyl compounds and ammonia, eq. 46). One of the striking features of these reactions is that a hydrogen is transferred

$$\underset{/\;\backslash}{\overset{\backslash\;/}{C}}\underset{OH}{\overset{H}{}} + NAD^+ \rightleftharpoons \overset{\backslash}{\underset{/}{C}}{=}O + NADH + H^+ \tag{45}$$

$$\underset{/\;\backslash}{\overset{\backslash\;/}{C}}\underset{H}{\overset{NH_2}{}} + NAD^+ \rightleftharpoons \overset{\backslash}{\underset{/}{C}}{=}NH + NADH + H^+ \tag{46}$$

$$\Big\| H_2O$$

$$\overset{\backslash}{\underset{/}{C}}{=}O + NH_3$$

directly and stereospecifically from the substrate to NAD$^+$, and vice versa in the reverse reaction. Also, there is no evidence that free radicals are intermediates. These facts, coupled with results from other extensive studies on the enzymic and model reactions, have led to the generally accepted view that these reactions proceed by a hydride transfer between the substrate and nicotinamide ring. This is illustrated in general terms for alcohol oxidations in equation 47 and is an example of the PHy mechanism. Several variations on this general mechanism have been suggested. One good possibility is that the proton is lost from the

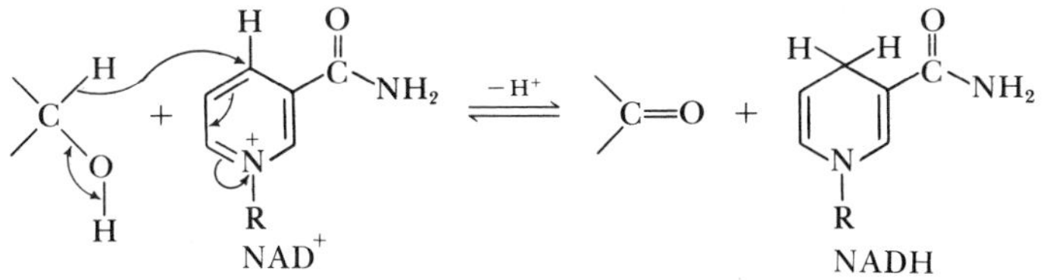

$$(47)$$

alcohol oxygen in a prior step and the alkoxide forms a salt bond with the positive charge on the nicotinamide ring, thus forcing the two molecules into close proximity for the subsequent hydride transfer.

Although the evidence seems strong that the nicotinamides react with most of their substrates by a hydride ion mechanism, a slightly different mechanism such as that shown in equation 48 deserves comment (the mechanism is given for the alcohol reaction, but amines would be expected to have similar properties). As written, this is the PPC mechanism applied to these systems: a covalent compound is formed between the NAD$^+$ and substrate and both hydrogens are lost from the substrate as protons. Certainly the first step should occur readily since it is similar to the addition of a nucleophile to an iminium compound. Also, compounds like NAD$^+$ are known to add nucleophiles readily [56–60]. Usually the addition occurs on the 4-position of the pyridine ring, but the 2- and 6-positions should have comparable reactivities towards nucleophiles. The mechanism of equation 48 could be written with

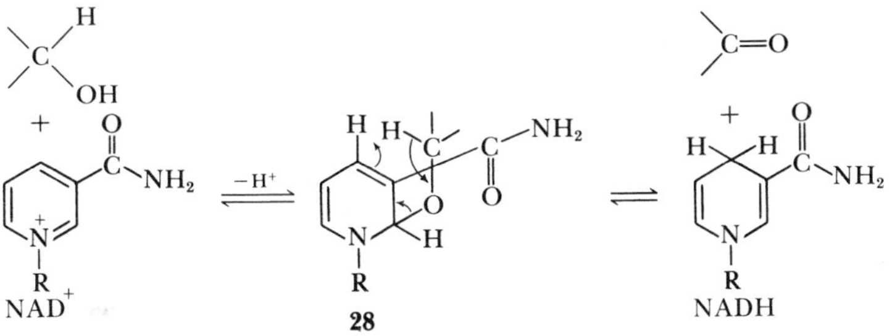

$$(48)$$

initial addition at the 6-position, but addition to the 4-position would give an intermediate which would not be expected to react readily to give NADH. A four-membered cyclic transition state would be required for such an interconversion and that is unlikely.

The key step in the mechanism of equation 48 is the second step—the conversion of **28** (or the 6-substituted isomer) to NADH and the carbonyl compound. An inspection of molecular models indicates that such an interconversion is sterically reasonable. Furthermore, six-membered cyclic transition states of this type are quite common in organic chemical reactions. However, in such a cyclic transition state the question as to whether a proton or a hydride is transferred loses much of its meaning. One could also imagine the decomposition of **28** to NADH and the carbonyl compound occurring as illustrated in equation 49. In this illustration the hydrogen is transferred as a hydride ion. In fact, it is

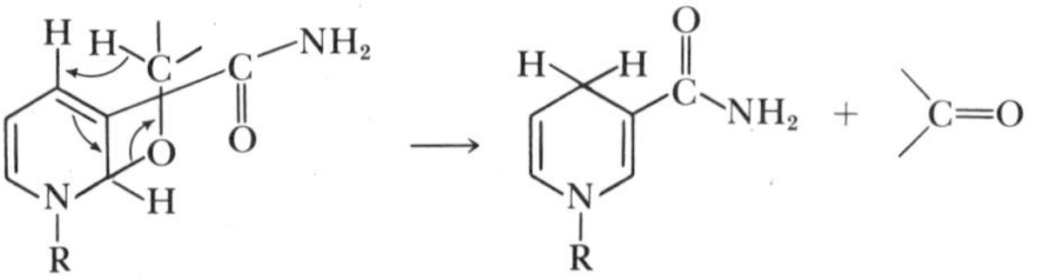

$$(49)$$

probably impossible to distinguish between the mechanisms for the decomposition of **28** shown in equations 48 and 49 (or even a mechanism involving a hydrogen atom transfer). Since electrons can move through overlapping orbitals much more rapidly than nuclei move, they will be arranged to give the most stable species at any stage along the reaction coordinate. In a cyclic transition state it is not possible to tell from which direction the electrons came; since they move so rapidly, they go back and forth many times while the nuclei are moving sufficiently to go from the ground to the transition state. It may be possible to determine whether the hydrogen which is transferred has no charge, a partial positive charge, or a partial negative charge in the transition state. Because of the presence of several more electronegative groups in the molecule it seems most likely that the transferred hydrogen would have a partial positive charge in the transition state. If such is the case then the mechanism as illustrated in equation 48 is probably conceptually better, but it should be emphasized that there is really no difference in the two mechanisms (eqs. 48 and 49) which can be experimentally tested.

Thus it appears that the enzymic reaction of NAD^+ with alcohols and amines may involve a proton removal and hydride ion transfer (the PHy mechanism), or possibly the hydrogen may be transferred in a cyclic transition state in which it is impossible to distinguish between the PHy and PPC mechanisms. If the reaction occurs by a direct hydride ion transfer then it is fairly unique in biological systems. Legitimate evidence for hydride transfers in other intermolecular redox reactions, catalyzed by enzymes which do not require an organometallic component (such as the cobamide coenzyme), is negligible. The distinctive role played by the nicotinamide coenzymes in metabolic pathways may be a result of their ability to transfer hydrogen directly and stereospecifically. However, in the previous section it was suggested that NADH reacts with flavins by the PPC mechanism. Therefore, perhaps the role of the nicotinamide coenzymes is also connected with their ability to react with various substrates by a hydride ion mechanism (the PHy mechanism) and then feed reducing equivalents into flavin by a proton mechanism (the PPC mechanism).

9 PROLINE REDUCTASE

A few enzymic reactions have been characterized in which the carbon–nitrogen bond of an amino acid is reductively cleaved. The enzyme in this group which has been studied sufficiently thoroughly to allow some speculation concerning its mechanism is proline reductase, characterized originally by Stadtman[61, 62] and investigated more recently by Hodgins and Abeles[63, 64]. The reaction catalyzed by proline reductase is shown in equation 50. Various dithiols can serve as reducing agents; possibly the actual reducing agent is an enzyme-bound dithiol. Hodgins

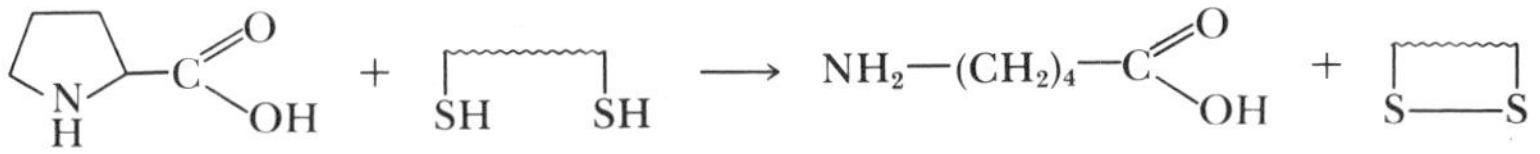

$$(50)$$

and Abeles have shown that, if the enzymic reaction is carried out in D_2O, the product, δ-aminovaleric acid, has one deuterium attached to a carbon, and this is in the α-position. Furthermore, they have found that a pyruvate residue is covalently attached to the enzyme probably by an amide bond, and this residue must be intact for the enzyme to be active. Reduction of the pyruvyl residue to a lactyl residue gives an inactive enzyme.

In considering mechanisms for this reaction one presumably must find some role for the enzyme-bound pyruvate. Hodgins and Abeles suggest that the amine group of the proline adds to the pyruvate carbonyl (eq. 51). Further, they suggest that in such an adduct the nitrogen will be a better leaving group, thus allowing a nucleophilic displacement to occur on the α-carbon. The positively charged nitrogen of **29** or **30** might serve this function, although it is not immediately

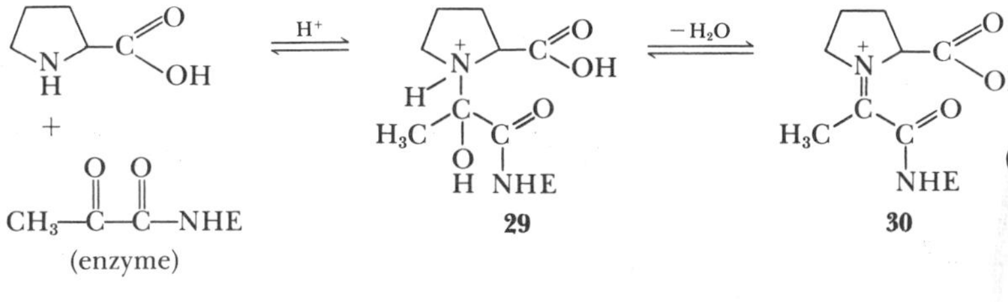

$$(51)$$

obvious why either would be much better as a leaving group than the positively charged nitrogen present in the zwitterionic form of proline. In general, amines are not good leaving groups in nucleophilic displacement reactions, but model compounds related to **29** or **30** have not been investigated to see if these would have special reactivity.

If **29** or **30** could undergo nucleophilic displacement at the α-carbon then the question becomes: what is the nucleophile? If it were a hydride ion, the observed product, δ-aminovaleric acid, would be obtained readily after hydrolysis of the initially formed Schiff base of this compound with the enzyme. However, it is very unlikely that this is the mechanism, firstly because it is not obvious what the hydride ion donor could be, and, secondly, hydride ion displacements of this type are not common even in organic chemistry. As the aforementioned experiment in D_2O indicates, the hydrogen which ends up on the α-carbon must come from water. This would be possible if the dithiol (which would exchange its thiol hydrogens rapidly with water) could donate a hydride ion and be oxidized to the disulfide. However, there is no chemical analogy for the donation of a hydride ion from a dithiol. Thus, a mechanism involving a hydride ion displacement on **29** or **30** seems very unlikely.

A mechanism for proline reductase, which does not involve a hydride ion displacement on the α-carbon, but a nucleophilic displacement by the dithiol, has been suggested by Hodgins and Abeles[64] and is illus-

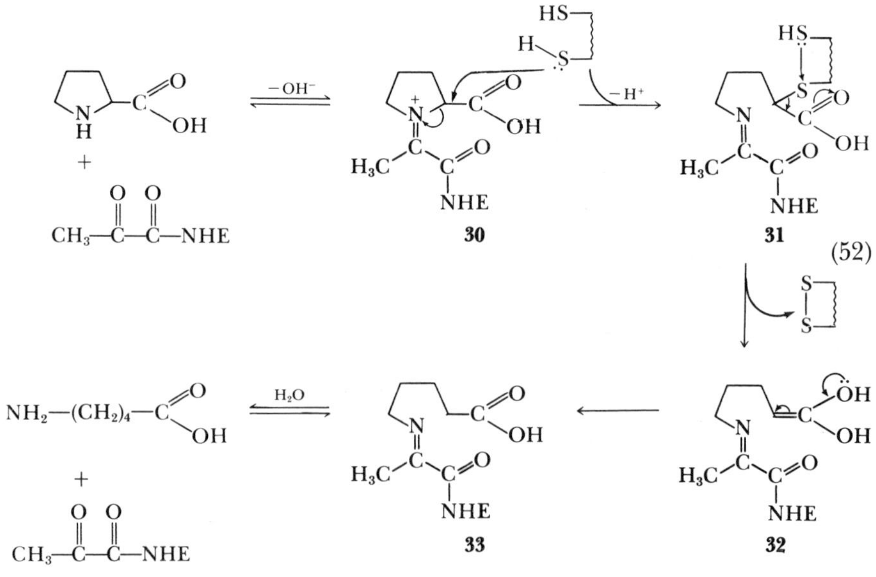

(52)

trated in equation 52. The mechanism is shown with the nucleophilic displacement occurring on the α-carbon of **30**, but a displacement on **29**[64] would be very similar mechanistically. The mercaptide ion is a good nucleophile, so the conversion of **30** to **31** is not unreasonable. The next step, the conversion of **31** to **32**, involves a nucleophilic displacement by a thiol on a sulfide with cleavage of a carbon–sulfur bond. Such reactions had not been studied in detail, but there are now strong indications that they occur readily in model systems. It has been reported that an α-acetoxy ketone can be reduced to acetic acid and the ketone by a dithiol[65]. The mechanism for this reaction is probably the same as in the conversion of **30** to **33**, except that acetate is the leaving group in the first step, rather than the imine. Recent work by Harkins in the author's laboratory[52] indicates that reduction of α-substituted ketones to ketones by thiols is a surprisingly facile reaction. Thus, α-bromoacetophenone reacts rapidly with dithiols (such as dithiothreitol or 1,3-dithiopropane), and even monothiols, at room temperature in organic solvents to give acetophenone and the disulfide. The reaction shows characteristics of an ionic reaction; for example, its rate is increased in more polar solvents. Presumably the mechanism for this redox reaction is the same as in the conversion of **30** to **33** (eq. 52), except that bromide ion is the leaving group in the first step. However, the main conclusion to be drawn from this model study is that the nucleophilic displacement by a thiol (or mercaptide ion) on a thioether, resulting in carbon–sulfur bond cleavage, occurs readily when a carbonyl-stabilized carbanion is formed. Thus, the conversion of **31** to **32** is a reasonable step. The subsequent reactions of **32** to regenerate the enzyme and give δ-aminovaleric acid are well known ionic reactions and should occur readily. The conversion of **32** to **33** involves a tautomerization in which a proton is put on the α-carbon. Thus, the overall redox reaction is an example of the PPC mechanism; both hydrogens are transferred as protons, and an intermediate covalent compound is formed between the oxidant and reductant to allow electron transfer. Several of the steps, including the conversion of **31** to **32**, would probably be subject to acid and/or base catalysis by suitably placed groups on the enzyme surface.

The most unlikely step in the mechanism shown in equation 52 is the conversion of **30** to **31**. As indicated above, a positively charged nitrogen is not a good leaving group in nucleophilic displacement reactions. Also, the mechanism of equation 52 does not ascribe a very unique role to the enzyme-bound pyruvate; there are many simpler ways to get a positive nitrogen, for example, merely by protonation. A mechanism which has characteristics related to that in equation 52, but which avoids both of these problems, is shown in equation 53. Here it is suggested that the thiol adds to the carbon of the iminium ion (**30**) rather than performing a nucleophilic displacement as in equation 52. The addition of the dithiol to **30** to give **34** should be very rapid; such iminium ions add a number of nucleophiles readily. The conversion of **34** to **35** is very similar to the previously discussed **31** to **32** reaction and should occur just as readily. In **35** the pyruvate part of the molecule has been reduced; **35** is just a

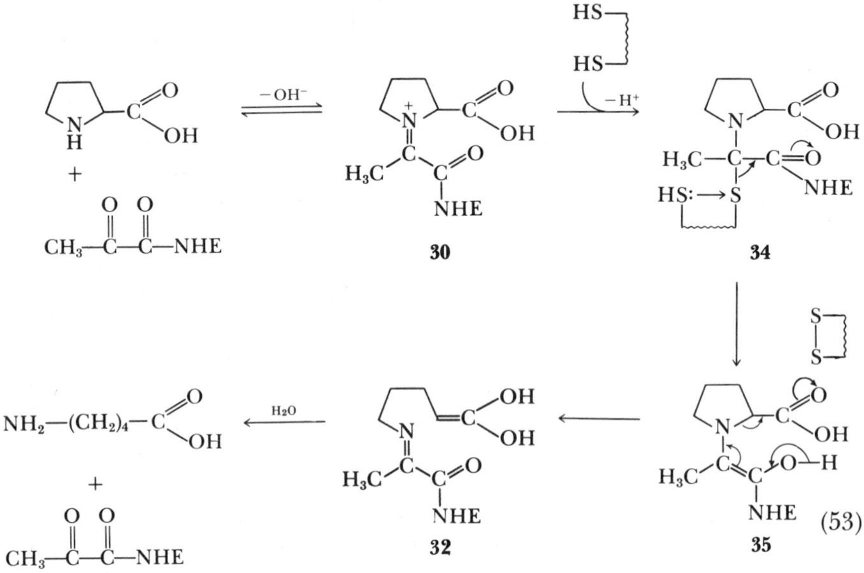

(53)

tautomer of a substituted α-amino acid amide. It is suggested that **35** (or possibly its tautomer) can be converted to **32** easily by a proton and electron migrations as shown. In this transformation the pyruvate part of the molecule is reoxidized and the proline part reductively cleaved. Actually, **35** may not be an obligatory intermediate; **34** could possibly be converted to **32** directly if the protonation occurs on the proline carboxylic oxygen. However, in such a step both a carbon–sulfur and a carbon–nitrogen bond would have to be broken more or less simultaneously. Therefore, it seems more likely that **35** would be an intermediate. As indicated in equation 52, **32** would give δ-aminovaleric acid and regenerate the enzyme by well known rapid ionic steps. In the overall reaction the hydrogens are again transferred as protons, and the electrons are transferred through a covalent compound formed between oxidant and reductant. Thus, it again falls in the category of the PPC mechanism, and several of the steps, including the conversion of **34** to **35** and **35** to **32**, should be catalyzed by acid and/or base groups on the enzyme. Because it avoids the nucleophilic displacement and gives a more direct role to the enzyme-bound pyruvate (it is reversibly reduced and oxidized), the mechanism shown in equation 53 for proline reductase seems more likely than that indicated in equation 52.

10 RIBONUCLEOTIDE REDUCTASES

There appear to be two distinct types of enzymes which convert ribonucleotides to deoxyribonucleotides; one type requires the cobamide coenzyme and the other does not [66]. Here, we will be mainly concerned with a possible mechanism for the latter type—those enzymes which do not require the vitamin B_{12} coenzyme. The reaction catalyzed by the ribonucleotide reductases is shown in equation 54. This reaction is not

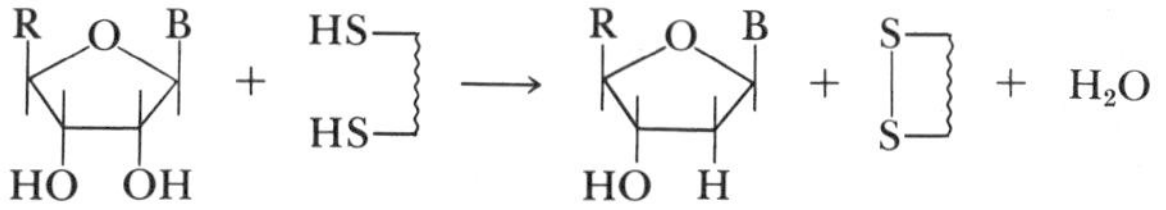

$$(54)$$

a dehydrogenation, but a deoxygenation reaction. However, since the mechanism seems related to that of proline reductase, it will be discussed here. The enzymes which do not require the cobamide coenzyme operate on the ribonucleotide diphosphates, while those requiring cobamide are specific for the triphosphates. In each case the enzyme will catalyze the reactions of nucleotides with either purine or pyrimidine bases at somewhat comparable rates, although the relative reactivity varies considerably, depending on the presence or absence of various allosteric effectors [66]. The dithiol reducing agent (eq. 54) is either the reduced form of thioredoxin or a reduced form of the enzyme ribonucleotide reductase, which can be formed from the disulfide form of the enzyme by reaction with reduced thioredoxin. Thioredoxin is a small protein (MW 12,000) whose structure varies somewhat depending on the species from which it is isolated. The oxidized form of thioredoxin is then reduced by one of the nicotinamides in a flavoprotein-catalyzed reaction.

It is known from isotopic studies that, with both types of enzymes, the hydrogen which ends up on the 2-position of the deoxyribose ring comes from water and, furthermore, that it replaces the original hydroxyl group with overall retention of configuration. None of the other hydrogens on the ribose carbons are exchanged during the enzymic reaction. It has been suggested that the mechanism involves a displacement of the hydroxy group by the enzyme, followed by a further displacement by hydride ion. However, there is no evidence for such a mechanism, and for the enzymes not requiring the cobamide coenzyme, it is not obvious how a dithiol could supply a hydride ion. More likely mechanisms for the enzymes which do not require cobamide are outlined in equations 55 and 56. In these equations the cytidine nucleotide is used as the example, but related mechanisms would be expected for other nucleotides. Actually these mechanisms are closely related. A requirement for this general type of mechanism is that the 1-position on the ribose ring acquire carbonyl character so that it becomes available for nucleophilic attack by the thiol. In the mechanism of equation 55 this is accomplished by cleaving the ribose ring to give 36, while in the other mechanism the cytosine is split off to give 40. If the 1-position of the ribose ring were hydrolyzed completely to give the aldehyde (not illustrated), again the same general type of mechanism could be written. The addition of the thiol to 36 or 40 to give 37 or 41, respectively, would be expected to occur very rapidly. The actual oxidation and reduction is accomplished by the conversion of 37 to 38 (or 41 to 42) with formation of a molecule of water and the disulfide. This step is closely related to the reaction discussed in the previous section; as indicated there, nucleophilic displacements of mercaptides on thioethers with cleavage of a carbon–sulfur bond appear to occur quite easily [52]. For such a displacement to occur, it is obvious that the negative charge on the carbon leaving group must somehow be dissipated. In the previous case (Sect. 9) this was possible by protonating on the carbonyl to form an enol. In the mechanisms of equations 55 and 56 this is accomplished by protonating on the 2-

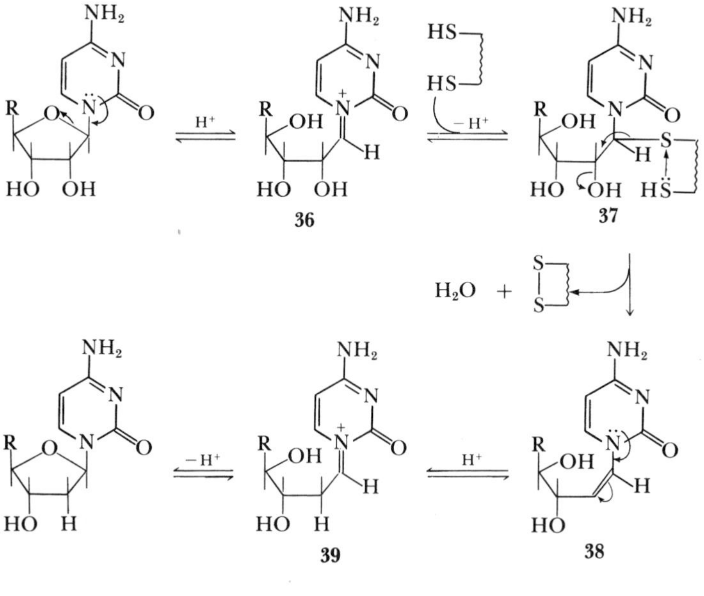

(55)

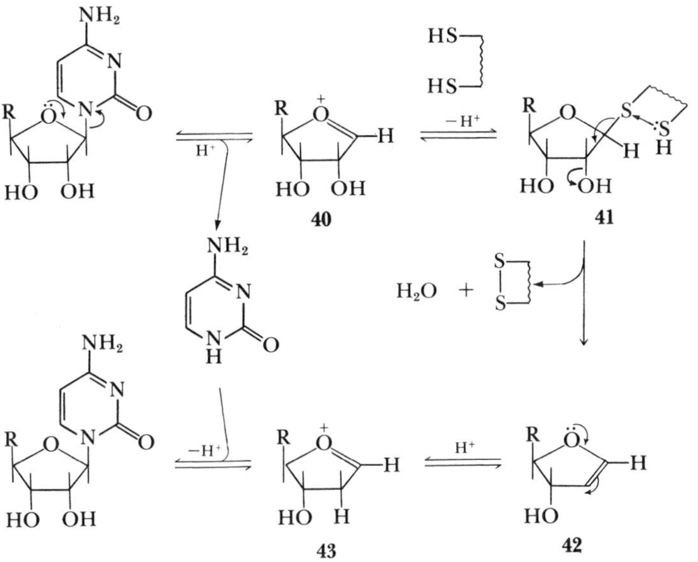

(56)

hydroxyl, breaking the carbon–oxygen bond, and thus forming an alkene. Although the author is unaware of a direct chemical analogy to this step, it seems reasonable as a logical extention of the reactions discussed in the previous section. One would expect that general acid (to protonate the 2-hydroxyl) and general base (to remove the proton from the thiol) groups on the enzyme could readily facilitate such a step.

The subsequent reactions of **38** or **42** to give the deoxyribonucleotide have numerous chemical analogies and should occur readily. The conversion of **38** to **39** (or **42** to **43**) merely involves the well known protonation of an enamine (or a vinyl ether) and the resulting ion should add a nucleophile easily; this last step is essentially just the reverse of the first.

A mechanism related to the above would involve nucleophilic displacement of the 2-hydroxy group of the ribose ring by the thiol, followed by ring opening to give **44**, which by proton and electron migrations as shown (eq. 57) would give **38** and disulfide. A similar mechanism (not

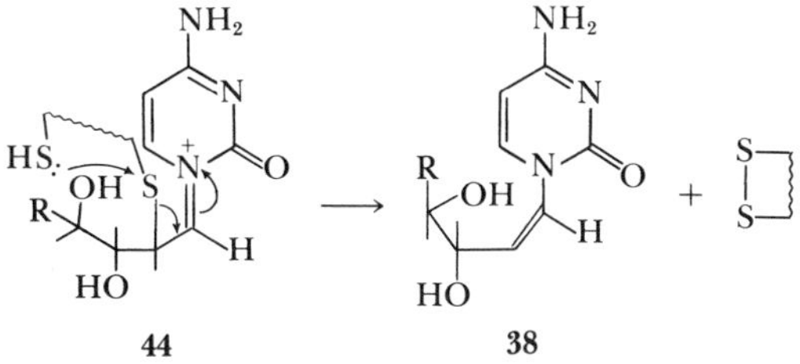

(57)

illustrated) would involve nucleoside cleavage (as in eq. 56) rather than ring cleavage. However, in either case the 1-position of the ribose ring plays a distinct role; a group attached to it acts as an electron sink for the redox step.

At this stage it is not clear which of the suggested mechanisms is more likely, but as indicated earlier, they are all quite similar. Regardless of which mechanism is correct, the important point is that the overall redox reaction probably proceeds by ionic steps involving proton transfers, and not a hydride ion transfer. Thus, this deoxygenation is closely related to the PPC mechanism for dehydrogenation; as in the dehydrogenation a covalent compound between the oxidant and reductant must be formed as an intermediate so that electrons can be transferred without a requirement for free radical intermediates.

It seems possible that the mechanism for cobamide-requiring ribonucleotide reductases is related to those discussed above. The similarity in the two reactions is striking: a thioredoxin is required as reducing agent, and the hydrogen which ends up in the deoxyribose ring comes from water and replaces the 2-hydroxyl with retention of configuration. Possibly the cobamide requirement is an evolutionary artifact and now acts merely for storage of reducing equivalents on the enzyme. Only further experiments will clarify this point.

11 SOME DEHYDROGENASES INVOLVING O_2 AND METAL IONS

Almost all enzymes which catalyze reactions involving molecular oxygen require either a transition metal ion or a highly conjugated coenzyme, such as a flavin, or possibly a pteridine, for activity (one notable exception is lipoxidase). Mechanisms for some flavoprotein reactions which are also dehydrogenations were considered in Section 7 and some oxygenations involving the flavin and pteridine coenzymes will be discussed in the following sections. Here, we will consider some aspects of dehydrogenases involving O_2 which do not contain a highly conjugated coenzyme, but do require a transition metal ion for activity. Possible mechanisms for such enzymes have been considered in detail elsewhere by the author[1]. In the present article only a few important points will be emphasized, and recent work on some nonenzymic systems will be discussed.

Included in the category of metal ion containing dehydrogenases, in which O_2 is the oxidant, are enzymes such as ascorbic acid oxidase, tyrosinase, laccase, uricase, amine oxidases which require pyridoxal phosphate (see Sect. 5), and galactose oxidase[1, 67]. In some of these, water is the product of the reduction of O_2, and in others, hydrogen peroxide is the product. Also, in some it is clear that free radicals are intermediates, while in others there is no evidence for free radical intermediates. In no case is there any evidence for hydride ion transfers, and from a mechanistic standpoint hydride ion transfers in such reactions seem unlikely. Also, there is no evidence for hydrogen atom transfers. The mechanisms of all these dehydrogenases probably involve only proton and electron transfers; all the steps in which the metal ion participates can be readily rationalized in terms of the PPR or PPM mechanism.

Ascorbic acid oxidase[68–70] is representative of the enzymes in this group in which free radicals are probably intermediates. This enzyme catalyzes the reaction shown in equation 58. Each mole of enzyme has eight atoms of copper associated, and during the catalysis apparently four of these undergo a reversible oxidation and reduction between the

$$2 \quad \text{[Ascorbic acid structure]} \quad + \quad O_2 \quad \longrightarrow \quad 2 \quad \text{[Dehydroascorbic acid structure]} \quad + \quad 2\,H_2O \qquad (58)$$

Ascorbic acid   Dehydroascorbic acid

Cu(I) and Cu(II) states. Electron spin resonance studies[71] indicate that the ascorbate semiquinone radical is formed during the enzymic reaction. The evidence indicates that the dehydroascorbic acid is formed by disproportionation of the ascorbate semiquinone to ascorbic acid and dehydroascorbic acid in a nonenzymic reaction. Such a reversible reaction is known to occur rapidly, probably by a mechanism involving proton and electron transfers. The enzyme, therefore, merely catalyzes the oxidation of ascorbic acid to the semiquinone. The most likely mechanism for this would involve electron transfer from an ascorbate ion (ascorbic acid minus a proton) to the enzyme. Such transfers would occur very readily if the ascorbate were complexed to the Cu(II)[1]. Four transfers of this type would lead to the reduction of four enzymic Cu(II)'s to Cu(I)'s and, after disproportionation, two molecules of dehydroascorbic acid would be formed from the four semiquinone molecules. If each of these steps occurs by proton and electron transfers as suggested, then this is an enzymic example of the PPR mechanism.

The mechanism for the reoxidation of the reduced enzyme by O₂ is not clear, but there is no evidence that free radicals (HOO· or OH·) formed from the reduction of O₂ are intermediates. Possibly the reaction proceeds as illustrated in equation 59. The formation of a complex

(45) of O_2 with one of the copper ions on the enzyme is reasonable. The protonation of such a complex would lead directly to 46, in which the O_2

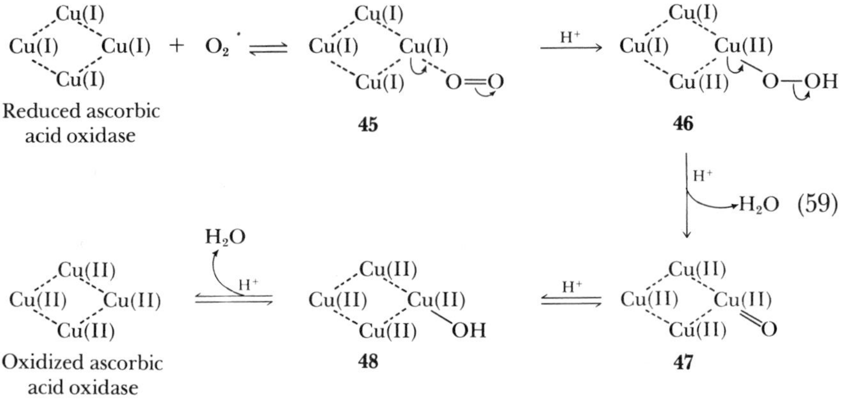

is reduced to the peroxide stage, if the copper ions in the enzyme are electronically connected so that two of them can be oxidized to the Cu(II) stage during the protonation. A bridging group, such as an oxygen, sulfur, and imidazole, between the copper ions would allow such a reaction. The important requirement is that orbitals of the metal ions overlap with one or more of those of the bridging group(s), so that one molecular orbital is formed which overlaps both metal ions. Electrons move within molecular orbitals much more rapidly than nuclei move. Thus, when a two-electron reduction of the O_2 occurs at one copper ion, an electron can be transferred from another copper ion in the same step so that each metal ion only undergoes a one-electron oxidation. Wang and Brinigar[72] have studied a model oxidation of an Fe(II)-heme-bipyridyl polymer which illustrates this principle. It is probable that all four of the active copper ions of ascorbic acid oxidase are so connected. If that is so, then 46 represents only one of the possible resonance hybrids. Other structures would have different copper ions in the Cu(I) and Cu(II) states. If one molecular orbital overlaps all the active copper ions, then electron transfer among the copper ions is very rapid, and one cannot say which of the coppers is in the oxidized or reduced state. If such were the case, then the O_2 reaction and all the ascorbate reactions could occur at only one of the coppers; the others could feed electrons to this site (or remove electrons) when required by the reaction.

The conversion of 46 to 47 (eq. 59) involves protonating on the oxygen of the peroxide complex, cleaving the oxygen–oxygen bond, and oxidizing two of the enzyme coppers. This type of cleavage of peroxide, when it is complexed to a metal ion, which also has a group (or groups) associated which can be readily oxidized, appears to occur readily[1, 73, 74, see also discussion later in this section]. The subsequent protonations of **47** to give **48** and eventually to give the free oxidized form of the enzyme should occur readily. In the mechanism of equation 59, four protons are added to the oxygen during the oxidation, and thus, as written, the enzyme would gain four positive charges. However, protons could be lost from other parts of enzyme (e.g., from ligands of other copper ions) during the oxidation, so that the enzyme might remain with essentially the same charge as it is oxidized or reduced.

The mechanism of equation 59 for the reduction of O_2 is an example of the PPM mechanism: all hydrogens are transferred as protons, a metal ion serves to connect the oxidant electronically to the reductant, and no free radicals (HOO· or OH·) are involved. A mechanism involving one-electron reductions of O_2 to give HOO· and OH· is very unlikely because these intermediate radicals (especially OH·) are very reactive and would be expected to attack other groups on the enzyme, thus destroying the enzyme. The mechanism of equation 59 avoids this problem.

Other enzymes (e.g., cytochrome oxidase) which include a step where a reduced form of the enzyme is reoxidized by O_2 probably proceed by a mechanism related to that in equation 59, and for the same reasons. In fact, cytochrome oxidase[75–77] probably has an overall mechanism closely related to that of ascorbic acid oxidase. The oxidized form of the enzyme apparently accepts electrons one at a time from the cytochrome (as ascorbic acid oxidase does from ascorbic acid) and is subsequently reoxidized by O_2. Other metal ions, in addition to copper, seem to be involved in the cytochrome oxidase, but if they were electronically connected by bridging groups, they should be able to react in the same manner as the coppers of ascorbate oxidase.

In the ascorbic acid oxidase catalyzed reaction, a free radical—the ascorbate semiquinone—is a reasonable intermediate because it is a resonance-stabilized species. It is interesting that in model systems some Cu(II)-catalyzed oxidations of ascorbate by O_2 involve the semiquinone as an intermediate and some do not[1, 78, 79]. It appears that the stability of the semiquinone puts the mechanism of ascorbic acid oxidation on the borderline, so that, depending on the conditions, it can be oxidized by the PPR or PPM mechanism. For some reason, which is not clear, the enzyme apparently chooses the PPR mechanism.

Other metal ion-containing oxidases which catalyze dehydrogenations like ascorbic acid oxidase, but whose substrates do not have the conjugation necessary to stabilize an intermediate radical, are apparently forced to carry out the reaction by the PPM mechanism. One such example is galactose oxidase[80, 81], which catalyzes the reaction shown in equation 60. The enzyme is apparently specific mainly for the 4-hydroxyl *cis* to the CH_2OH because many carbohydrate derivatives with this configuration are substrates[82, 83]. Also, it has recently been found

$$\begin{array}{c} CH_2OH \\ HO \underset{OH}{\overset{O}{\diagup}} OH \\ OH \end{array} \;+\; O_2 \;\longrightarrow\; \begin{array}{c} CHO \\ HO \underset{OH}{\overset{O}{\diagup}} OH \\ OH \end{array} \;+\; H_2O_2 \qquad (60)$$

that simpler compounds, such as dihydroxy acetone[84] and glycerol[85], are good substrates. The enzyme has one atom of Cu(II) per mole (MW approx. 55,000), and apparently the Cu(II) does not change its valence during the enzymic reaction[86]. Other evidence indicates that a ternary complex of the enzyme, substrate, and O_2 is the species which reacts to give products[87]. A mechanism consistent with all these results has been suggested by the author[1] and is illustrated in equation 61. It is suggested that the actual redox step, the conversion of **49** to **50**, involves a proton transfer through the solvent and electron transfer through the complex as shown. As illustrated in equation 61, suitably placed acid and base groups on the enzyme surface would be expected to accelerate this step.

A nonenzymic reaction, recently studied in detail by Adolf in the author's laboratory[88], appears to proceed by a mechanism closely related to that suggested for galactose oxidase. The reaction studied is the oxidation of acetoin to biacetyl by O_2 with Fe(III) as a catalyst. Some results obtained with this system are the following: from pH 1.5 to 2.5 the rate of O_2 uptake follows the rate law, rate $= k[Fe(III)][acetoin]/[H^+]$; the stoichiometry is: two acetoin molecules react with one O_2 molecule to give two molecules of biacetyl and water; under the reaction conditions, H_2O_2 reacts more rapidly than O_2 with acetoin to give biacetyl and water (this reaction is also catalyzed by Fe(III)); mono-deuterated acetoin (3-deuterio-3-hydroxy-2-butanone) reacts at about one-fifth the rate of acetoin; Fe(II) is not an intermediate. Evidence that Fe(II) is not an intermediate is the following: Fe(II) alone is not a catalyst for the reaction, and it does not react with O_2 under the reaction conditions; when Fe(II) is present in the Fe(III)-catalyzed reaction the stoichiometry changes so that in this case one mole of O_2 is required to oxidize one mole of acetoin to biacetyl [apparently the Fe(II) intercepts the H_2O_2 when it is formed]. Mechanisms for the O_2 and H_2O_2 reactions, which are consistent with all the data, are shown in equations 62 and 63. In these reactions the Fe(III) in solution and in the complexes

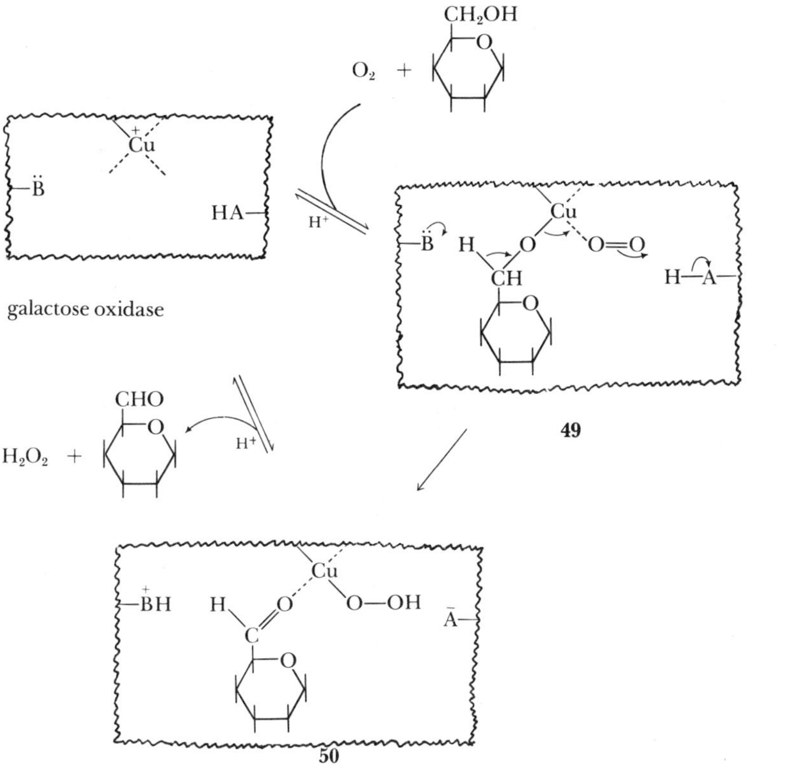

O_2 + [structure: CH₂OH sugar ring]

Cu⁺

—B̈

HA— , H⁺

galactose oxidase

—B̈ H, O, CH, O=O, H—A, Cu [structure]

49

H_2O_2 + [structure: CHO sugar ring] , H⁺

—B̈H H, O, C, O—OH, Ā, Cu [structure]

50

will have additional ligands (some of which will be negatively charged), but these have been omitted for simplicity. The kinetics indicate that the rate-determining step for the O_2 reaction is the conversion of **51** to the enediol complex **52** (the effect of pH shows that the transition state has one less positive charge than the reactants). Since Fe(II) is not an intermediate, it is suggested that O_2 complexes with **52** to give **53**, which, by a protonation and electron reorganization, gives **54**. This is the actual redox step and it should occur readily by the ionic mechanism indicated. One would expect **54** to be in equilibrium with biacetyl, free ferric ion, and hydrogen peroxide.

As indicated earlier, the oxidation of acetoin by H_2O_2 in the presence of Fe(III) occurs more rapidly than the O_2 reaction. The H_2O_2 reaction can apparently occur by two mechanisms[88], one of which is a free radical reaction when small amounts of Fe(II) are present. However, in the strict absence of Fe(II), the reaction shows no radical character, and the kinetics indicate a mechanism as shown in equation 63. The rate-determining step is apparently the conversion of **55** to **56** because the

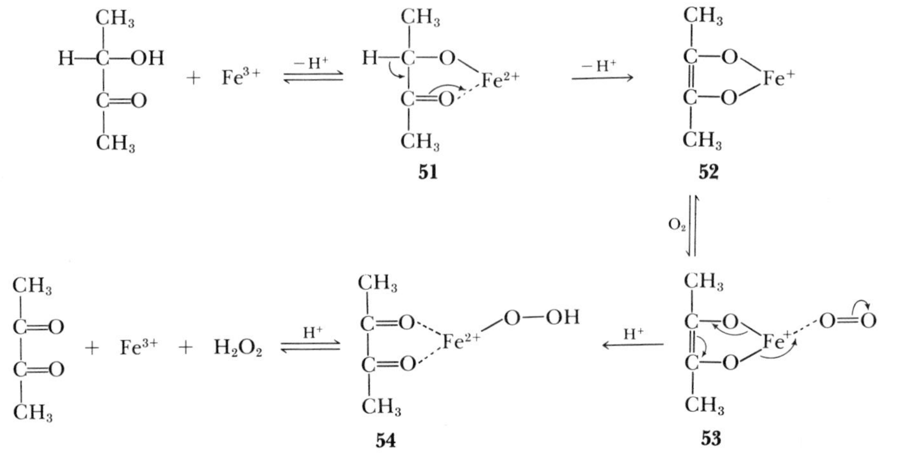

$$
\begin{array}{c}
\text{CH}_3 \\
\mid \\
\text{H}-\overset{\mid}{\text{C}}-\text{OH} \\
\mid \\
\text{C}=\text{O} \\
\mid \\
\text{CH}_3
\end{array}
\;+\; \text{Fe}^{3+}
\;\underset{-\text{H}^+}{\rightleftharpoons}\;
\mathbf{51}
\;\overset{-\text{H}^+}{\longrightarrow}\;
\mathbf{52}
\tag{(6}
$$

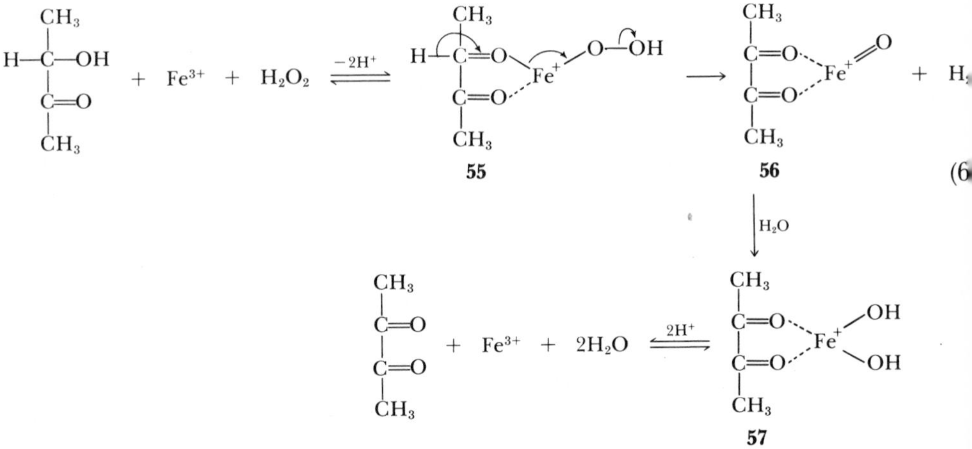

$$
\mathbf{52} \;\xrightarrow{\;\text{O}_2\;}\;
$$

$$
\begin{array}{c}
\text{CH}_3 \\
\mid \\
\text{C}=\text{O} \\
\mid \\
\text{C}=\text{O} \\
\mid \\
\text{CH}_3
\end{array}
\;+\; \text{Fe}^{3+} \;+\; \text{H}_2\text{O}_2
\;\overset{\text{H}^+}{\rightleftharpoons}\;
\mathbf{54}
\;\overset{\text{H}^+}{\longleftarrow}\;
\mathbf{53}
$$

$$
\begin{array}{c}
\text{CH}_3 \\
\mid \\
\text{H}-\overset{\mid}{\text{C}}-\text{OH} \\
\mid \\
\text{C}=\text{O} \\
\mid \\
\text{CH}_3
\end{array}
\;+\; \text{Fe}^{3+} \;+\; \text{H}_2\text{O}_2
\;\overset{-2\text{H}^+}{\rightleftharpoons}\;
\mathbf{55}
\;\longrightarrow\;
\mathbf{56} \;+\; \text{H}
\tag{(6}
$$

$$
\mathbf{56} \;\xrightarrow{\;\text{H}_2\text{O}\;}\;
$$

$$
\begin{array}{c}
\text{CH}_3 \\
\mid \\
\text{C}=\text{O} \\
\mid \\
\text{C}=\text{O} \\
\mid \\
\text{CH}_3
\end{array}
\;+\; \text{Fe}^{3+} \;+\; 2\text{H}_2\text{O}
\;\overset{2\text{H}^+}{\rightleftharpoons}\;
\mathbf{57}
$$

rate of the reaction is first order in the concentrations of each of acetoin, Fe(III), and H_2O_2. Furthermore, monodeuterated acetoin reacts at only about one-half the rate of acetoin. The conversion of **55** to **56** is the redox step, and it apparently occurs by an ionic mechanism as shown: a proton and hydroxide ion are removed from **55** with cleavage of the peroxide bond. The subsequent addition of water to **56** to give **57** and eventually biacetyl should occur rapidly.

Both the model dehydrogenations shown in equations 62 and 63 occur by ionic mechanisms and are examples of the PPM mechanism. However, the O_2 reaction (eq. 62) does not occur by the exact mechanism suggested for galactose oxidase (eq. 61). With acetoin one has the possibility of going to an enol complex (**52**), which is not possible with simple alcohols, such as the one being oxidized by galactose oxidase. However, the actual oxidation step (**53** to **54**) does require a ternary complex as in the enzymic reaction. The conversion of **55** to **56** in the model H_2O_2 reaction is very similar to the conversion of **49** to **50** in the enzymic reaction, except that different oxidizing agents are involved. Thus, the work with the model system indicates that metal ion catalyzed ionic oxidations of alcohols by O_2 and H_2O_2 are possible and suggests that the proposed mechanism for galactose oxidase is reasonable.

The ionic cleavage of a peroxide bond, as in the conversion of **55** to **56**, is similar to that suggested for the conversion of **47** to **48** (eq. 59) in the oxygen reaction with the reduced form of ascorbic acid oxidase. In the model system acetoin is the reducing agent, whereas two Cu(I)'s perform this function in the ascorbic acid oxidase reaction. This type of ionic cleavage of peroxides in the presence of reducing agents and metal ions apparently occurs quite frequently in both model and enzymic reactions. It is believed to occur in the formation of the aromatic hydroxylating agent in a system composed of Fe(III), H_2O_2, catechol, and an aromatic compound[1, 73, 74]. Furthermore, several oxidations by H_2O_2 in the presence of Cu(II) complexes[89–91] can be readily rationalized in terms of this mechanism. These include the oxidation of NH_2—NH_2, NH_2—OH, and H_2O_2. As suggested previously by the present author[1], this mechanism is also the most likely one for the disproportionation of H_2O_2 catalyzed by the complex of Fe(III) with triethylenetetramine[92, 93]; the mechanism suggested by Wang[92–94] is without chemical analogy. In addition to the enzymic reactions catalyzed by ascorbic acid oxidase and cytochrome oxidase (see earlier), this mechanism is probably also involved in the catalase, peroxidase, and tyrosinase reactions[1]. One thus expects that it will occur in many other enzymic redox reactions as well.

As indicated earlier, mechanisms related to the ones discussed above have been suggested for a number of other oxidases which require metal ions[1]. In each case the reaction proceeds either by the PPR mechanism (if free radical intermediates are reasonable and are indicated by experiments) or by the PPM mechanism.

12 XANTHINE AND ALDEHYDE OXIDASES

Closely related and complex enzymes which contain Mo, Fe, and flavin catalyze a variety of reactions of the type shown in equation 64, where X is O or NR[95, 96]. These enzymes are in general not very specific; for example, xanthine oxidase catalyzes the oxidation of purines readily,

$$\underset{R}{\overset{X}{\diagdown}}C-H + H_2O + O_2 \longrightarrow \underset{R}{\overset{X}{\diagdown}}C-OH + H_2O_2 \tag{64}$$

but will also catalyze the oxidation of aldehydes and N-methylnicotin-amides. Other oxidizing agents, such as cytochrome C and 2,6-dichloro-phenol indophenol, can replace O_2 as the ultimate electron acceptor. It is known that the oxygen atom which ends up in the substrate comes from water. Thus, these enzymic reactions are quite distinct from the mixed function oxidases or monooxygenases to be discussed in the following sections.

Recent work [97–99 and references therein] with xanthine and aldehyde oxidases has clearly shown that the substrate (purine or aldehyde) reacts at a molybdenum site on the enzyme. In the initial steps the substrate is oxidized and two molybdenums are reduced from Mo(VI) to Mo(V). Subsequently, the electrons are transferred to flavin and the bound iron atoms and ultimately to the electron acceptor. Although these subsequent reactions most certainly involve one-electron transfers, there is no evidence for purine or aldehyde free radicals at any stage during the reaction. Fridovich [100] has shown that it is the unhydrated aldehyde which is the substrate for the enzyme.

Several people have suggested that the oxidation of the purine or aldehyde involves a hydride transfer to the enzyme followed by hydroxide attack on the resultant carbonium ion. However, there is no evidence for such a mechanism, and intermediate carbonium ions of this type, especially in the oxidation of ring systems such as purines or N-methylnicotinamides, are very unlikely. A mechanism which is more reasonable, and which is closely related to those discussed previously, is shown in equation 65. It is suggested that the molybdenum site on the enzyme (58) has a bound hydroxide which adds by nucleophilic addition to the carbon–nitrogen or carbon–oxygen double bond of the substrate to give 59. Such an addition should occur readily. The oxidation of the substrate occurs in the conversion of 59 to 60 and 61. In this step it is suggested that the hydrogen is lost from the substrate as a proton and that both molybdenums are reduced to Mo(V). This can occur all in one step [i.e., a two-electron oxidation of the substrate and 2 one-electron reductions of 2 Mo(VI)'s] if the molybdenums are connected electronically (see previous section). The subsequent reoxidation of 61 to 58 presumably occurs by electron transfer to the flavin and iron, and ultimately to an electron acceptor; the OH on 58 comes from a solvent water.

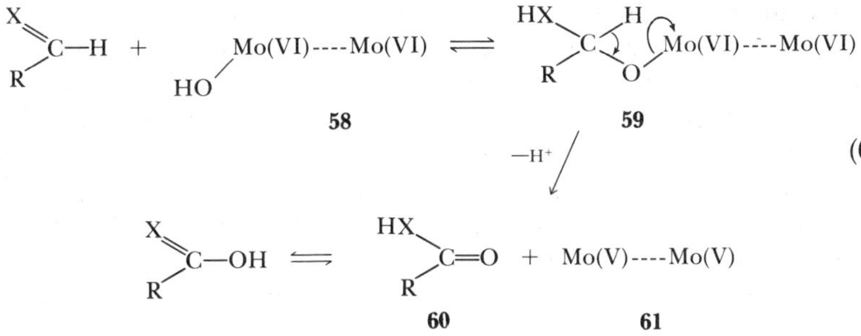

$$
\underset{\substack{\\ \textbf{58}}}{\overset{X}{\underset{R}{>}}C{-}H} \;+\; \underset{\substack{HO \\ }}{Mo(VI){-}{-}{-}{-}Mo(VI)} \;\rightleftharpoons\; \underset{\substack{\\ \textbf{59}}}{\overset{HX}{\underset{R}{>}}C{\diagdown}\overset{H}{\underset{O}{\diagup}}Mo(VI){-}{-}{-}{-}Mo(VI)} \tag{65}
$$

$$\xrightarrow{-H^{+}}$$

$$
\underset{\substack{\\ \textbf{60}}}{\overset{X}{\underset{R}{>}}C{-}OH} \;\rightleftharpoons\; \underset{\substack{\\ \textbf{61}}}{\overset{HX}{\underset{R}{>}}C{=}O} \;+\; Mo(V){-}{-}{-}{-}Mo(V)
$$

In the proposed mechanism for aldehyde or purine hydroxylation, the hydrogen is transferred as a proton. Thus the mechanism is closely related to the dehydrogenations discussed previously. The conversion of **59** to **60** and **61** is very similar to that which occurs in the generally accepted mechanism for the chromic acid oxidation of aldehydes or alcohols [1, 101]. The main difference is that in the chromate oxidation, Cr(VI) is converted to Cr(IV), whereas in the present case it is suggested that two Mo(VI)'s are converted to Mo(V)'s. However, it is possible that Mo(IV) may also be a fleeting intermediate in the enzymic cases. In any event, the important point is that the hydrogen is apparently transferred from the substrate as a proton in both cases.

13 SOME OXYGENATION–DECARBOXYLATIONS CATALYZED BY FLAVOENZYMES

Several enzymes have been characterized which catalyze reactions of the type shown in equation 66 [102]. In such reactions X can be either O as in lactic acid oxygenase [103] or NH as, for example, in lysine oxygenase [104], arginine oxygenase [105], and tryptophan oxygenase [106]. In each case the enzyme requires a flavin coenzyme in the oxidized

$$
\underset{\substack{| \\ XH \\ \textbf{62}}}{R{-}CH{-}COOH} + O_2 \longrightarrow \underset{\substack{\diagdown \\ XH}}{R{-}\overset{\overset{\textstyle O}{\parallel}}{C}} + CO_2 + H_2O \tag{66}
$$

form, none of the enzymes requires a metal ion, and the oxygen which ends up on the product comes from O_2. Although these enzymes are included in the general class of monooxygenases, it is clear that their mechanism is considerably different from that of most monooxygenases, which apparently proceed by an "oxene" mechanism (1, see Sects. 14 and 15). A reasonable overall pathway for the enzymic reactions illustrated in equation 66 is shown in equation 67. The first step, involving the dehydrogenation of **62** by the oxidized form of enzyme-bound flavin (F) to give **63** and the reduced form of the flavin (FH$_2$), has been discussed in Section 7 and certainly occurs readily in a number of other enzymic reactions. There is also evidence that it occurs in these monooxygenase reactions[102, 103, 105, 107]. Similarly, the ready reoxidation

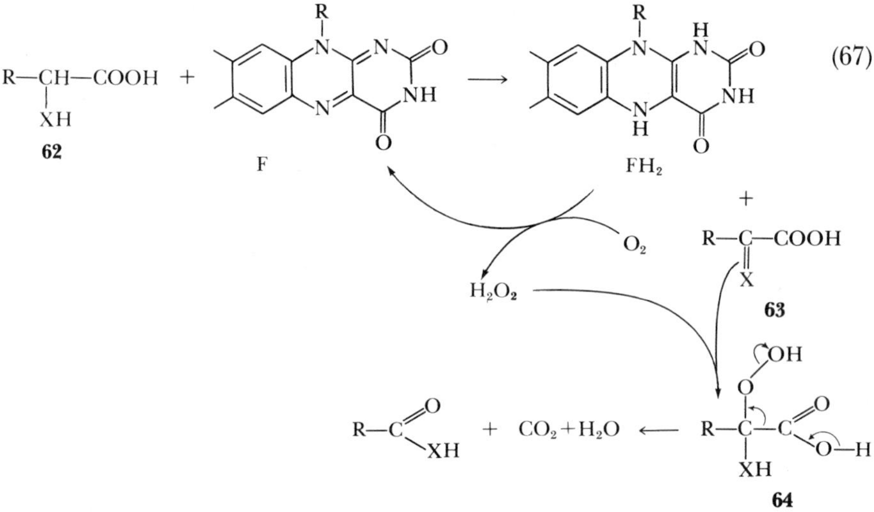

$$(67)$$

of FH$_2$ by O$_2$ to give F and H$_2$O$_2$ was discussed in Section 7. It has been known for some time that H$_2$O$_2$ reacts nonenzymically with α-ketoacids (**63**, X = O) to give CO$_2$ and the carboxylic acid[108–110], and this presumably occurs with **64** as an intermediate. A similar reaction of **63** when X is NH would be expected to give the amide. In the enzymic reactions there is no direct evidence that H$_2$O$_2$ is an intermediate – possibly the H$_2$O$_2$ is bound in some way to the enzyme or flavin. However, such a hydroperoxide (R'OOH) would be expected to react with **63** in the same way as H$_2$O$_2$.

Although it has been known for some time that H$_2$O$_2$ reacts readily at room temperature in aqueous solution with α-ketoacids to give the carboxylic acid and CO$_2$, this reaction until recently had not been studied mechanistically and no evidence for an intermediate such as **64** had been obtained. However, Rohrbaugh in the author's laboratory[111] has now looked at the kinetics of this reaction [using benzoylformic acid

(ϕCOCOOH) as the α-ketoacid] over a wide pH range and the data can only be explained on the basis of a peroxide–carbonyl adduct being an intermediate. At low pH's the formation of the adduct is rapid and the decarboxylation step is rate determining, while at high pH's the decarboxylation is rapid and the slow step is the formation of the intermediate. From additional effects of pH one can determine which of the various ionized forms of the reactants and intermediates are reacting in kinetically important steps. These results are summarized in the mechanism shown in equation 68. In this mechanism only one of the tautomers of the various intermediates is drawn, but it should be remembered that

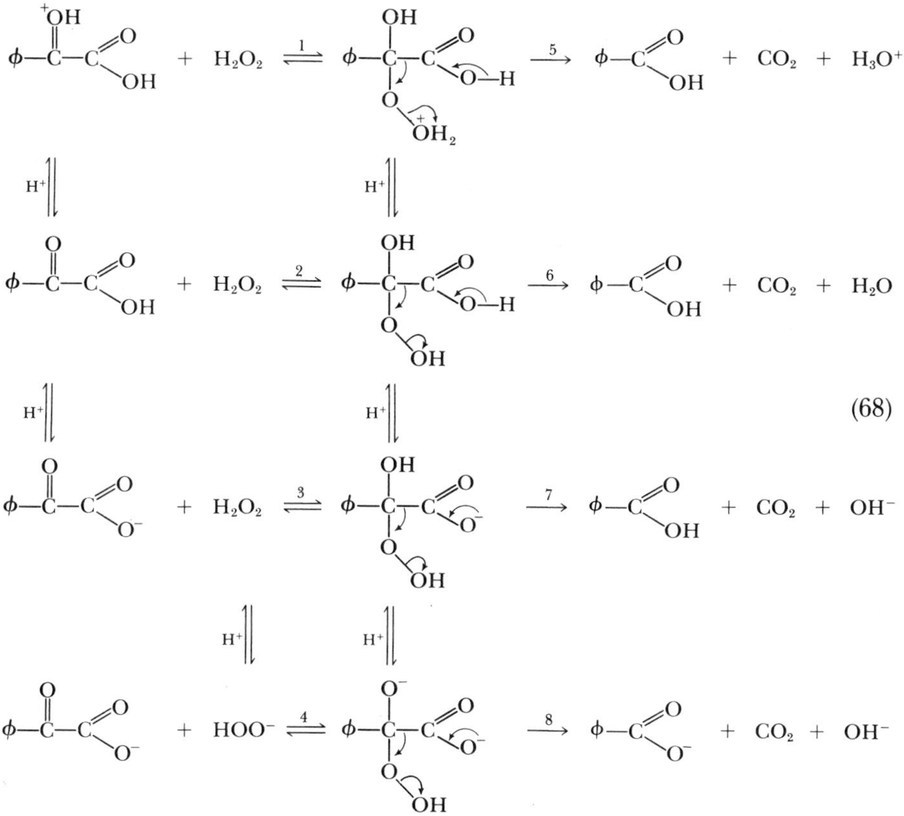

$$
\text{(68)}
$$

a reaction involving a different tautomer is kinetically indistinguishable. The rate of the oxidative decarboxylation is low at pH's 2 to 4 where steps 6 and 7 (eq. 68) are rate determining. At lower pH's the rate increases because step 5 occurs more rapidly. Also, from pH 4 to 6 the rate increases sharply with pH because step 8 is rapid. However, from pH 8 to 11 the rate is constant and this can only be explained if the rate-determining step changes so that step 3 now is the slow step. Above pH 11 the rate increases somewhat because of the contribution of step 4.

At physiological pH then, the formation of the carbonyl–peroxide adduct and its decarboxylation occur at comparable rates. The addition of H_2O_2 to carbonyl compounds is known to be an ionic reaction and is catalyzed by acids and bases[112]. Thus, one can understand how the enzyme could speed up this step. The decarboxylation of the peroxide-carbonyl adduct is also most certainly an ionic reaction; the mechanism is akin to that for the decarboxylation of β-ketoacids. Since proton transfers are probably involved in the decarboxylation of some of the various ionized forms of the intermediate, one would expect an enzyme to be able to increase the rate of this step also by having suitably placed acid and base groups on the enzyme.

Thus, although the enzymic reactions shown in equation 66 involve an oxygenation and decarboxylation as well as a dehydrogenation, it appears that again the reaction occurs mainly by an ionic mechanism involving proton transfers. The only step where free radicals are probably involved is in the oxidation of FH_2 by O_2 (see Sect. 7). However, free radicals formed from the substrate (**62**), or any of the intermediates involving this substrate, are almost certainly not formed, because they could not be stabilized by resonance.

14   ENZYMIC "OXENE" REACTIONS INVOLVING FLAVINS AND TETRAHYDROPTERIDINES

The reaction shown in equation 69 is catalyzed by a large number of different enzymes which have been termed mixed function oxidases by Mason[113] and monooxygenases by Hayaishi[114] (for reviews see 102, 115–117). In equation 69, S refers to the substrate and AH_2 to some

$$S + O_2 + AH_2 \longrightarrow SO + A + H_2O \qquad (69)$$

reducing agent. In crude systems NADH or NADPH is usually the reducing agent which is consumed, but with purified enzyme preparations it has been found that the reducing agents which interact directly with the enzymes are compounds such as tetrahydropteridines, reduced flavins, reduced metal ions, ferredoxins, and α-ketoacids. The oxygen atom which ends up in SO comes from O_2. The various enzymes catalyze a wide variety of oxidations, including hydroxylation of aromatic compounds, hydroxylation of aliphatic compounds, epoxidation of olefins, oxidative decarboxylation, oxidative dehydroxymethylation, desaturation, lactonization, and formation of amine oxides from amines.

As has been discussed in detail previously[1], there is considerable evidence that monooxygenases (except those discussed in Sect. 13) catalyze the oxidation of the various substrates by an "oxene" mechanism. The actual oxidizing agent is apparently able to transfer an oxygen atom, with six electrons in its outer shell, to the substrate. This general mechanism was suggested in 1964 by the present author[118] and has been termed the "oxene" mechanism because it is analogous to carbene and nitrene reactions[119–124].

Although several groups of workers, interested in the mechanisms of such monooxygenase-catalyzed reactions, have concluded that the "oxene" mechanism is a general mechanism for these enzymic reactions [1, 125–129], it is still not clear what the actual oxidizing agent is and how it is able to transfer the oxygen atom. One thing which is clear, however, is that the oxidizing agent is not exactly the same for each enzyme. In addition to the variety of reducing agents which have been found to function with the different enzymes, it is now known that some of the enzymes, which presumably function by an "oxene" mechanism, require a transition metal ion while others do not. In the previous article[1], possible mechanisms for some monooxygenase reactions which require a metal ion were discussed in some detail, and the important role the proton would have in such reactions was mentioned. In this section, the discussion is focussed on mechanisms for those enzymes which do not require a metal ion. In the following section possible mechanisms for monooxygenases which use an α-ketoacid as a reducing agent are considered. These latter enzymes probably require a metal ion, but the oxidizing agent is most likely quite different from those discussed previously[1].

One group of monooxygenases, for which it is quite clear no metal ion is necessary[130], requires a flavin as a cofactor. Several enzymes in this group have been purified to homogeneity and crystallized[102, 131]. These include salicylate decarboxylase[132, 133], imidazoleacetate hydroxylase[134], and p-hydroxybenzoate hydroxylase[135, 136]. In each case the enzyme, with the reduced form of flavin attached, interacts with O_2 and the substrate to give the oxidized form of flavin, the oxidized substrate, and water. The reduced form of the enzyme-bound flavin is regenerated for subsequent turnovers by reaction with a reducing agent.

In Section 11 it was emphasized that transition metal ions are required for reactions involving O_2 if free radicals cannot be intermediates. Since the monooxygenases which require flavin do not have a transition metal ion, it is clear that free radicals must be intermediates at some stage in the mechanism. It is unlikely that substrate radicals are involved because these would not be very stable. It seems more likely that O_2 reacts with the reduced form of flavin to give an intermediate which then reacts by an ionic mechanism with the substrate. As discussed in Section 7, the O_2 reaction with reduced flavin involves free radicals as intermediates; but singlet species, H_2O_2 and the oxidized form of flavin are the ultimate products. Perhaps a spin-paired compound formed on reaction of O_2 with reduced flavin can be altered by the enzyme so that it can act as an oxygen atom transfer or "oxene" reagent.

Mager and Berends and co-workers[137–140] have observed that aromatic compounds are hydroxylated by O_2 in a nonenzymic system when various reduced flavins are present. They suggest that the hydroxylating agent in this system is a flavin peroxide which they formulate as **65**. A

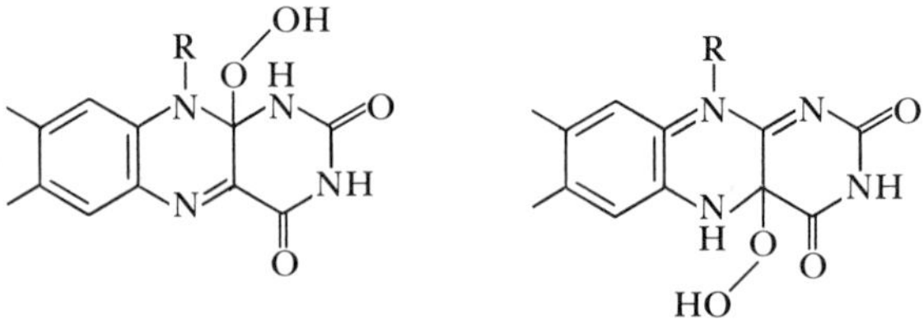

flavin peroxide is certainly a reasonable spin-paired intermediate in the reaction of O_2 with reduced flavin. However, both **65** and **66** should be considered as possible structures for such a peroxide. Mager and Berends' suggestion that a flavin peroxide such as **65** is the hydroxylating agent is unlikely, because alkyl hydrogen peroxides are not hydroxylating agents in the absence of metal ion catalysts.

The suggestion is made here that the actual hydroxylating agent is not the flavin peroxide, but a compound formed from one of the peroxides **65** or **66**. Since the reagent formed from **66** seems like the most likely candidate, it will be discussed first. It is suggested that **66** reacts by a proton transfer to give **67** (eq. 70) and that **67** is the actual

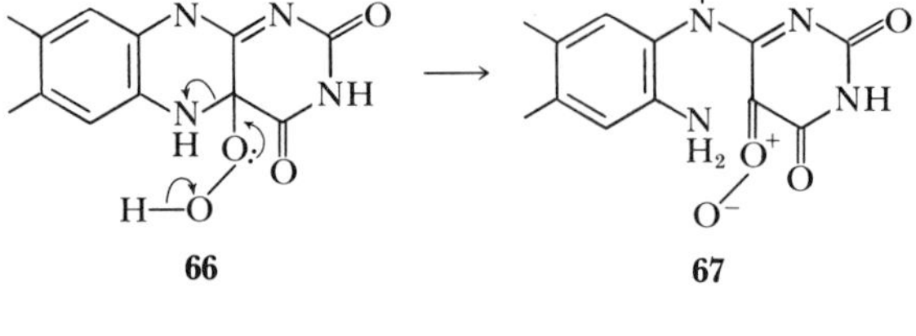

(70)

hydroxylating agent in the enzymic reactions. **67** is a carbonyl oxide; such compounds are known to be intermediates in the reaction of ozone with alkenes[141, 142]. In ozonolyses the intermediate carbonyl oxides usually react with carbonyl compounds to give ozonides. However, in some cases carbonyl oxides apparently transfer oxygen atoms to organic compounds, for example, alkenes are converted to epoxides[143]. Also, Giacin, working in the author's laboratory[144], found that the carbonyl oxide formed on reacting certain carbenes with O_2 would oxidize saturated hydrocarbons to alcohols and ketones. However, this carbonyl oxide may not be directly related to **67** because the carbene–oxygen adduct is a triplet species and reacts by a radical mechanism. It appears that oxygen atom transfer occurs more readily the more electronegative the groups attached to the carbon of the carbonyl oxide are. In the extreme case where the carbon of the carbonyl oxide is replaced by the electronegative element oxygen, one has ozone, and this is known to be an oxygen atom transfer reagent in some reactions. In the reaction of ozone with pi-bonds (olefins or aromatic compounds) a different mechanism operates because a 1,3-dipolar addition is possible[141], but with alkanes, ozone apparently transfers an oxygen atom ("oxene" mechanism) to give alcohols mainly with retention of configuration[145, 146]. In summary, then, carbonyl oxides are probably good oxygen atom transfer reagents if the groups attached to the carbon of the carbonyl group are able to stabilize a negative charge.

It is fairly obvious why the electronegative groups are necessary. As written in **67** (and as usually written), the terminal oxygen of the carbonyl oxide has a negative charge and would be expected to be a nucleophilic reagent. However, for attacking an alkene, aromatic compound, alkane,

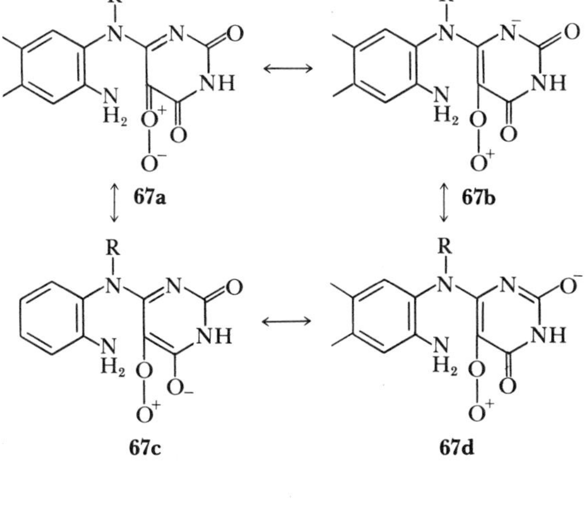

(71)

etc. so that an oxygen atom transfer can occur, an electrophilic species is required. Carbenes and nitrenes are electrophilic reagents. The terminal oxygen will be more electropositive the more a negative charge can be stabilized on the rest of the molecule. The carbonyl oxide (**67**) should be able to make the terminal oxygen quite electrophilic because the resonance forms **67b–67d** (eq. 71) all put the negative charge on an electronegative element. If one of the negatively charged sites illustrated in **67b**, **67c**, or **67d** were protonated by an acid on the enzyme, the terminal oxygen would be even more electrophilic. Therefore, in the transition state the terminal oxygen can develop an electron deficiency, and **67**, or a protonated form of **67**, should be a good oxygen atom transfer reagent. It is thus a likely candidate as the actual hydroxylating agent in the monooxygenase reactions involving flavin.

The suggested overall mechanism for these enzymic reactions is illustrated in equation 72. As discussed in Section 7, the reaction of FH_2 with O_2 to give **66** would almost certainly have to proceed by a free

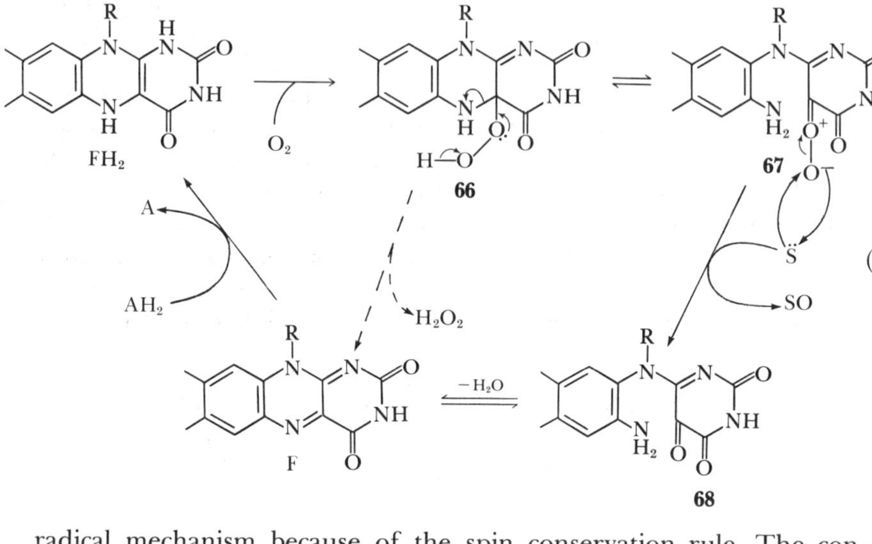

$$(72)$$

radical mechanism because of the spin conservation rule. The conversion of **66** to **67** involves a proton transfer and cleavage of a nitrogen–carbon bond of the flavin. This should not be a difficult process, especially if appropriately placed acid and base groups on the enzyme assist the protonation and deprotonation. If **67** transfers an oxygen atom to the substrate as shown, then **68** would be formed; **68** should give F very readily, even nonenzymically, because it is just a Schiff base formation. To complete the cycle, F would have to be reduced back to FH_2 by some reducing agent (frequently NADH).

It seems less likely, but possible, that a compound formed from the flavin peroxide **65** is the hydroxylating agent. A reaction of **65** related to the suggested conversion of **66** to **67** would lead to **69** or **70** (eq. 73). Again, the terminal oxygen of these carbonyl oxides should be electrophilic because of resonance forms such as **69b** (a similar resonance structure for **70** is not illustrated) or possibly if the nitrogen originally in the 5-position of the flavin is protonated. If **69** or **70** transferred an oxygen atom to a substrate then **71** or **72** would be formed, respectively. The regeneration of the flavin ring system (F) from these again involves only the loss of water, but contrary to the conversion of **68** to F, the conversion of **71** or **72** to F would not be expected to occur readily. Such a conversion involves the attack of a nitrogen nucleophile on an amide, and amides are quite unreactive toward nucleophilic attack under physiological conditions. It is possible that the enzymes could catalyze such a reaction, but this would be an extra requirement for the enzyme. In a nonenzymic system, in which a reduced flavin derivative is oxidized by O_2 (no other substrate present), Mager, Berends, and Addink[138]

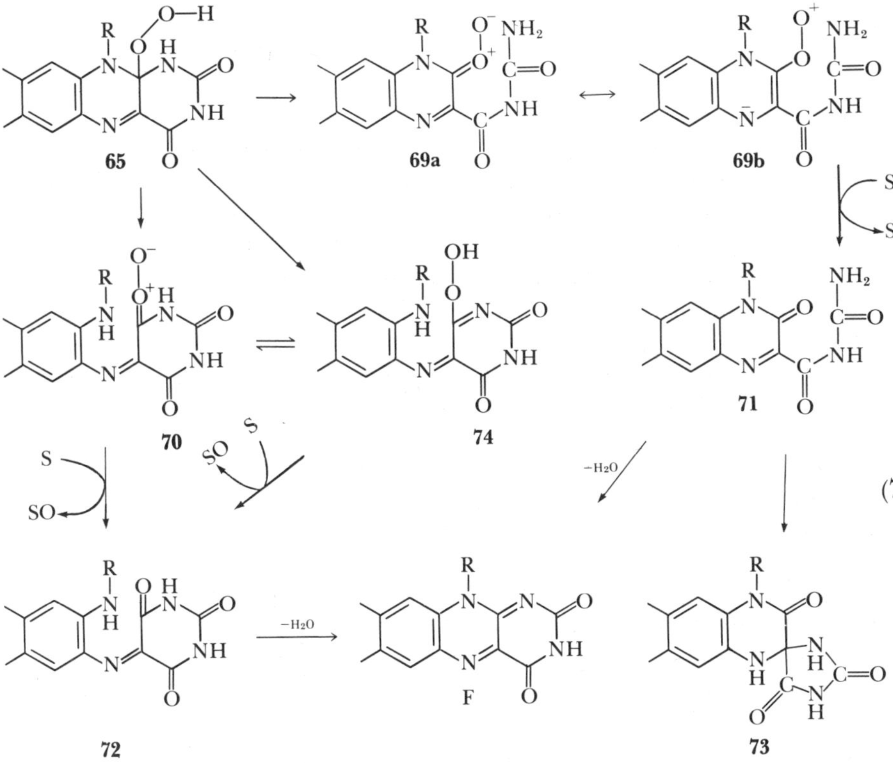

$$(73)$$

observed that a product is a compound related to **73**. This strongly indicates that a compound related to **71** is in fact formed in their system; **71** would be expected to give **73** readily, whereas it would not be expected to regenerate the flavin ring system at neutral pH's. In Mager and Berends' case[138] the 1-position of the flavin has a methyl. Therefore the intermediate would have a positive charge if a peroxide related to **66** were formed, while it would not if the peroxide structure was similar to **65**. Perhaps this forces the peroxide related to **65** to form in this particular case. However, this work does not prove that in the enzymic reaction a similar peroxide would be formed.

Another alternative for flavin peroxide (**65**) is that it reacts to give **74**, either directly or through **70** as an intermediate. If **74** was formed, one would expect it to be a good oxygen atom transfer reagent. It is similar to the intermediate formed on reacting a nitrile with H_2O_2, and it is known that such compounds transfer oxygen ,atoms readily. For example, this reagent is sometimes used to epoxidize olefins[147]. The mechanism for such an oxygen atom transfer is probably closely related to that for the transfer from peracids (147, see also next section). If **74** were the hydroxylating agent in the enzymic reactions, then **72** would be the product after "oxene" transfer, and again the enzyme would have to catalyze the formation of F from **72** so that the catalytic reaction could continue.

Very recently it has been reported[148] that benzoate is a "pseudosubstrate" for a salicylate decarboxylase from a soil bacterium. With this enzyme benzoate is not oxidized by O_2, but in the presence of benzoate, H_2O_2 is formed. Thus, benzoate uncouples the hydroxylation reaction. These results are quite consistent with the suggested mechanisms. Benzoate would be less reactive than salicylate; thus, **67** (eq. 72) might not be able to hydroxylate it. However, **66** and **67** should be in rapid equilibrium, and **66** would be expected to give F and H_2O_2 readily. Thus the reaction would be short circuited to give H_2O_2 as observed. Similarly **65** (eq. 73) would also be expected to give F and H_2O_2 if the hydroxylation could not occur.

In summary then, it seems very likely that the "oxene" reagent in monooxygenase-catalyzed reactions, which involve flavin and no metal ion, is a carbonyl oxide, formed from a flavin peroxide by ring cleavage of the flavin ring system. In organic reactions, compounds of this type are known to be oxygen atom transfer reagents. This author believes that the most likely such carbonyl oxide in the enzymic reactions is **67**. However, at this stage one cannot eliminate the possibility that **68**, **70**, or **74** is the actual "oxene" reagent in the enzymic reactions; the regeneration of the flavin ring system (**71** or **72** to F) is the most difficult step to rationalize by this mechanism. In any event it is very unlikely that a flavin peroxide itself (**65** or **66**) is the "oxene" reagent because alkyl peroxides are not good oxygen atom transfer reagents. Since many of the steps leading to the formation of the possible carbonyl oxide reagents involve proton transfers, the enzyme could presumably direct the pathway to give the most effective reagent by having appropriately placed acid and base groups on the enzyme surface.

The transfer of an oxygen atom from the "oxene" reagent to the aromatic compounds, which are substrates for the enzymes that require flavin, would probably lead to a benzene epoxide as the initial intermediate. This is the expected initial product by analogy to related carbene and nitrene reactions[119–124], and the elegant work on the "NIH shift" (129, and references therein) indicates that benzene epoxides are intermediates in a number of monooxygenase-catalyzed oxidations of aromatic compounds. Where phenols are the ultimate products they are presumably formed by rearrangement of the aromatic epoxide[129, 149]. In the case of salicylate hydroxylase[132, 133, 148] it seems likely that the initially formed aromatic epoxide (**75**) reacts as shown in equation 74[1]. The conversion of **75** to catechol and CO_2

$$\qquad\qquad\qquad\qquad\qquad\qquad\qquad (74)$$

75

should occur very readily because the aromatic resonance can be regenerated by a facile proton shift and bond cleavages as shown.

The monooxygenases, such as phenylalanine hydroxylase and tyrosine hydroxylase, which require a tetrahydropteridine[102, 150] may generate an "oxene" reagent very closely related to that suggested above for the flavin enzymes. For some time it has been unclear whether the tetrahydropteridine enzymes require a metal ion or not. If they do, then the "oxene" mechanism suggested earlier by this author[1, 118] seems more likely. However, the tetrahydropteridines (**76**) have structures very closely related to reduced flavin (FH_2); the main differences are that a nitrogen replaces oxygen on carbon-2, and the aromatic ring is missing in the tetrahydropteridines. Nevertheless, the tetrahydropteridines have chemical properties similar to those of reduced flavin; they would be expected to react with O_2 to give a peroxide analogous to either **65** and **66** and thus could generate a carbonyl oxide analogous to those discussed above for the flavins. In equation 75 such an overall mechanism for the enzymic reactions is illustrated. For the same reasons discussed above, the carbonyl oxide (**78**) generated from the peroxide (**77**) is the most likely "oxene" reagent. The product (**79**) following oxygen atom transfer would cyclize to **80** very readily, while the product which would have formed if the "oxene" reagent were similar to **69** or **70** would have the same unreactivity as **71** or **72**.

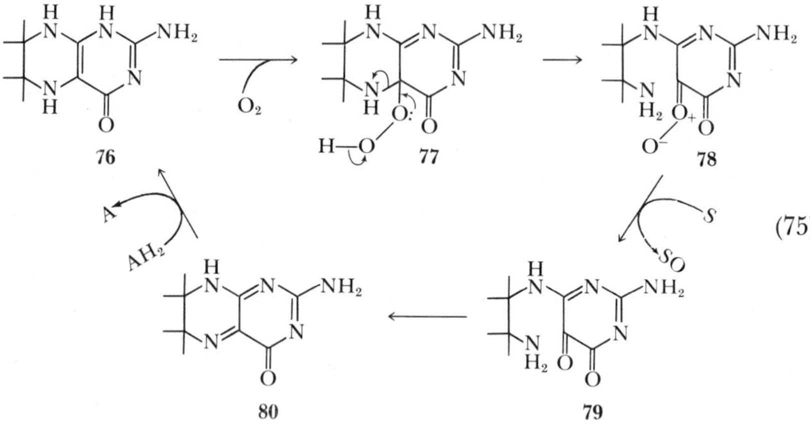

(75)

There is some evidence that this type of mechanism for the tetrahydropteridine monooxygenases is more likely than that involving a metal ion[1]. The fact that a dihydropteridine such as **80** is the product following oxygen atom transfer is consistent with either mechanism. However, Kaufman and co-workers[151, 152] have recently reported that, with phenylalanine hydroxylase, the O_2 reduction can be partially uncoupled from the hydroxylation reaction so that some H_2O_2 is formed. This occurs when *p*-fluorophenylalanine is substituted for phenylalanine[151] or when 7-methyltetrahydropterin is substituted for the natural cofactor tetrahydrobiopterin[152]. This is consistent with the mechanism of equation 75; the peroxide (**77**) would be expected to give H_2O_2 and **80** if oxidation of the aromatic substrate did not occur readily (one would expect **78** and **77** to be in equilibrium with one another). Kaufman's results cannot be readily rationalized in terms of a mechanism involving the O_2, enzyme-bound tetrahydropteridine, and substrate all reacting at once to give the dihydropteridine (**80**) and oxidized substrate [1].

15 ENZYMIC "OXENE" REACTIONS REQUIRING α-KETOACIDS

Very recently it has become apparent that, in several monooxygenase reactions, an α-ketoacid is used for taking up the two oxidizing equivalents not used in the oxidation of the other substrate by O_2 [153, and references therein]. Such enzymes usually use α-ketoglutarate as the α-ketoacid, and they catalyze reactions such as the hydroxylation of proline derivatives and betaines. In the overall reaction (eq. 76) the

$$
S + O_2 + R-\underset{\underset{O}{\|}}{C}-C\overset{\nearrow O}{\underset{\searrow OH}{}} \longrightarrow SO + R-C\overset{\nearrow O}{\underset{\searrow OH}{}} + CO_2 \qquad (76)
$$

α-ketoacid is oxidatively decarboxylated. Recent experiments with isotopic oxygen [153] indicate that one atom of oxygen from O_2 ends up in SO and the other in the carboxyl group of the product carboxylic acid. An intramolecular reaction, apparently of the same type, has been known for some time. This is the reaction catalyzed by *p*-hydroxyphenylpyruvate oxidase (eq. 77). For this enzymic reaction it is known

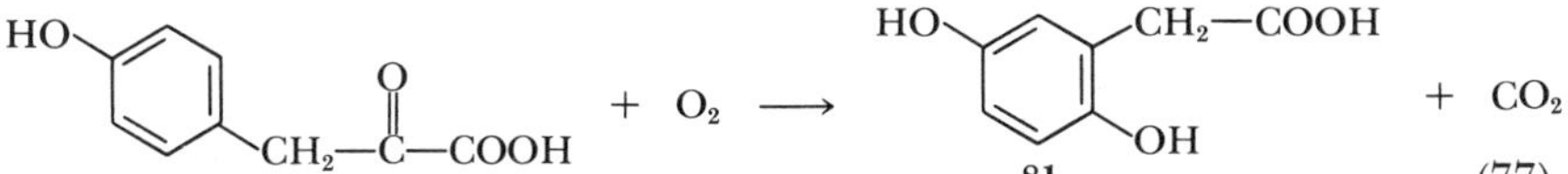

$$(77)$$

[153, 154] that the side chain migrates to an *ortho* position, the new oxygen on the aromatic ring comes from O_2, and one atom of O_2 ends up in the carboxyl group of homogentisic acid (**81**). Clearly the reactions of equations 76 and 77 are related, but up to this time their mechanisms have not been obvious.

Witkop and co-workers some time ago [155, 156] suggested a mechanism (eq. 78) for *p*-hydroxyphenylpyruvate oxidase involving the peroxide **82** as an intermediate. This was an ingenious and reasonable mechanism which fit the data then available, but later work has indicated

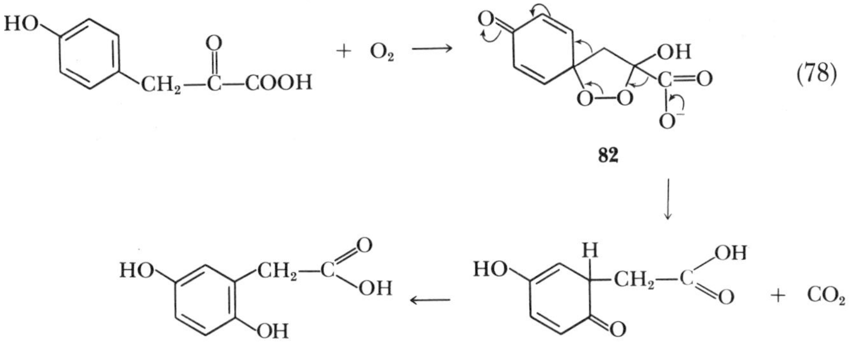

$$(78)$$

that it is probably not correct. It has been shown[157] that the same enzyme which catalyzes the reaction of equation 77 will also catalyze the conversion of phenylpyruvate to o-hydroxyphenylacetic acid, albeit at a slower rate. Since the mechanism of equation 78 requires the *para*-hydroxy group, it cannot be the correct mechanism, at least for the phenylpyruvate reaction, and probably not for the p-hydroxyphenyl-pyruvate reaction either.

Lindblad et al.[153] suggest that the initial step in enzymic reactions using α-ketoglutarate is the formation of a substrate peroxide which subsequently reacts with the α-ketoacid to give SO, CO_2, and RCOOH. The second step would certainly be expected to occur readily; as discussed in Section 13, peroxides oxidatively decarboxylate α-ketoacids under physiological conditions even in the absence of an enzyme. The problem with the suggested mechanism of Lindblad et al. is that it is not obvious how the substrate peroxide could be formed. The positions on the substrate (proline, betaine, etc.) which get hydroxylated during the enzymic reaction are completely unactivated, and enzymic formation of a peroxide at such a position is without analogy in enzymic reactions. Furthermore, it is not obvious how an enzyme could catalyze such peroxide formation. An additional argument against the suggested mechanism of Lindblad et al. is that the enzymic reactions requiring α-ketoglutarate and that catalyzed by phenylpyruvate oxidase appear similar, and it is not clear what kind of substrate peroxide could be formed in the latter case.

In fact, both reactions appear to be examples of "oxene" mechanisms. An oxene being analogous to carbenes and nitrenes, would be expected to insert into a completely unactivated carbon–hydrogen bond to give an alcohol[1], as occurs in some of the enzymic reactions requiring α-ketoglutarate. Also, the rearrangement of the side chain in the phenyl-pyruvate oxidase catalyzed reaction is very similar to rearrangements ("NIH shifts") observed in other enzymic reactions involving benzene epoxides as intermediates[129, 158]. In these cases the benzene epoxide is almost certainly formed by an oxene mechanism. Therefore, the question concerning the mechanisms of the reactions requiring α-keto-acids is: What is the oxene reagent? The suggestion is made here that the oxene reagent is a peracid formed by interaction of O_2 with the α-ketoacid (eq. 79). Peracids are known to be oxygen atom transfer

$$\underset{\text{OH}}{\overset{\overset{\displaystyle O}{\|}\quad\overset{\displaystyle O}{/\!/}}{R\!-\!C\!-\!C}} \quad +\,O_2 \longrightarrow \underset{\text{O\!-\!OH}}{\overset{\overset{\displaystyle O}{/\!/}}{R\!-\!C}} \quad +\,CO_2 \qquad (79)$$

reagents; they epoxidize olefins readily [147]. Furthermore, it has been shown that some peracids give the highest amount of the NIH shift of any reagent in a model system [159]. This latter test apparently measures the amount of benzene epoxide formed by an "oxene" transfer to the aromatic ring.

In order for the reaction of equation 79 to occur, it is most likely that a transition metal ion would be necessary as a catalyst. Again because of the spin conservation rule the direct reaction of the singlet *α*-ketoacid with the triplet O_2 to give all singlet products is spin forbidden. Therefore, free radicals would have to be intermediates or a metal ion must be present. It is not likely that a free radical formed from the *α*-ketoacid could be an intermediate because no obvious relatively stable free radical can be envisioned. There are hints, but no proof, that a transition metal ion is required by *p*-hydroxyphenylpyruvate oxidase. In the presence of certain complexing agents the enzymic reaction is inhibited, and those which complex Cu(II) well are the best inhibitors [154]. If a transition metal ion (M^{2+}) is present on the enzyme, then a reasonable mechanism for the formation of the peracid from the *α*-ketoacid and O_2 is that shown in equation 80. It is suggested that the enzyme-bound metal ion, O_2, and the *α*-ketoacid are in equilibrium with a complex such as **83**. If the orbitals of the O_2 overlap with those of the transition metal one cannot say where the unpaired electrons are, and one can get an ionic reaction such as the conversion of **83** to **84** if **84** has the same number of unpaired electrons as **83**. One would expect **84** to lose CO_2 readily to

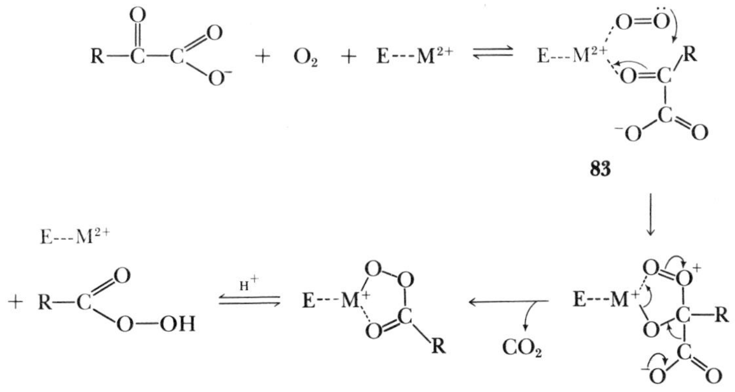

$$(80)$$

give **85**; in fact, a tetrahedral intermediate such as **84** might only have transitory existence. **85** is clearly just a complex of the peracid and enzyme-bound metal ion.

In order for the peracid to react with the other substrate by an "oxene" mechanism one expects that the terminal oxygen of the peracid would have to be free and not complexed to the metal ion. In the case of p-hydroxyphenylpyruvate oxidase, a complex like **85** could definitely not transfer an oxygen atom intramolecularly to the aromatic ring because it is stereochemically impossible. If the peracid were formed by the mechanism of equation 80, then **85** would have to give free peracid or possibly be protonated and isomerize to **86**. However, it is possible that **86** might be formed directly, for example, by the mechanism shown in equation 81. In **87** the O_2 is assumed to be enzyme bound, but not

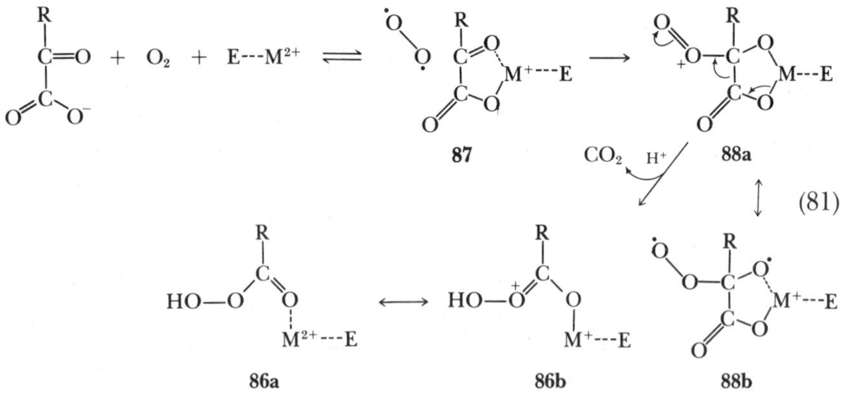

coordinated to the metal ion. The conversion of **87** to **88** is spin allowed if **88** has the same number of unpaired electrons as the total on **87** (on the enzyme-bound metal ion and O_2), or if **88** has two more or two less unpaired electrons than **87**. One could think of this conversion in terms of the initial formation of **88b**, but this is only a resonance form of **88a** if they have the same number of unpaired electrons. In the transition state for conversion of **87** to **88** the orbitals on O_2 will overlap with those of the ligand and metal ion, so that again one cannot say where the unpaired electrons are. The conversion of **88** to **86b**, which is just a resonance form of **86a**, should occur readily. In summary then it is not necessary for the O_2 to be directly bound to the transition metal ion in order to get the formation of a peracid from O_2 and the α-ketoacid (this would also be true for other metal ion catalyzed reactions of O_2). However, the metal ion is necessary and must somehow be in electronic contact with the transition state in order to circumvent the spin conservation rule. For the enzymic reaction, the mechanism of equation 81 has advantages because the enzyme-bound peracid (**86**) is in a state where it can react directly as an oxene reagent with the other substrate.

As far as the author is aware, there appear to be no nonenzymic examples of the formation of a peracid from an α-ketoacid and O_2. However, the mechanisms suggested in equations 80 and 81 seem reasonable. Also, the data on the enzymic reactions which utilize α-ketoacids can be rationalized very easily on the basis of a peracid intermediate and cannot be readily rationalized in terms of other mechanisms. Therefore, it seems like a very probable intermediate.

The details of the mechanism of oxygen atom transfer from peracids is known to some extent. In the epoxidation of olefins the peracid acts as an electrophilic reagent[147]. The best description of the mechanism is that shown in equation 82. A proton transfer as shown either directly

$$\tag{82}$$

precedes or occurs concurrently with the oxygen atom transfer. For the reaction catalyzed by p-hydroxyphenylpyruvate oxidase such a proton transfer in a cyclic transition state could not occur because the oxygen atom sterically could not then be transferred to the aromatic ring. A reasonable mechanism for the oxene transfer in this enzyme is outlined in equation 83. The species which transfers the oxene intramolecularly in the enzymic reaction could be either the peracid **89** or metal-complexed peracid **86**. In equation 83 the mechanism is illustrated for the peracid. In the enzymic conversion of **89** to **90** (or **91**) a proton transfer from one oxygen to the other still occurs, as in the nonenzymic epoxidation mechanism (eq. 82), but now it does not occur in a cyclic

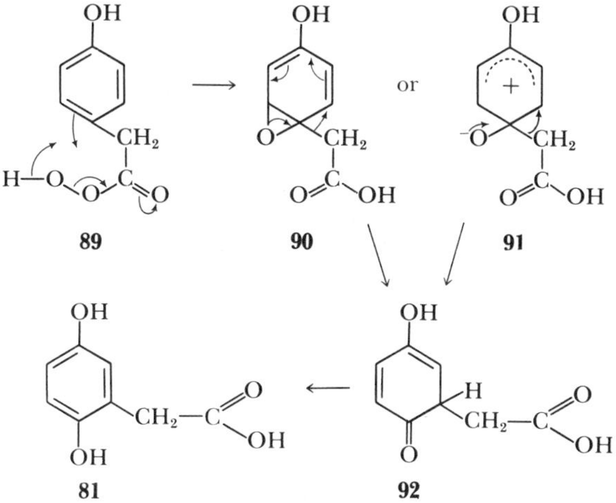

$$\tag{83}$$

transition state. Presumably, suitably placed acid and base groups on the enzyme would catalyze such a step. On oxygen atom transfer, either **90** or **91** could be formed. Most of the enzymic mechanisms involving oxygen atom transfer to aromatic rings have been interpreted in terms of a benzene epoxide such as **90** as an intermediate [1, 129], but it should be pointed out that nearly all the work on model systems, and on the NIH shift with enzymic systems, can be rationalized in terms of **91** as the important intermediate. The difference between **90** and **91** is that **91** is a zwitterion and has one less bond. However, **91** should be the most stable of all possible electrophilic aromatic substitution intermediates because the opposite charges would tend to stabilize one another. Furthermore, the rearrangement of **91** to **92** should be a more facile process than that of **90** to **92**. For some time it has bothered this author that the NIH shift occurs so readily. It is not obvious that a benzene epoxide should rearrange so completely when it is possible to go to the phenol merely by a proton shift and epoxide ring cleavage. However, if a zwitterion like **91** is the intermediate actually formed in the many enzymic hydroxylation reactions, it is easy to understand why the NIH shift is so general. In any event, the difference between **90** or **91** being the intermediate is a small point.

The insertion of oxygen atoms into saturated hydrocarbons by peracids has not been studied extensively, but a recent report [160] indicates that it does occur. Thus it is reasonable that the enzymes which utilize α-ketoglutaric acid could catalyze such a reaction of persuccinic acid (the intermediate peracid expected in these cases). The mechanistic details of carbon–hydrogen insertion reactions of carbenes and nitrenes are not too clear. Therefore, the details of carbon–hydrogen insertions by enzymic oxene reagents, such as peracids, will have to await further study.

16   CONCLUSION – WHY PROTON TRANSFERS RATHER THAN HYDRIDE ION OR HYDROGEN ATOM TRANSFERS?

The main thesis of this Chapter is that, in nearly all biological redox reactions in which hydrogen is transferred, the hydrogen is transferred as a proton and not a hydride ion or a hydrogen atom. Furthermore, it is the author's contention that basically the mechanisms of these redox reactions are not too different from the mechanisms of nonredox enzymic reactions. In particular, general acid and base catalysis by suitably placed groups on the enzymes probably plays as large a role in the enzymic redox reactions as in nonredox enzymic reactions. Using these principles one can develop reasonable mechanisms for a large number of enzymic redox reactions and relate many of them to non-enzymic reactions. However, one may legitimately inquire: (1) why are proton transfers favored so greatly over hydride ion and hydrogen atom transfers in biological redox reactions and (2) why is general acid and base catalysis (proton transfers) so effective in catalyzing reactions? This latter question can be asked concerning nonredox enzymes as well, and in the author's opinion has never been suitably answered.

Before considering these questions a comment should be made about possible experimental techniques for determining whether a hydride ion, hydrogen atom, or proton is transferred in a given reaction. There is essentially no experimental method which can distinguish among these possibilities. Usually a particular mechanism for transfer has been proposed on the basis of the stoichiometry of a given reaction. If, for example, a neutral molecule gives after hydrogen transfer a positively charged product, then it is usually assumed a hydride ion was transferred; if it gives a neutral product, a hydrogen atom transfer is postulated, and if it gives a negatively charged product, a proton transfer is proposed. Various chemical and physical methods are available for identifying such charged or radical products. However, these products could also have arisen by mechanisms involving electron and other hydrogen species transfers. Therefore, the overall stoichiometry of a hydrogen transfer reaction cannot be used to distinguish the various possible mechanisms for transfer. Consequently, evidence for a given transfer mechanism must come from chemical intuition and from the correlation of large numbers of reactions with a given transfer mechanism and with other types of reactions. This is what has been attempted in this article.

One reason why one might expect hydride ion transfers to occur infrequently is because hydrogen is not a very electronegative element compared to many other elements, such as oxygen, nitrogen, and sulfur, commonly found in biological molecules. Consequently, in order to transfer a hydride ion it would be necessary to remove electrons from elements of comparable or greater electronegativity and localize them somewhat on the hydrogen. One might expect this to require considerable energy and thus not occur too readily.

Contrary to the general consensus, hydride ion transfers are not commonly found in organic reactions which bear any resemblance to biological ones[161]. It seems fairly clear that the metal hydrides ($NaBH_4$, $LiAlH_4$, etc.) reduce compounds by hydride ion transfer, but in these cases one starts with a high energy species (the metal hydride). Since hydrogen is the most electronegative element in these compounds, one might expect that it would carry the electrons (hydride transfer) to another molecule. However, species comparable to the metal hydrides are not encountered very frequently in biological systems (the cobamide coenzyme may be one exception).

Overall hydride transfer to a carbonium ion either intermolecularly or intramolecularly (for example, 1,2-shifts) certainly occurs readily in many organic reactions[161]. However, in these cases the system is very electron deficient, and one would expect that in the transition state the hydrogen being transferred has a partial positive charge. Therefore, the question arises whether one can consider this as a hydride ion transfer or as the transfer of electrons and a proton. Certainly the overall effect is that a hydride ion is transferred, but it is also just as certain that in the transition state the hydrogen does not have a negative or even a partial negative charge. In any event, there are probably few cases in biological systems where an electron deficiency comparable to that present on a free carbonium ion (to which such transfers occur readily) is encountered. In the reactions discussed in this article, the sites where reduction occurs have much more modest electron deficiencies.

Another type of reaction where hydride ion transfers have been suggested frequently[161] is the reduction of aldehydes and ketones by alcohols or aldehydes in the presence of base. Again in these cases the question arises whether the electron transfer accompanies the hydrogen or not. In many such examples it is probable that the transition state is cyclic, as for example in the Oppenauer oxidation or Meerwein-Pondorf-Verley reduction[162]. One could illustrate the mechanism as shown in equation 84 or equation 85. In the mechanism of equation 84 the hydrogen is shown being transferred as a hydride ion, while in 85 it is shown being transferred as a proton. In fact, one cannot distinguish between these two possibilities.

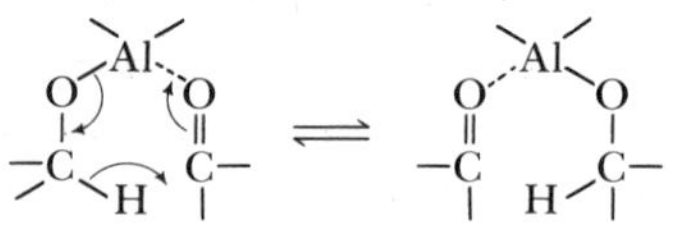

$$\tag{84}$$

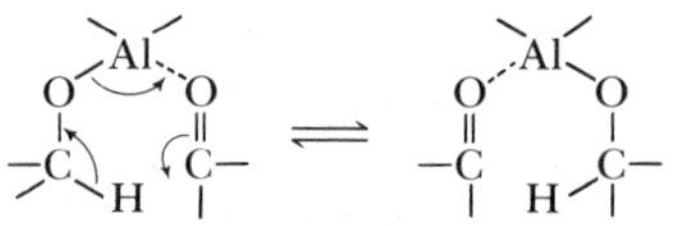

$$\tag{85}$$

The author does not want to give the impression that hydride ion transfers never occur. However, the point which needs stressing is that hydride transfers are not common in organic chemical reactions; they occur only in special cases and probably in fewer cases than is generally believed. Hydride transfers in a few select biological systems will undoubtedly be found to occur, but this is probably not a general mechanism for biological redox reactions, and other possibilities, such as those discussed in this article, should be considered.

The hydrogen is probably not transferred as a hydrogen atom very frequently in biological redox reactions because this would require free radicals as intermediates. Many of the substrates which are oxidized or reduced in the enzymic reactions discussed here could not stabilize such free radicals, and thus their formation is unlikely. Hydrogen atom transfers undoubtedly occur in many organic reactions which occur at high temperatures or in the gas phase, where ionic intermediates cannot be stabilized. However, under physiological conditions, hydrogen atom transfers are again not prevalent in organic reactions, unless an extremely reactive radical reagent (for example, the chlorine atom) is used or the product radical is very stable.

Probably the main reason that hydrogens are usually transferred as protons rather than hydride ions or hydrogen atoms is because proton transfers usually occur rapidly. There can be little doubt that this is the case for proton transfers from one electronegative element (oxygen, nitrogen, sulfur, etc.) to another. The extensive work of Eigen[6] indicates that such transfers are frequently diffusion controlled. When the proton is transferred to or from carbon the reactions are slower, probably because there is a geometry change of the carbon usually required[163]. However, proton transfers to or from carbon are still relatively rapid if stable products are formed.

What is special about the proton which allows it to be transferred so readily relative to other species? The proton is unique in that it is the only species in which a nucleus can be transferred without having to drag along any electrons either in an inner or outer shell. Since the nucleus is so much smaller than any species with electrons, one might expect this to impart special properties to proton transfers. One can think of this crudely in frictional terms. The transfer of a proton from one atom to another is to the transfer of a species with electrons associated, as the transfer of a small steel ball is to the transfer of a balloon through a somewhat viscous medium. The analogy is not completely accurate because it does not include attractive and repulsive forces. However, these should also aid proton transfers relative to other species. When a proton is being transferred from one molecule to another it is under an attractive force at all stages until it gets close to the other nuclei. Of course, the attractive force during transfer may not be as great as in the product or reactant molecule itself, and thus there will usually be an activation energy for transfer. For the transfer of species which have electrons associated, these electrons will have to penetrate the electron cloud around the nuclei to which the species is being transferred. Since electrons repel one another, this will not be easy unless the electrons can arrange themselves into stable orbitals overlapping the reactant, product, and transferring group in the transition state. This therefore puts an additional requirement on groups transferring with electrons associated. In essence, the electrons act as a shield, minimizing the effectiveness of the attraction of the nucleus for the electrons of the new molecule. The net effect is that a repulsion can occur as the transferring group with electrons approaches the reacting molecule.

There is now mounting evidence that proton transfers are sometimes so rapid that they occur before the environment around the product has reached its equilibrium state [164, 165]. For example, proton transfers are more rapid than solvent relaxation [166], so that immediately after the transfer, the products are in a higher energy state than their equilibrium state. This gives a hint as to why general acid and base catalysis is so effective in catalyzing reactions. If the protons are transferred to give a high energy state, which can be relieved most readily by a specific reaction occurring, then one would have a special catalysis. Presumably this is one of the reasons proton transfers and acid and base catalysis are so prevalent in enzymic reactions, including enzymic redox reactions.

ACKNOWLEDGEMENTS

The author would like to thank the following agencies for support of the research discussed in this chapter: the National Institute of Arthritis and Metabolic Diseases, Public Health Service (AM 13448); the National Science Foundation; and the Alfred P. Sloan Research Foundation.

REFERENCES

1. G. A. Hamilton, *Advances in Enzymology*, Vol. 32, F. F. Nord, Ed., Interscience, New York, 1969, p. 55.
2. J. March, *Advanced Organic Chemistry: Reactions, Mechanisms, and Structure*, McGraw-Hill, New York, 1968, p. 865.
3. W. I. Taylor and A. R. Battersby, *Oxidative Coupling of Phenols*, Dekker, New York, 1967.
4. F. H. Westheimer, in *The Mechanism of Enzyme Action*, W. D. McElroy and B. Glass, Eds., The Johns Hopkins Press, Baltimore, 1954, p. 321.
5. I. A. Rose, *Ann. Rev. Biochem.*, **35**, 23 (1966).
6. M. Eigen, *Angew. Chem. Intern. Ed. Engl.*, **3**, 1 (1964).
7. E. E. Snell, *Vitamins. Hormones*, **16**, 77 (1958).
8. A. E. Braunstein, *Enzymes*, **2**, 113 (1960).
9. E. E. Snell and W. T. Jenkins, *J. Cellular Comp. Physiol.*, **54**, Suppl. 1, 161 (1959).
10. T. C. Bruice and S. J. Benkovic, *Bioorganic Mechanisms*, Vol. 2, Benjamin, New York, 1966, p. 267.

11. T. C. Bruice and R. M. Topping, *J. Amer. Chem. Soc.*, **85**, 1480 (1963).
12. J. W. Thanassi, A. R. Butler, and T. C. Bruice, *Biochemistry*, **4**, 1463 (1965).
13. H. Yamada and K. T. Yasunobu, *J. Biol. Chem.*, **238**, 2669 (1963).
14. F. Buffoni, *Pharmacol. Rev.*, **18**, 1163 (1966).
15. E. V. Goryachenkova and E. A. Ershova, *Biokhimiya*, **30**, 165 (1965).
16. J. M. Hill and P. J. G. Mann, *Biochem. J.*, **91**, 171 (1964).
17. B. Mondovi, M. T. Costa, A. Finazzi-Agro, and G. Rotilio, *Arch. Biochem. Biophys.*, **119**, 373 (1967).
18. B. Mondovi, M. T. Costa, A. Finazzi-Agro, E. Chiancone, R. E. Hansen, and H. Beinert, *J. Biol. Chem.*, **242**, 1160 (1967).
19. Y. Yamada, O. Adachi, and K. Ogata, *Agr. Biol. Chem. (Tokyo)*, **29**, 912 (1965).
20. E. E. Snell, A. E. Braunstein, E. S. Severin, and Y. M. Torchinsky, Eds., *Pyridoxal Catalysis*, Interscience, New York, 1968, pp. 339–423.
21. H. Yamada, K. Yasunobu, T. Yamano, and H. S. Mason, *Nature*, **198**, 1092 (1963).
22. F. Buffoni, L. D. Corte, and P. F. Knowles, *Biochem. J.*, **106**, 575 (1968).
23. E. V. Goryachenkova, L. J. Stcherbatiuk and C. I. Zamaraev, in *Pyridoxal Catalysis*, E. E. Snell, A. E. Brannstein, E. S. Severin, and Y. M. Torchinsky, Eds., Interscience, New York, 1968, p. 391.
24. G. A. Hamilton and A. Revesz, *J. Amer. Chem. Soc.*, **88**, 2069 (1966).
25. G. A. Hamilton, in *Pyridoxal Catalysis*, E. E. Snell, A. E. Braunstein, E. S. Severin, and Y. M. Torchinsky, Eds., Interscience, New York, 1968, p. 375.
26. J. M. Hill and P. J. G. Mann, *Biochem. J.*, **99**, 454 (1966).
27. L. Schirch, private communication.
28. V. A. Gillis and G. A. Hamilton, unpublished results.
29. F. Buffoni, in *Pyridoxal Catalysis*, E. E. Snell, A. E. Braunstein, E. S. Severin, and Y. M. Torchinsky, Eds., Interscience, New York, 1968, p. 363.
30. C. S. Foote, *Accounts Chem. Res.*, **1**, 104 (1968).
31. Various authors, *Enzymes*, **7**, 275–648 (1963).
32. E. C. Slater, Ed., *Flavins and Flavoproteins*, Elsevier, New York, 1966.
33. K. Yagi, Ed., *Flavins and Flavoproteins*, University Park Press, Baltimore, 1968.
34. T. P. Singer, Ed., *Biological Oxidations*, Interscience, New York, 1968.
35. H. Beinert, *Enzymes*, **2**, 339 (1960).
36. A. Ehrenberg and P. Hemmerich, in *Biological Oxidations*, T. P. Singer, Ed., Interscience, New York, 1968, p. 239.
37. P. Hemmerich, C. Veeger, and H. C. S. Wood, *Angew. Chem. Intern. Ed. Engl.*, **4**, 671 (1965).
38. G. Palmer and V. Massey, in *Biological Oxidations*, T. P. Singer, Ed., Interscience, New York, 1968, p. 263.
39. T. P. Singer and E. B. Kearney, *J. Biol. Chem.*, **183**, 409 (1950).
40. C. H. Suelter and D. E. Metzler, *Biochim. Biophys. Acta*, **44**, 23 (1960).
41. G. K. Radda and M. Calvin, *Biochemistry*, **3**, 384 (1964).
42. J. L. Fox and G. Tollin, *Biochemistry*, **5**, 3865, 3873 (1966).
43. I. M. Gascoigne and G. K. Radda, *Biochim. Biophys. Acta*, **131**, 498 (1967).
44. M. J. Gibian and D. V. Winkleman, *Tetrahedron Letters*, **1969** 3901.
45. G. D. Weatherby and D. O. Carr, *Biochemistry*, **9**, 351 (1970).
46. L. E. Brown and G. A. Hamilton, *J. Amer. Chem. Soc.* **92**, 7225 (1970).
47. A. H. Neims, D. C. DeLuca, and L. Hellerman, *Biochemistry*, **5**, 203 (1966).
48. D. J. T. Porter and H. J. Bright, *Biochem. Biophys. Res. Commun.*, **36**, 209 (1969).
49. T. T. Tchen and H. van Milligan, *J. Amer. Chem. Soc.*, **82**, 4115 (1960).
50. J. Retey, J. Seibl, D. Arigoni, J. W. Cornforth, G. Ryback, W. P. Zeylemaker, and

C. Veeger, *Nature*, **216**, 1320 (1967); D. Arigoni, private communication; G. R. Drysdale, unpublished results.

51. O. Gawron, A. J. Glaid, K. P. Mahajan, G. Kananen, and M. Limetti, *J. Amer. Chem. Soc.*, **90**, 6825 (1968).
52. J. L. Harkins and G. A. Hamilton, unpublished observations.
53. P. Hemmerich, V. Massey, and G. Weber, *Nature*, **213**, 728 (1967).
54. D. E. Guttman and T. E. Platek, *J. Pharm. Sci.*, **56**, 1423 (1967).
55. V. Massey, S. Strickland, S. G. Mayhew, L. G. Howell, P. C. Engel, R. G. Matthews, M. Schuman, and P. A. Sullivan, *Biochem. Biophys. Res. Commun.*, **36**, 891 (1969).
56. E. M. Kosower, *Molecular Biochemistry*, McGraw-Hill, New York, 1962, p. 166.
57. N. O. Kaplan, *Enzymes*, **3**, 105 (1960).
58. Various authors, *Enzymes*, **4**, 1–274 (1963).
59. T. C. Bruice and S. J. Benkovic, *Bioorganic Mechanisms*, Vol. 2, Benjamin, New York, 1966, p. 301.
60. H. Sund, in *Biological Oxidations*, T. P. Singer, Ed., Interscience, New York, 1968, pp. 603, 641.
61. T. C. Stadtman, *Biochem. J.*, **62**, 614 1956).
62. T. C. Stadtman and P. Elliot, *J. Biol. Chem.*, **228**, 983 (1957).
63. D. S. Hodgins and R. H. Abeles, *J. Biol. Chem.*, **242**, 5158 (1967).
64. D. S. Hodgins and R. H. Abeles, *Arch. Biochem. Biophys.*, **130**, 274 (1969).
65. D. J. Cram and M. Cordon, *J. Amer. Chem. Soc.*, **77**, 1810 (1955).
66. A. Larsson and P. Reichard, *Progr. Nucleic Acid Res. Mole. Biol.*, **7**, 303 (1967).
67. E. Frieden, S. Osaki and H. Kobayashi, *J. Gen. Physiol.*, **49**, 213 (1965).
68. G. R. Stark and C. R. Dawson, *Enzymes*, **8**, 297 (1963).
69. Z. G. Penton and C. R. Dawson, in *Oxidases and Related Redox Systems*, T. E. King, H. S. Mason and M. Morrison, Eds., Wiley, New York, 1965, p. 222.
70. C. R. Dawson, in *The Biochemistry of Copper*, J. Peisach, P. Aisen and W. E. Blumberg, Eds., Academic Press, New York, 1966, p. 305.
71. I. Yamazaki and L. H. Piette, *Biochim. Biophys. Acta.*, **50**, 62 (1961).
72. J. H. Wang and W. S. Brinigar, *Proc. Natl. Acad. Sci. U.S.*, **46**, 958 (1960).
73. G. A. Hamilton, J. P. Friedman and P. M. Campbell, *J. Amer. Chem. Soc.*, **88**, 5266 (1966).
74. G. A. Hamilton, J. W. Hanifin, Jr., and J. P. Friedman, *J. Amer. Chem. Soc.*, **88**, 5269 (1966).
75. Q. H. Gibson, in *Biological Oxidations*, T. P. Singer, Ed., Interscience, New York, 1968, p. 379.
76. Various authors, in *Oxidases and Related Redox Systems*, T. E. King, H. S. Mason, and M. Morrison, Eds., Wiley, New York, 1965, pp. 549–812.
77. Various authors, in *The Biochemistry of Copper*, J. Peisach, P. Aisen, and W. E. Blumberg, Eds., Academic Press, New York, 1966, pp. 211–292.
78. M. M. T. Kahn and A. E. Martell, *J. Amer. Chem. Soc.*, **89**, 4176 (1967).
79. M. M. T. Kahn and A. E. Martell, *J. Amer. Chem. Soc.*, **89**, 7104 (1967).
80. J. A. D. Cooper, W. Smith, M. Bacila and H. Medina, *J. Biol. Chem.*, **234**, 445 (1959).
81. D. Amaral, L. Bernstein, D. Morse and B. L. Horecker, *J. Biol. Chem.*, **238**, 2281 (1963).
82. G. Avigad, D. Amaral, C. Asensio, and B. L. Horecker, *J. Biol. Chem.*, **237**, 2736 (1962).
83. R. A. Schlegel, C. M. Gerbeck and R. Montgomery, *Carbohydrate Res.*, **7**, 193 (1968).
84. G. T. Zancan and D. Amaral, *Biochim. Biophys. Acta*, **198**, 146 (1970).
85. J. De Jersey and G. A. Hamilton, unpublished observations.
86. W. E. Blumberg, B. L. Horecker, F. Kelly-Falcoz, and J. Peisach, *Biochim. Biophys. Acta*, **96**, 336 (1965).

87. T. L. Fabry, J. P. Kim and L. M. Titcomb, Abstracts 156th National Meeting, American Chemical Society, Atlantic City, New Jersey, September, 1968, Biol-231.
88. P. K. Adolf and G. A. Hamilton, *J. Amer. Chem. Soc.*, in press.
89. H. Sigel, C. Flierl and R. Griesser, *J. Amer. Chem. Soc.*, **91**, 1061 (1969); H. Erlenmeyer, C. Flierl, and H. Sigel, *J. Amer. Chem. Soc.*, **91**, 1065 (1969).
90. R. Griesser, B. Prijs and H. Sigel, *J. Amer. Chem. Soc.*, **91**, 7758 (1969).
91. J. Schubert, V. S. Sharma, E. R. White, and L. S. Bergelson, *J. Amer. Chem. Soc.*, **90**, 4476 (1968).
92. J. H. Wang, *J. Amer. Chem. Soc.*, **77**, 4715 (1955).
93. R. C. Jarnigan and J. H. Wang, *J. Amer. Chem. Soc.*, **80**, 6477 (1958).
94. J. H. Wang, *Accounts Chem. Res.*, **3**, 90 (1970).
95. R. C. Bray, *Enzymes*, **7**, 533 (1963).
96. K. V. Rajagopalan and P. Handler, *Biological Oxidations*, T. P. Singer, Ed., Interscience, New York, 1968, p. 301.
97. M. P. Coughlan, K. V. Rajagopalan and P. Handler, *J. Biol. Chem.*, **244**, 2658 (1969).
98. K. V. Rajagopalan, P. Handler, G. Palmer and H. Beinert, *J. Biol. Chem.*, **243**, 3784 (1968).
99. V. Massey, P. E. Brumby, H. Komai and G. Palmer, *J. Biol. Chem.*, **244**, 1682 (1969).
100. I. Fridovich, *J. Biol. Chem.*, **241**, 3126 (1966).
101. K. B. Wiberg, Ed., *Oxidation in Organic Chemistry*, Academic Press, New York, 1965, p. 69.
102. O. Hayaishi, *Ann. Rev. Biochem.*, **38**, 21 (1969).
103. W. B. Sutton, *J. Biol. Chem.*, **226**, 395 (1957).
104. H. Takeda, S. Yamamoto, Y. Kojima, and O. Hayaishi, *J. Biol. Chem.*, **244**, 2935 (1969).
105. A. Olomucki, D. B. Pho, R. Lebar, L. Delcambe, N. V. Thoai, *Biochim. Biophys. Acta*, **151**, 353 (1968).
106. O. Hulzinger and T. Kosuge, *Biochim. Biophys. Acta*, **136**, 389 (1967).
107. T. Nakazawa, S. Yamamoto, Y. Maki, H. Takeda, Y. Kujita, M. Nozaki and O. Hayaishi, in *Flavins and Flavoproteins*, K. Yagi, Ed., University Park Press, Baltimore, 1968, p. 214.
108. *Methods Enzymol.*, **1**, 483 (1955); **2**, 204 (1955); **3**, 414 (1957); **4**, 611 (1957).
109. A. Meister and D. Wellner, *Enzymes*, **7**, 610 (1963).
110. G. J. Moody, *Advan. Carbohydrate Chem.*, **19**, 149 (1964).
111. D. K. Rohrbaugh and G. A. Hamilton, unpublished results.
112. E. G. Sander and W. P. Jencks, *J. Amer. Chem. Soc.*, **90**, 4377 (1968).
113. H. S. Mason, in *Advances in Enzymology*, Vol. 19, F. F. Nord, Ed., Interscience, New York, 1957, p. 128.
114. O. Hayaishi, *Proc. 6th Intern. Congr. Biochem., Plenary Sessions, New York*, **33**, 31 (1964).
115. H. S. Mason, *Ann. Rev. Biochem.*, **34**, 595 (1965).
116. T. E. King, H. S. Mason and M. Morrison, Eds., *Oxidases and Related Redox Systems*, Vols. 1 and 2, Wiley, New York, 1965.
117. K. Bloch and O. Hayaishi, Eds., *Biological and Chemical Aspects of Oxygenases*, Maruzen, Tokyo, Japan, 1966.
118. G. A. Hamilton, *J. Amer. Chem. Soc.*, **86**, 3391 (1964).
119. J. Hine, *Divalent Carbon*, Ronald Press, New York, 1964.
120. W. Kirmse, *Carbene Chemistry*, Academic Press, New York, 1964.
121. J. S. McConaghy, Jr. and W. Lwowski, *J. Amer. Chem. Soc.*, **89**, 4450 (1967).
122. D. S. Breslow, T. J. Prosser, A. F. Marcantonio and C. A. Genge, *J. Amer. Chem. Soc.*, **89**, 2384 (1967).
123. A. G. Anastassiou and H. E. Simmons, *J. Amer. Chem. Soc.*, **89**, 3177 (1967).

124. G. Kobrich, *Angew. Chem. Intern. Ed. Engl.*, **6**, 41 (1967).

125. R. O. C. Norman and J. R. Lindsay Smith, in *Oxidases and Related Redox Systems*, T. E. King, H. S. Mason, and M. Morrison, Eds., Wiley, New York, 1965, p. 131.

126. M. B. Dearden, C. R. E. Jefcoate and J. R. Lindsay Smith, *Advan. Chem. Ser.*, **77**, 260 (1968).

127. V. Ullrich and H. Staudinger, in *Biological and Chemical Aspects of Oxygenases*, K. Bloch and O. Hayaishi, Eds., Maruzen, Tokyo, Japan, 1966, p. 235.

128. V. Ullrich, *Z. Naturforsch.*, **24b**, 699 (1969).

129. D. M. Jerina, J. W. Daly, B. Witkop, P. Zaltman-Nirenberg, and S. Udenfriend, *Biochemistry*, **9**, 147 (1969).

130. S. Yamamoto, H. Takeda, Y. Maki, and O. Hayaishi, *J. Biol. Chem.*, **244**, 2951 (1969).

131. D. Wellner, *Ann. Rev. Biochem.*, **36**, 669 (1967).

132. S. Yamamoto, M. Katagiri, H. Maeno, and O. Hayaishi, *J. Biol. Chem.*, **240**, 3408 (1965).

133. M. Katagiri, H. Maeno, S. Yamamoto, O. Hayaishi, T. Kitao, and S. Oae, *J. Biol. Chem.*, **240**, 3414 (1965).

134. Y. Maki, S. Yamamoto, M. Nozaki and O. Hayaishi, *J. Biol. Chem.*, **244**, 2942 (1969).

135. K. Hosokawa and R. Y. Stanier, *J. Biol. Chem.*, **241**, 2453 (1966).

136. K. Yano, M. Morimoto, N. Higashi, and K. Arima, in *Biological and Chemical Aspects of Oxygenases*, K. Bloch and O. Hayaishi, Eds., Maruzen, Tokyo, Japan, 1966, p. 329.

137. H. I. X. Mager and W. Berends, *Rec. Trav. Chim.*, **84**, 1329 (1965).

138. H. I. X. Mager, R. Addink, and W. Berends, *Rec. Trav. Chim.*, **86**, 833 (1967).

139. W. Berends, J. Posthuma, J. S. Sussenbach, and H. I. X. Mager, in *Flavins and Flavoproteins*, E. C. Slater, Ed., Elsevier, New York, 1966, p. 22.

140. H. I. X. Mager and W. Berends, *Biochim. Biophys. Acta*, **118**, 440 (1966).

141. P. S. Bailey, *Chem. Rev.*, **58**, 925 (1958).

142. J. March, *Advanced Organic Chemistry, Reactions, Mechanisms, and Structure*, McGraw-Hill, New York, 1968, p. 871.

143. J. P. Heicklen, in *Advances in Photochemistry*, Vol. 7, W. A. Noyes, Jr., G. S. Mammond, and J. N. Pitts, Jr., Ed., Interscience, New York, 1969, p. 57.

144. G. A. Hamilton and J. R. Giacin, *J. Amer. Chem. Soc.*, **88**, 1585 (1955).

145. G. A. Hamilton, B. S. Ribner, and T. M. Hellman, *Advan. Chem. Ser.*, **77**, 15 (1968).

146. T. M. Hellman and G. A. Hamilton, unpublished results.

147. D. Swern, *Org. Reactions*, **7**, 378 (1953); H. O. House, *Modern Synthetic Reactions*, Benjamin, New York, 1965, p. 109; J. March, *Advanced Organic Chemistry; Reactions, Mechanisms, and Structure*, McGraw-Hill, New York, 1968, p. 618.

148. R. H. White-Stevens and H. Kamin, *Federation Proc.*, **29**, Abstract 890 (1970).

149. E. Vogel and H. Gunther, *Angew. Chem. Intern. Ed. Engl.*, **6**, 385 (1967).

150. S. Kaufman, *Ann. Rev. Biochem.*, **36**, 171 (1967).

151. C. B. Storm and S. Kaufman, *Biochem. Biophys. Res. Commun.*, **32**, 788 (1968).

152. D. B. Fisher and S. Kaufman, *Biochem. Biophys. Res. Commun.*, **38**, 663 (1970).

153. B. Lindblad, G. Lindstedt, M. Tofft, and S. Lindstedt, *J. Amer. Chem. Soc.*, **91**, 4604 (1969).

154. S. E. Hager, R. I. Gregerman, and W. E. Knox, *J. Biol. Chem.*, **225**, 935 (1957).

155. S. Goodwin and B. Witkop, *J. Amer. Chem. Soc.*, **79**, 179 (1957).

156. J. W. Daly and B. Witkop, *Angew. Chem. Intern. Ed. Engl.*, **2**, 421 (1963).

157. K. Taniguchi, T. Kappe, and M. D. Armstrong, *J. Biol. Chem.*, **239**, 3389 (1964).

158. J. Daly and G. Guroff, *Arch. Biochem. Biophys.*, **125**, 136 (1968).

159. D. M. Jerina, J. W. Daly, W. Landis, B. Witkop, and S. Udenfriend, *J. Amer. Chem. Soc.*, **89**, 3347 (1967).

160. V. Ullrich and H. Staudinger, in *Microsomes and Drug Oxidations*, J. R. Gillette, A. H.

Conney, G. J. Cosmides, R. W. Estabrook, J. R. Fouts, and G. J. Mannering, Eds., Academic Press, New York, 1969, p. 199.

161. N. C. Deno, H. J. Peterson and G. S. Saines, *Chem. Rev.*, **60**, 7 (1960).

162. R. Stewart, *Oxidation Mechanisms*, Benjamin, New York, 1964, p. 19.

163. R. P. Bell, *The Proton in Chemistry*, Cornell University Press, Ithaca, New York, 1959, p. 171.

164. J. M. Williams and M. M. Kreevoy, in *Progress in Physical Organic Chemistry*, Vol. 6, A. Streitwieser, Jr., and R. W. Taft, Eds., Interscience, New York, 1968, p. 63.

165. R. P. Bell, *Disc. Faraday Soc.*, **39**, 16 (1965).

166. E. Grunwald and E. Price, *J. Amer. Chem. Soc.*, **86**, 2970 (1964).

METAL IONS IN ENZYMATIC CATALYSIS

JOSEPH E. COLEMAN

Department of Molecular Biophysics and Biochemistry, Yale University, New Haven, Connecticut

1 INTRODUCTION

Participation of metal ions in the structure and function of biological systems has now been documented in a great many systems. The first examples were provided by the oxidative enzymes containing heme-iron (the peroxidases, catalases, and cytochomes). Their early recognition in the late 1800's was due in no small measure to their intense visible absorption spectra[1].

With the advent of more sophisticated analytical techniques applicable to biological systems, a large number of other enzymatic reactions and cellular functions have now been found to require metal ions as one of the essential components[2,3]. When surveying the periodic table for those metal ions most frequently encountered in biological systems, two rather restricted groups stand out, a light group consisting of the alkali metals Na and K together with the alkaline earths Ca and Mg, and a heavier group comprising the second half of the first transition elements, Mn, Fe, Co, Ni, and Cu. The lightest IIB element, Zn, is properly included to complete the latter series since it completes the filling of the $3d$ electronic orbitals for this series and is quite commonly found in metalloenzymes[2,4]. Ni(II) is not known to be significantly involved in biological catalysis, although it is found in procarboxypeptidase A[5]. There are two known biologically important metals outside these restricted groups, vanadium and molybdenum, but they occur under rather restricted circumstances. Vanadium occurs in the vanodocytes of the sea squirt in the presence of $1.8 N$ H_2SO_4 to maintain it in reduced form, and molybdenum occurs in a small group of metalloflavoproteins (see Sect. 3.2.2).

The alkali metals and alkaline earths have received the greatest study in terms of their contribution to the maintenance of the ionic environment necessary for cell function, although there are many instances where these metals are known to participate in enzymatic reactions as activators. Of particular note are the many reactions involving transfer of high energy phosphate where magnesium is an activator[6] and the activation of the trypsinogen–trypsin[7] and prothrombin–thrombin[8] conversions by Ca(II).

If consideration is limited to purified well characterized metalloenzymes, the participating metal ions are almost entirely limited to the series of Mn to Zn, with the single exception of molybdenum. One may speculate on the evolutionary significance of this restriction. Biologically important metals occur in the earth's crust in the following relative abundance: Fe > Ca > Na > K > Mg > Mn > V > Ni > Cu > Zn > Co[9]. Iron, calcium, sodium, potassium and magnesium are the fourth, fifth, sixth, seventh, and eighth most abundant elements in the earth's crust, making up 2–4% each of the total composition[9]. The others, however, are distinctly trace elements making up from 0.003 or 0.004% (Co and Zn) to 0.1% (Mn) of the earth's crust. Zinc is 28th in relative abundance. Thus other reasons aside from their relative abundance would appear to have determined the incorporation of certain functional metals into biological systems.

It is likely that there are chemical reasons why these particular metals were picked up as functioning groups by the evolving biological system. All these ions form a variety of complexes with organic molecules. Space precludes a detailed discussion of this coordination chemistry and only brief discussions of certain features relating to metalloenzymes will be given. Complete discussions of transition metal chemistry are found in references 10–13.

Much of this interesting chemistry arises from the nature of the orbitals in which the electrons interacting with the ligands are located, namely the $3d$-orbitals. Many electronic properties of these ions and their complexes can be attributed to a partially filled set of d orbitals (Fe, Co, Ni, and Cu), although the two ions that remain spherically symmetrical in their complexes [Mn(II) with five d-electrons, and Zn(II) with ten d-electrons] also form varied and interesting coordination compounds [13]. It may be inferred that the extraordinarily varied chemistry of the complexes of these metals with organic ligands is the general property that leads to their participation in catalysis by biological systems.

The chapter is divided into two major sections. The first describes various physicochemical properties related to the metal ion present in metalloprotein catalysts. Although many of these properties cannot as yet be related directly to the catalytic mechanism, it seems likely that a significant proportion of the mechanistic features of these enzymes will ultimately be related to particular chemical features of the coordination complex present at the active site.

Specific catalytic mechanisms may often be relatively unique to a particular enzyme, hence the second major section of the chapter is devoted to a discussion of a number of specific metalloenzymes in terms of the information that can be brought to bear on their mechanism of catalysis. A great many enzymes have been described which are activated by metal ions or contain a metal ion as an integral part of the structure. The present discussion is not meant to be exhaustive, but includes from this large number of enzymes those which best illustrate the various physico-chemical features of metal–protein interaction. Because it has been a great deal easier to characterize those enzymes which contain a stoichio-metric amount of firmly incorporated metal ion [2], the metalloenzymes have received the most extensive and detailed investigation. Precise details of the metal ion coordination have not been as extensively investi-gated in the enzyme systems activated by metal ions. Hence, the examples used in this chapter are largely from the metalloenzyme literature.

2 PHYSICOCHEMICAL PROPERTIES OF METALLOENZYMES

2.1 Stabilities, Metal Ion Exchange Reactions, pH, pM, and Apo-enzymes

2.1.1 Metal–Protein Equilibria. Catalytic properties of metal ions in biological systems almost certainly derive in large measure from the properties of the metal ion after it has formed coordinate bonds with the ligand. The hydrated metal ions have not been observed to catalyze reactions at rates that even approach by several orders of magnitude the rates observed for catalysis by metalloenzymes. The fact that the ligands are macromolecules with certain rigid internal structure of their own complicates the picture, but certain basic features of coordination chem-istry may still be expected to apply, albeit in modified form.

Complex formation in aqueous solution between a metal cation and a negatively charged (or dipolar) donor atom on the ligand, L, may be formally described by equation 1. Equation 1 is written for a metal ion with a charge of n^+ and a ligand with a single negative charge in the

$$\text{Me}^{n+}(\text{H}_2\text{O})_p + q\text{HL} \rightleftharpoons \text{Me}^{n+}(\text{L}^-)_q + p\text{H}_2\text{O} + q\text{H}^+ \tag{1}$$

unprotonated form. The hydrated metal ion is almost always involved in aqueous solution, and p equals 6 for most commonly occurring divalent metal ions [14]. In most cases of complex formation, displacement of a proton from the ligand accompanies complex formation over at least part of the pH range[10].

The most easily accessible thermodynamic parameter associated with the reaction formalized by equation 1 is the dissociation constant for the metal–ligand complex. In coordination chemistry these constants are commonly expressed as the reciprocal of the dissociation constant or stability constant[10,15]. In general, such constants must be written as step-wise constants, since unless the ligand is multidentate such that it completely fills the coordination sphere, several ligands can be added to the complex in a stepwise manner. Hence,

$$k_1 = \frac{[MeL_1]}{[Me][L]} \quad k_2 = \frac{[MeL_2]}{[MeL_1][L]} \cdots \quad k_n = \frac{[MeL_n]}{[MeL_{n-1}][L]} \tag{2}$$

and $K = (k_1 \cdot k_2 \cdots k_n)$ is the overall stability constant for the complex containing n ligands.

Most proteins are multidentate ligands, and in the examples discussed in this chapter the ligand does appear to fill the coordination sphere at a 1:1 stoichiometry per site (i.e., no additional protein ligands can be added). Hence, only one stability constant, K, applies for the combination of the metal ion with the apoprotein. Much recent data, however, shows that in many metalloenzymes at least one coordination site remains occupied by a water molecule which can be displaced by a substrate or inhibitor donor atom or can participate in the catalytic mechanism (see p. 244).

The concentration of ligand, $[L]$, in equation 2 refers to the unprotonated ligand, hence a second set of simultaneous equations (eq. 3) must be used to completely describe the process in equation 1. In equation 3 n is the number of protonatable coordinating groups on the ligand

$$\frac{[LH_{n-1}][H^+]}{[LH_n]} = K_{a_n} \tag{3}$$

and K_{a_n} is the usual acid dissociation constant for the nth group. Since $[L]$ in equation 2 is the concentration of unprotonated ligand, the k_n's thus defined are pH independent and describe the intrinsic affinity of the metal ion for the donor atom. This should be remembered if stability constants are measured by some equilibrium procedure at a given pH. Such constants are not pH independent unless they can be corrected for the amount of protonated ligand present. This correction requires a knowledge of the pK_a's involved, which is often not available (see below).

The proton can be considered the simplest metal cation, and if the ligand has affinity for Me^{n+} it usually has at least some and often rather great affinity for H^+. This general feature of metal complex formation and the ease and accuracy of modern methods of measuring hydrogen ion concentration have resulted in the use of the changes in hydrogen ion equilibria accompanying complex formation as the most widely applied method for study of the quantitative aspects of metal coordination[10]. A solution of the ligand is titrated in the presence and absence of the metal ion. From the known pK_a's of the ligand and the shifts in the proton dissociations caused by the metal ion the concentrations of the various species required for the solution of simultaneous equations 2 and 3 can be determined. The resultant k_n's relating the two titration curves can then be calculated. The detailed methods for setting up and solving these equations can be found in reference 10. The general phenomenon is that the presence of the metal shifts the apparent dissociation of the proton of the ligand to much lower pH (see Fig. 1).

Unfortunately this relatively simple procedure cannot be applied with ease to proteins, since the ligand contains a large number of dissociating groups that are not involved in the metal ion coordination. The number of these on the average protein is usually so great that it is difficult, if not impossible, to separate the few proton dissociations shifted by the presence of the metal ion from the general mass of proton dissociations that make up a continuous titration curve[16–18].

In the case of the coordination of Zn(II) by insulin, however, a continuous titration was applied with successful results[19]. Comparison of the continuous titration curve in the absence of Zn(II) with the titration curve of Zn(II) insulin showed the shift to acid pH of 3 groups with pK_a's of ~ 6 in the Zn-free insulin. These pK_a's were assigned to the imidazole groups of histidyl residues as the coordinating groups[19]. Although later chemical evidence in solution tended to implicate the N-terminal α-amino groups in Zn(II) coordination[20], the recent X-ray data on crystalline Zn(II) insulin show that the coordination is to the imidazole nitrogens of the histidyl residues[21]. Metalloenzymes in which the pH stability of the metal complex has been studied in any detail have been much larger than insulin, and continuous data have not been obtained on the apoenzyme. Often such phenomena as acid or alkaline denaturation also complicate the picture by uncovering buried groups which may have nothing to do with metal ion binding[17,18].

Individual parts of the equilibria described in equations 1–3, however, can be examined, and if enough information from different approaches is collected, the complete $Me^{n+} \rightleftharpoons H^+$ equilibria can be reconstructed. One of the easiest measurements is the pH dependence of metal binding to the apoenzyme. The pH dependence of ^{65}Zn binding to apocarboxypeptidase is shown in Figure 1. $^{65}Zn(II)$ dissociates between pH 6 and 4 and is half dissociated at pH 5. This dissociation coincides with the loss

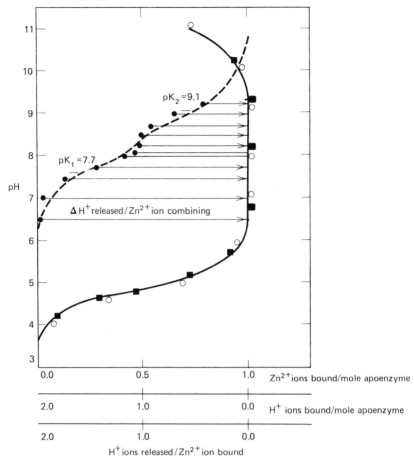

Figure 1   Relationship between the complexometric titration curve and the pH-formation curve for zinc carboxypeptidase. The experimental points (●) are the complexometric titration curve for zinc carboxypeptidase as described in the text. (-----) The theoretical curve for two independent ionizations of pK 7.7 and pK 9.1; (———) the formation curve for zinc carboxypeptidase. (○) ^{65}Zn bound to the enzyme; (■) peptidase activity assayed at pH 7.5*. Between pH 6.5 and pH 9.5 the metalloenzyme normally contains its full complement of metal, 1.0 g-atom per mole, and the metal binding groups bind no hydrogen. On the other hand, the apoenzyme binds two hydrogen ions per mole at pH 6.5, the average number decreasing along the curve given by the complexometric titration. The addition of Zn^{2+} ions to the apoenzyme serves to transform the protein from the point on the apoenzyme titration curve to the corresponding point on the formation curve. The bound hydrogen ions are displaced as indicated by the arrows. Below pH 6.5, the hydrogen ion concentration is high enough to begin to displace the more firmly bound Zn^{2+} ion. The two curves pictured in this graph are quantitatively related by the stability constant for the metal–protein chelate and the pK_a's of the donor groups. Data plotted from Coleman and Vallee [22, 28].

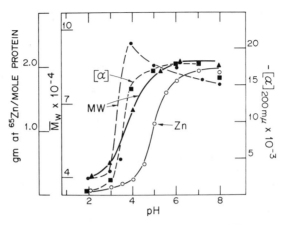

Figure 2   pH-dependence of Zn(II) binding and structure changes of alkaline phosphatase. (○) Protein bound ^{65}Zn for biosynthesized ^{65}Zn-enzyme, gram atoms per mole; (●) molecular weight $\bar{M}_w$, determined for samples adjusted from neutral pH to each indicated pH by dialysis; (▲) molecular weight, $\bar{M}_w$, determined for samples prepared at pH 2, then dialyzed to the appropriate pH; (■) specific rotation [α] at 200 mμ for samples prepared at each pH for both dissociation and reassociation. Data taken from reference 26.

of activity,* and both activity and Zn binding are completely restored as the pH is raised[22]. There is no evidence of major change in protein structure over this pH range. Hence, this curve has been interpreted as the uncomplicated competition of H$^+$ for the donor atoms of the protein ligand[22,23]. Thus the ^{65}Zn binding curve can be considered completely analogous to the titration of a small ligand in the presence of a metal ion. Such an interpretation for a protein must be approached with great caution because of possible structural changes in the protein as a function of pH which might change the nature of the metal binding site. Carbonic anhydrase and liver alcohol dehydrogenase both undergo apparently irreversible changes in structure as the pH is lowered sufficiently to displace the metal ion[24,25]. These problems have no counterpart in simple coordination chemistry.

A somewhat intermediate case is represented by alkaline phosphatase [26]. In Figure 2 two assays of protein structure, the optical rotation and the weight average molecular weight, $\bar{M}_w$, are included along with ^{65}Zn binding as functions of pH. In this case the protein undergoes some major changes in secondary, tertiary, and quaternary structure as the pH is lowered from 7 to 2[26]. The protein dissociates into monomers, MW = 40,000, and the polypeptide folding is radically altered as shown

*Activity of carboxypeptidase A from which Zn(II) has been removed by acid follows the pH dependence of Zn(II) binding, even though the enzyme is assayed at pH 7.5, because the substrate prevents the rebinding of zinc.

by the change in $[\alpha]_{200}$. The two $^{65}Zn^{2+}$ ions of the protein have completely dissociated by pH 4, before the major structural changes have become well developed. Hence, the actual dissociation of the metal ion probably represents primarily H^+ ion competition for the donor atoms. Both the structural changes and the metal dissociation are completely reversible and a fully active dimeric protein can be reconstituted by returning the pH to 7 [26]. As will be shown below the metal ion does influence the monomer → dimer equilibrium, and hence in this case independent structural changes in the ligand cannot be entirely separated from $Me^{2+} \rightleftharpoons H^+$ equilibria.

While complete titration curves cannot be used to determine stability constants for metalloenzymes, apparent equilibrium constants can often be determined under rather specific conditions. Measurements of stability constants for metalloenzymes have usually been carried out by equilibrium dialysis procedures where the concentrations of aquated metal ions, protein-complexed metal ions, and, in certain cases, competing small metal–chelate complexes have been allowed to come to equilibrium. The concentration of each species is then determined and the unknown metal–protein stability constant calculated. Use of radioactive metal ions has been a very convenient method of determining free and protein-bound metal ion concentrations [22]. The properties of the proteins restrict the conditions rather severely. In order for the method to be applicable the apoenzyme must be stable, since all species must be present in reversible equilibrium, a condition that is not met by liver alcohol dehydrogenase, for example [24]. It is often not possible to examine a wide pH range because the protein is unstable.

The stability constants for the complexes of bovine carboxypeptidase A with Mn(II), Co(II), Ni(II), Cu(II), Zn(II), Cd(II), and Hg(II) were the first to be determined [27–29]. Certain properties of this enzyme make it uniquely satisfactory for studying the metal–protein binding. The apoenzyme is stable [30], one divalent metal ion at the active site readily exchanges with another in the above series [22, 28], and Co(II) carboxypeptidase partially dissociates to apoenzyme and free cobalt at reasonably high concentrations of free cobalt, $\sim 10^{-6} M$ [28]. The latter property makes it possible to calculate the dissociation constant for the Co(II) enzyme directly from an equilibrium dialysis experiment. At pH 8.0 log K_{Co} is 5.0 in $0.02 M$ Tris–$1M$ NaCl [28]. The stability constant for Mn(II) carboxypeptidase can also be determined in this manner [28]. Since the metal ion at the active center of carboxypeptidase readily exchanges with its radioactive species or another species of metal ion (see below), the constants for more stable metallocarboxypeptidases in the series can be determined by competing these metal ions for either

Co(II) or Mn(II) at the active site and calculating the unknown constants from the ratio of the two types of metallocarboxypeptidase present at equilibrium. Constants determined in this manner are listed in Table 1A.

TABLE 1A STABILITIES OF METALLOCARBOXYPEPTIDASES [28]

Metal	Log $K_{apparent}$ (pH 8.0)	Log $K_{corrected}$ (pH 8.0 for Cl⁻, Tris)	pH-Independent Stability Constant Log K
Mn(II)	5.6	5.6	–
Co(II)	5.8	7.0	10.1
Ni(II)	5.7	8.2	–
Cu(II)	5.1	10.6	–
Zn(II)	8.3	10.5	12.6
Cd(II)	7.9	10.8	12.8
Hg(II)	6.7	21.0	–

TABLE 1B STABILITIES OF METALLOCARBONIC ANHYDRASES [35]

Metal	Log K, pH 5.5	pH-Independent Stability Constant Log K
Mn(II)	3.8	–
Co(II)	7.2	–
Ni(II)	9.5	–
Cu(II)	11.6	–
Zn(II)	10.5	15.0
Cd(II)	9.2	–
Hg(II)	21.5	–

In marked contrast to carboxypeptidase, the central Zn(II) ion in carbonic anhydrase will *not exchange* with metal ions in the medium at a measurable rate [25, 31, 32], although the metal ion can be removed by chelating agents, most readily between pH 5 and 6. Fortunately the resultant apoenzyme is stable [25, 33, 34]. Thus by using various chelating agents and apocarbonic anhydrase as competing ligands for the central metal ion, the stability constants for Mn(II), Co(II), Ni(II), Cu(II), Zn(II), Cd(II), and Hg(II) carbonic anhydrase have been calculated from the final equilibrium concentrations of the two liganded species, the concentration of free metal ion, and the known stability constants of the metal complexes of the small chelating ligands[35]. This particular investigation used 1,10-phenanthroline, 8-OH-quinoline sulfonic acid, and ethylenediamine tetraacetate as competing ligands[35]. Log K values for the series of metallocarbonic anhydrases determined at pH 5.5 are given in Table 1B. The data given in Tables 1A and 1B represent the two sets of complete data that are available at the present time. Fe(II) is missing because of the complication introduced by the Fe(II) → Fe(III) transformation in aqueous solution – the ferric ion precipitates as ferric hydroxide if equilibrium proceeds for any length of time.

A determination of the stability constants applying for the two active Zn(II) ions in the dimeric alkaline phosphatase of E. coli has been determined by the method of using a chelating agent to compete as a ligand for the central metal ion[36]. The level of alkaline phosphatase activity, however, rather than a direct analytical technique was used to assay for the amount of Zn(II) alkaline phosphatase present. The resulting data appeared to be best analyzed if it was assumed that two stability constants characterize the system, one with a $\log K = 10.2$ and the other with a $\log K = 7.66$ at pH 8.0. The precise interpretation of this is unclear at present, since the metal ion has been shown to have a distinct affect on the dimer–monomer equilibrium[26] (see p. 181). Absence of the metal ion favors dissociation, hence the nature of the chelating site may change in the apoenzyme, and equilibria in addition to simple metal–donor interactions may be involved. Keeping in mind the above qualifications, it is possible to interpret these findings as indicating the presence of negative cooperativity such that the presence of one Zn(II) ion causes a second Zn(II) ion to be bound less tightly than the first.

The required experimental conditions for the proteins has tended to limit the determination of the stability constants for these metallo-enzymes to a single pH. Knowledge of the state of protonation of the ligand donor atoms is generally not known and thus the constants are not pH independent, since no distinction could be made between protonated and unprotonated ligands. Thus these apparent stability constants would be expected to rise with pH. An experiment showing this has been carried out by Nyman and Lindskog[35] on Zn(II) carbonic anhydrase and the results are shown in Figure 3. The constant determined by the ligand competition method rises from $\log K = 7$ at pH 4.5 to $\log K = 15.0$ at pH 10, where the $\log K$ value appears to become relatively constant with pH. Although precise interpretation in the case of a protein is difficult, it could be postulated that at least one of the donor groups of the ligand has a relatively high pK_a, perhaps 9 or above. pH-Dependent changes in protein structure not directly related to the state of ligand protonation, however, cannot be ruled out.

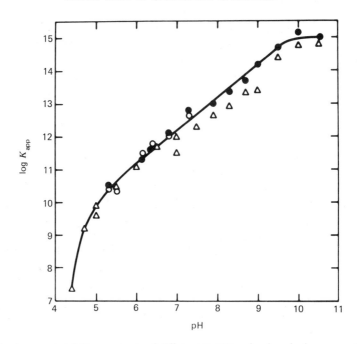

Figure 3 Apparent stability constants of different Zn(II) carbonic anhydrases as a function of pH: ($\triangle$) bovine CA; ($\bigcirc$) human high activity CA; ($\bullet$) human low-activity CA. After Nyman and Lindskog [35].

In the case of carboxypeptidase A a slightly different approach to this problem may be used. The feature lacking for the complete analysis of the pH–formation curves of these metalloenzymes is the continuous titration curve of the ligand and, hence, the pK_a values of the donor atoms. While a complete titration curve cannot be interpreted to such a fine point that the coordinating groups can be picked out, a discontinuous titration can be performed taking advantage of the fact that according to equation 4 the protons released at any given pH upon combination of Zn(II) with the apoenzyme will reflect the state of protonation of the protein donor atoms. If the proton release accompanying this reaction is measured at enough different pH values a complete titration curve

$$\text{EnzH}_n + \text{Zn}^{2+} \longrightarrow \text{EnzZn} + n\text{H}^+ \tag{4}$$

of the ligand (in the absence of metal ion) can be constructed [23, 28]. By this means the ionization of the groups combining with the metal ion can be separated from all other protein ionizations. Once again, there are some restrictions introduced by the presence of a protein. The protein must be stable over a wide pH range and the metal complex must be of high stability such that it is 100% formed over the pH range of the titration. Theoretically one could correct for partial formation at low pH, but considering the experimental difficulties of the method, such corrections would be difficult. For carboxypeptidase A the workable range is pH 6–10. The results of such a complexometric titration of apocarboxypeptidase A are included in Figure 1 from the data of Coleman and Vallee[28] expressed as the ratio of H^+ ions released to Zn^{2+} ions combining with the protein. Two protons are released at pH 6.0 which titrate as the pH rises with pK_a's of approximately 7.7 and 9.1. Thus two of the groups binding the metal ion appear to have alkaline pK_a values. Both Cd(II) and Co(II) also release the two protons, further indication that the same site is involved in binding this series of cations[23, 29].

With this data on the pK_a values of the ligand groups, one can construct the complete titration data needed for the calculation of a pH-independent stability constant, since the pH–formation of the metalloenzyme is known from the pH dependency of ^{65}Zn-binding. This construction has been completed in Figure 1.

A restriction must once again be introduced in that if a ligand is present which has a pK_a below the pH limits over which the complexometric titration can be performed there will be an error in the final calculations. In carboxypeptidase A the X-ray data (see p. 271) do show that a third ligand is present, a carboxyl group of aspartic acid which apparently has a pK_a below pH 5. Since the metalloenzyme is completely dissociated at pH 4.0 this does not introduce a large error. The log of the pH-independent stability constant for Zn(II) carboxypeptidase A calculated from the data in Figure 1 is 12.6, two orders of magnitude higher than the apparent constant determined at pH 8.0 by equilibrium dialysis[29]. Similar calculations from the data on Co(II) and Cd(II) are given in Table 1.

While these features of stability and hydrogen–metal ion equilibria are relevant to the mechanism of action of these metalloenzymes in that they help define the nature of the coordination complex present at the active center, they are probably not in many cases directly involved in the mechanism of catalysis. The pH stability of metal binding must be considered of course, in interpretation of pH–rate profiles for metalloenzymes (see p. 262).

Shifts in hydrogen ion equilibrium accompanying addition of ligands (substrates of inhibitors) to open coordination sites on metal ions already incorporated at the active site, however, may provide information directly relevant to the mechanism of catalysis. Carbonic anhydrase provides an example of this approach. Metal binding anions such as cyanide and sulfide, are powerful inhibitors of carbonic anhydrase[31]. If the anions like CN^- and HS^- inhibit carbonic anhydrase by co-ordinating the metal ion (see Sect. 3.1.3), the reaction should be accompanied by the displacement of a proton from the inhibitors at pH values below the pH region for the dissociation of HCN and H_2S, the former described by a pK_a of 9.3, the latter by a pK_a of 6.9 at about $23°$[15]. The net hydrogen ion release will depend on other hydrogen ion equilibria altered by the binding of anions. In order to investigate the hydrogen ion equilibria that accompany the reaction of carbonic anhydrase with cyanide and sulfide, a set of equilibrium measurements between pH 6 and 10 were made by the direct measurement of H^+ or OH^- release with a difference titration method[37]. The method consists of adjusting both the concentrated anion solution and the protein solution to an identical pH, adding a small equimolar aliquot of the anion to the protein, and measuring the uptake or release of protons on a pH stat at a precision of 1.0 ± 0.02 μmole H^+. The experiment can be done over a pH range from 6 to 10, since cyanide forms a firm $1:1$ complex with carbonic anhydrase over this pH range.

The results are best described by formulating one of the possible models that may apply to the anion–carbonic anhydrase reaction and comparing the predicted results with the observed findings. If a coordinated water molecule is displaced from the metal ion by the anions, the various proton equilibria may be pictured as in equation 5.

$$CAZn \cdot H_2O \quad + \quad HCN \overset{(a)}{\rightleftharpoons} CAZn \cdot CN^- + H_2O + H^+$$

$$pK_a=(8) \,\Big\Updownarrow\, -(H^+) \qquad pK_a=9.3 \,\Big\Updownarrow\, -(H^+)$$

$$CAZn \cdot OH^- \quad + \quad CN^- \overset{(b)}{\rightleftharpoons} CAZn \cdot CN^- + OH^- \tag{5}$$

At low pH, where the metal ion is in the hydrated form and the inhibitor is in the acid form, the reaction is described by the release of a water molecule and a proton (a). At very high pH the inhibitor is in the anion form and if the coordinated water molecule has a pK_a within the experimental pH range, the metal ion will be in the hydroxide form (b). As seen in Figure 4, the reaction of cyanide with the Zn(II) enzyme does show a biphasic titration curve; one mole H^+ is released at pH 6.0, while 0.7 mole OH^- (H^+ uptake) is released at pH 10.0. These alterations in proton equilibria are a function of the metal ion, since the apoenzyme solution does not show any change in pH upon the addition of cyanide. The Co(II) enzyme shows a very similar titration curve to the Zn(II) enzyme (Fig. 4). The results can be explained if a single additional

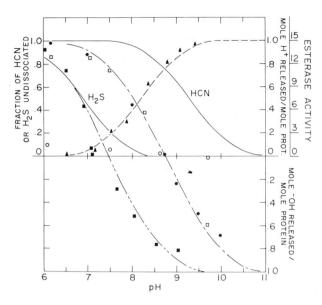

Figure 4   Displacement of H^+ and OH^- from human carbonic anhydrases B by cyanide and sulfide. The solid lines represent the continuous titration curves for the dissociation of HCN to H^+ and CN^- and H_2S to H^+ and SH^-, expressed as the fraction of inhibitor undissociated (left-hand ordinate). (– – –) Theoretical titration curve for an acidic group of $pK_a = 8.1$, expressed as moles H^+ dissociated per mole of protein (upper right-hand ordinate); (△) esterase activity of human carbonic anhydrase B, micromoles of substrate hydrolyzed per minute per micromole of enzyme; (– – – and – – – –) theoretical difference titration curves expected for the displacement of H^+ (upper right-hand ordinate) and OH^- (lower right-hand ordinate) from Zn(II) carbonic anhydrase by cyanide and sulfide, respectively, as a function of pH according to equation 5. Moles of H^+ or OH^- released per mole of protein on the addition of cyanide to (●) Zn(II) carbonic anhydrase, (○) apocarbonic anhydrase, and (□) Co(II) carbonic anhydrase. Maximum H^+ and OH^- release from the Co(II) enzyme was not reached until the addition of ca. 1.5 equivalents of cyanide. Moles of H^+ or OH^- released per mole of protein on the addition of sulfide to (■) Zn(II) carbonic anhydrase. Data from Coleman [37].

proton equilibrium associated with the protein, coupled to the metal ion, and described by a pK_a of approximately 8 is added to the titration curve (dashed line, Fig. 4). The results are compatible with equation 5 if the coordinated water molecule is assumed to have a pK_a of 8. As the pH rises more enzyme is in the OH^- form. The release of this OH^- by cyanide progressively neutralizes the proton from HCN along the theoretical curve shown in the figure. Finally the CN^- species will displace only the OH^-. An additional feature of the system predicted from equation 5 is the dependence of the difference titration on the pK_a of the inhibitor. This can be tested by using sulfide, since H_2S has a pK_a of 6.3. As expected the biphasic titration curve shifts about 1.5 units toward acid pH. Once again a theoretical curve can be generated by the same additional ionization with a pK_a of 8.

These results are compatible with the presence of a coordinated water molecule at the active center of carbonic anhydrase according to equation 5; they do not prove the identity of this ionization. An alternative is possible by assuming that the anion displaces a metal-bound protein ligand with a very high pK_a which subsequently takes up a proton. This group cannot be initially bound to the metal ion, but must bind along a pH function with a midpoint of 8. The evidence is less convincing for this alternative. The interpretation that the pK_a observed is that for a coordinated water molecule is strongly supported by nmr data (see Sect. 2.4).

The ionization described by the pK_a of 8 does occur in the region of the inflection point describing the esterase pH–rate profile for the enzyme. The esterase activity of human carbonic anhydrase B as a function of pH is plotted in Figure 4 and follows the ionization revealed by the difference titration. The mechanisms that can be postulated for

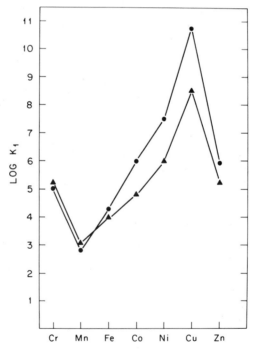

Figure 5 Logarithms of the stability constants for the 1:1 complexes of first transition metals and zinc with (●) ethylenediamine and (▲) alanine [15].

carbonic anhydrase on the basis of a coordinated H_2O or OH^- are discussed in detail in Sect. 3.1.3.

The stabilities of the complexes among the metals Mn(II), Fe(II), Co(II), Ni(II), Cu(II), and Zn(II) with the same nitrogen- or oxygen-containing ligands have been observed to follow a fairly consistent pattern, shown in Figure 5 for the complexes of ethylenediamine and alanine[15]. The Mn(II) complex generally has the lowest stability, except when the donor atoms are exclusively oxygen[3]. Mn(II) complexes are also less stable than the corresponding chromium complexes. Stability then rises to the Cu(II) complexes and falls sharply to Zn(II) complexes. The existence of this series is evident from a large amount of comparative data[3, 15]; however, the relative stability of these complexes can be predicted from consideration of ligand-field theory, which is discussed briefly in Section 2.6. If comparisons are made between this series and the relative stability constants presently available for metalloenzymes (Table 1), the general trend is apparent. In contrast to the small ligand case, the Zn(II) metalloenzymes (the native state) appear to have unexpectedly high stability. The possible reasons for this are discussed in Section 2.6.

2.1.2 Kinetics of Metal Binding and Exchange. Data discussed thus far have been based on equilibrium measurements. Such data tell only part of the story, since the kinetic phenomena underlying these equilibria may actually tell more about a given process than the final equilibrium constants. A large amount of information is now available on the kinetics of formation and dissociation of coordination complexes[38–41]. Information is far from complete, however, as to the actual mechanisms that may be involved in central metal ion or ligand exchange. Many of the studies have involved the measurement of exchange rates for the central metal ion or the ligand in a complex by using an isotopically labeled ligand or metal ion. These data have been summarized and discussed by Stranks[38]. In general, it has been found that formation of a stable complex is very fast (often immeasurably fast), while dissociation is slow. Exchange rates are not necessarily a good measure of the stability of a complex. Rapid exchange of a central metal ion with metal ion in the medium is, however, a direct indication of lability. Exchange rates are affected by a number of factors, including charge on the central metal ion, as well as the charge, the electronegativity, the polarizability, the size, and particular stereochemical features of the ligand[38]. Geometry often has a major effect on exchange, e.g., the central metal ion in tetrahedral complexes of Co(II) exchange relatively rapidly, while exchange in square-planar, low-spin complexes of Co(II) is very slow [42]. In general, tetrahedral complexes are more labile than octahedral complexes [42, 43].

Unfortunately only spotty information is available on metalloproteins, but some interesting phenomena have been discovered. Metal ions exchange readily for each other at the active site of carboxypeptidase A [22, 28], and the exchange of Co(II) and Zn(II) and the reverse is illustrated in Figure 6. The rate of exchange depends on the concentration of free metal ion. For example, greater concentrations of Co(II) are needed to replace Zn(II) than the reverse (Fig. 6). Such exchange rates probably depend largely on the dissociation rate of the given metal–protein complex. One would guess that the dissociation rate of Co(II) carboxypeptidase is more rapid than that of the Zn(II) enzyme; however, no direct information is available as to the magnitude of the on constants.

Binding of the inhibitor β-phenylpropionate markedly slows the exchange of ^{65}Zn(II) at the active center of carboxypeptidase A for cold

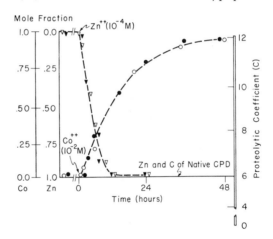

Figure 6   Replacement of radiocobalt bound to carboxypeptidase by stable zinc and the replacement of radiozinc bound to carboxypeptidase by stable cobalt. The ^{60}Co or ^{65}Zn enzymes, containing 1 g-atom of either metal, were dialyzed against the zinc or cobalt ion concentrations indicated below and contained in $1 M$ NaCl, $0.05 M$ Tris, $0.05 M$ sodium acetate solutions at pH 8.0 and 4° for the periods indicated on the abscissa. The proteolytic coefficients and enzyme-bound metal concentration were measured simultaneously as a function of time. Assays were carried out at pH 7.5 and 0° with $0.02 M$ CGP (carbobenzoxyglycyl-L phenylalanine) in $0.02 M$ sodium Veronal containing $0.1 M$ NaCl. (▼). Proteolytic coefficient (C) of $1 \times 10^{-4} M$ [(CPD)^{60}Co] exposed to $1 \times 10^{-4} M$ Zn^{2+} at 0 time; (▽) radiocobalt bound to $1 \times 10^{-4} M$ [(CPD)^{60}Co] exposed to $1 \times 10^{-4} M$ Zn^{2+} at 0 time; (●) proteolytic coefficient (C), of $1 \times 10^{-4} M$ [(CPD)^{65}Zn] exposed to $1 \times 10^{-2} M$ Co^{2+} at 0 time; (○) radiozinc bound to $1 \times 10^{-4} M$ [(CPD)^{65}Zn] exposed to $1 \times 10^{-2} M$ Co^{2+} at 0 time. From Coleman and Vallee [22], by permission.

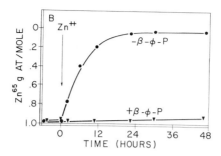

Figure 7   Prevention of the exchange $^{65}Zn^{2+} \rightleftharpoons Zn^{2+}$ at the active site by β-phenylpropionate. [(CPD)^{65}Zn], $1 \times 10^{-5} M$, was exposed to $1 \times 10^{-5} M$ Zn^{2+} as indicated, and ($\bullet$) the enzyme-bound $^{65}Zn^{2+}$ was measured as a function of time. ($\blacktriangledown$) Enzyme-bound $^{65}Zn^{2+}$ measured as a function of time in the presence of $2 \times 10^{-2} M$ β-phenylpropionate. From Coleman and Vallee[44], by permission.

Zn(II) in the medium (Fig. 7)[44]. The half-time for exchange goes from 5 to 576 hr[44]. β-Phenylpropionate thus radically slows the dissociation rate. This data, along with additional information on inhibitor complexes (see Sect. 3.1), suggests that the inhibitor adds a ligand, probably the carboxylate, to the central metal ion[44]. Cd(II) and Co(II) exchange in the enzyme have half-times of 320 and 96 hr, respectively, in the presence of β-phenylpropionate, suggesting that these mixed complexes are more labile than with Zn(II)[44]. Other inhibitors and substrates can also be shown to produce slowing of the exchange rate [44–46]. The opposite phenomenon is seen with certain of the copper oxidases. The active copper ions in tyrosinase do not exchange with ^{64}Cu in the medium[47]. Exchange will take place, however, if the enzyme is actively catalyzing the oxidation of substrate[47]. Thus the substrate appears to labilize the metal complex.

The above exchange studies reveal interesting features of metal complexes at the active centers of these enzymes, but they give only a rather limited and qualitative assay of the kinetic parameters involved in their formation and dissociation. The only thorough and reasonably complete study of the kinetics of formation of a metalloprotein has been carried out by Henkens and Sturtevant on bovine carbonic anhydrase[32]. The Zn(II) ion in carbonic anhydrase does not exchange at any measurable rate with external metal ions except at pH values near 5[25, 33]. Spectral studies of the Co(II) enzyme show that the geometry of the complex has shifted rather markedly at pH 5.5 from the geometry present at alkaline pH, where no exchange is observed (see p. 189). A stable apoenzyme can be prepared by removing the metal ion at pH 5.5[25, 33]. Reactivation of this apoenzyme by Zn(II) is accompanied by small changes in the ultraviolet spectrum of the protein[32]. Either the absorptivity change or the appearance of enzymatic activity can be used to measure the rate of combination of Zn(II) with the apoenzyme in a stopped-flow apparatus[32]. The reaction is first order with respect to both Zn(II) and apoprotein. The rate constants observed are $\sim 10^4$ $M^{-1}\,\text{sec}^{-1}$[32]. While the combination of Zn(II) with the protein is rapid, the rate constant is two orders of magnitude smaller than those observed for the formation of small Zn(II) chelate compounds. A number of studies report second-order rate constants in the range 10^6–$10^8\,M^{-1}\,\text{sec}^{-1}$ [39–41]. The latter rates are believed to be controlled primarily by the dissociation of H_2O from the coordination positions to be occupied by the incoming ligand[39]. The rate of combination of Zn(II) with the apocarbonic anhydrase rises from pH 5.5 to pH 7.5 (Table 2), but does not approach the rate observed in model systems. A study of the temperature dependence of the rates satisfactorily yielded Arrhenius plots, and the calculated energies and entropies of activation are given in Table 2. The energy of activation observed is considerably higher than that characteristic of model systems, but is partially compensated for by a large increase in entropy of activation. Values of these parameters observed for Zn(II) chelation to small molecules are 7–8 kcal mole^{-1} for E_a and -4 to -8 cal deg^{-1} mole^{-1} for $\Delta S\ddagger$ [40, 41].

The second order rate constant observed in the model systems has

TABLE 2   RATE OF RECOMBINATION OF Zn(II) WITH APOCARBONIC
ANHYDRASE AT 25°. HEAT AND ENTROPY OF ACTIVATION[a]

			Method				
			Ultraviolet Difference Spectrum	Recovery of Hydration Activity	Recovery of Esterase Activity		$\Delta S\ddagger$
pH	Ionic Strength, M	H$^+$ Released by Reaction	Apparent Second Order Rate Constant, $M^{-1}\,\text{sec}^{-1} \times 10^{-4}$			E_a, kcal mole^{-1}	cal deg^{-1} mole^{-1}
5.5	0.051	—	—	—	—	21.2	27.7
5.90	0.001	0.18	0.27	—	—		
5.90	0.021	0.42	0.54	—	—		
6.27	0.056	0.85	—	—	0.81		
7.0	0.056	1.5	—	1.36	—		
7.40	0.020	1.6	1.33	—	—		
7.56	0.056	2.2	—	—	2.0		
7.5	0.051	—	—	—	—	20.8	30.0

[a]Data taken from Henkens and Sturtevant[32].

been formulated as the product of an outer sphere association constant (K_0) and a first order rate constant (k_1) for the exchange of water for the ligand. Hence, the observed rate constant is $K_0 k_1$ [32, 39]. The various factors that may account for the slow rate of Zn(II) binding in the enzyme can only be speculated upon. The general magnitude of the rate would indicate that the binding site is relatively intact before combination of the apoprotein with the metal ion. Thus complex formation does not appear to involve major rearrangements of protein structure [32]. The Zn(II) atom, however, is bound at the bottom of a large cavity in the protein (see p. 286) and occupies what is probably a rather unusual geometry [25]. Considerable rearrangement of water bound to the protein as well as to the metal ion may have to take place [32]. Small conformational changes in the protein may also occur.

By using the equilibrium constant previously determined for Zn(II) carbonic anhydrase, Henekens and Sturtevant [32] have calculated an off rate for the Zn(II) ion of 1.5×10^{-9} sec^{-1}. Some interesting comparisons have been made to carboxypeptidase A. Since the half-time for the replacement of Zn(II) by Cd(II) in carboxypeptidase A is ~20 hr [23] at pH 8.0, 4°, the rate constant for dissociation has been estimated to be 10^{-6} sec^{-1} [32]. Because of some difference in stability of the Zn(II) and Cd(II) enzymes the ^{65}Zn–Zn exchange probably gives a better figure. This half-time of exchange is 5 hr [44], hence $k_{off} = $ ~2.5×10^{-5} sec^{-1}, 4–5 orders of magnitude faster than the dissociation of Zn(II) from carbonic anhydrase. If, as seems likely, the on rates for the two enzymes don't differ by more than an order of magnitude, then the very slow dissociation of Zn(II) from carbonic anhydrase explains the failure to observe any exchange at pH 8.0. The half-time for dissociation of Zn(II) from carbonic anhydrase at pH 8.0, 0.1 M ionic strength, 25°, can be calculated to be 5–6 years [32].

2.1.3 Apoenzymes. The *in vitro* preparation of the apoenzymes by removal of the native metal ion from metalloenzymes by various chelating agents has formed an important part of the study of these catalysts. Preparation of an inactive metal-free apoenzyme which can be totally reactivated by readdition of the metal ion answers the fundamental question about the function of the metal ion. It is a necessary part of the active centers of these enzymes. This concept should be qualified in the sense that the metal ion could bind some distance from the actual substrate combining site and exert its influence via structural, electronic, or allosteric effects on the "catalytic site." Such effects could result from changes in quaternary structure in the case of multimeric metalloenzymes. It is true that in the two monomeric enzymes for which detailed X-ray data on the metal binding site is available, carbonic anhydrase and carboxypeptidase (see Sect. 3), the metal ion is within the radius of the substrate or inhibitor binding site. What specific features of chemical structure cause the metal ion to exert functional control remains a partially answered question. The properties of the various apoenzymes that have been prepared do contribute some information to the answer to this question.

For three Zn(II) metalloenzymes, carboxypeptidase A, carbonic anhydrase, and alkaline phosphatase, stable apoenzymes have been prepared and their physicochemical properties studied in detail[25, 26, 30]. In the case of the two monomeric proteins (carboxypeptidase A and carbonic anhydrase) crystals of the apoenzymes have been prepared and X-ray data shows them to be isomorphous with the native enzyme [30, 48]. These findings imply that there must not be major conformational changes induced by removing the metal ion, a conclusion supported by optical rotatory dispersion measurements[25, 30]. While minor changes around the metal binding site are not ruled out, it is of interest that the only major feature in the difference map of the electron density between apocarbonic anhydrase and the native enzyme at 5.5 Å resolution is a density corresponding to the Zn(II) ion (Fig. 8) [48].

The situation in the case of the dimeric protein alkaline phosphatase is more complex. The Zn(II) ions can be removed slowly with high concentrations of chelating agents like 1,10-phenanthroline or 8-hydroxyquinoline sulfonate [26, 49]. Removal is much more rapid if the enzyme is gently shaken with Chelex resin [50]. The resulting apoenzyme can be

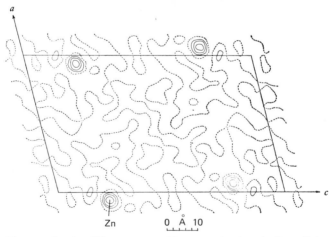

Figure 8   Electron density difference map. Human carbonic anhydrase C–human apocarbonic anhydrase C. After Tilander et al.[48].

completely reactivated by Zn(II) ions [26, 49–51]. Circular dichroism of the apoenzyme in the far ultraviolet shows some small, but significant, changes [51], suggesting that loss of the metal ions is associated with some structural change. A change in quaternary structure is apparent from sedimentation ultracentrifugation of the apoenzyme. Approximately 25% of the apoenzyme is present as the monomer, while 75% is dimer [26, 51] (Fig. 9). Thus, the metal ion appears to favor dimerization, but is not the sole force holding the dimer together. Another significant feature of the Zn(II)–subunit equilibria is revealed by studies of ^{32}P-phosphate binding. If one Zn(II) ion per dimer is added back to the apoenzyme in the presence of $H^{32}PO_4^{2-}$, very tight binding of phosphate (equal to the affinity of phosphate for the native enzyme) is induced, but at a stoichiometry of 0.5 mole phosphate/dimer [52]. The most reasonable explanation of this finding is that phosphate stabilizes the 2Zn(II) dimer over the 1Zn(II) form of the enzyme (Fig. 9). These results are discussed further in Section 2.6.

Despite probable effects of Zn(II) on the secondary, tertiary, and quaternary structure of the alkaline phosphatase dimer, it has been

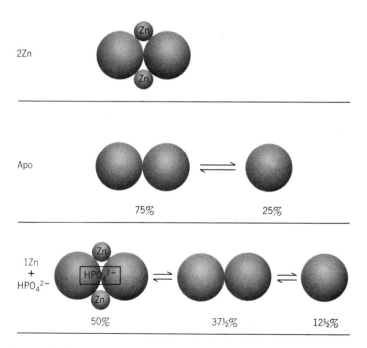

Figure 9   Postulated subunit structure for *E. coli* alkaline phosphatase containing 2Zn(II), apoalkaline phosphatase, and apoalkaline phosphatase + 1Zn(II) + HPO_4^{2-}.

shown the *E. coli* grown under Zn-free conditions synthesize a completely metal-free inactive apoenzyme[51]. When isolated in pure form, this apoenzyme has physicochemical properties identical to the apoenzyme prepared *in vitro*[26, 51]. It can be completely reactivated by Zn(II) ions [51]. Thus it appears that the metal ion need not be involved in the synthesis of this metalloenzyme, but that the polypeptide chain assumes a conformation in which the donor atoms at the metal coordination sites are in a position to assume the final active coordination geometry if the metal ion is inserted[51].

Zn(II) appears to play a much more direct role in the structure of yeast alcohol dehydrogenase than in the Zn(II) enzymes discussed above. If the enzyme (MW = 151,000) is exposed to chelating agents there is both an instantaneous reversible inhibition due to mixed complex formation and a slow time-dependent irreversible inhibition[53, 54]. The latter can be shown to be accompanied by slow removal of the four Zn(II) ions and the concomitant dissociation of the tetrameric molecule into four subunits of MW = 36,000[24]. The coenzyme, NAD, protects the enzyme against this loss of the metal ion in the presence of chelating agents. Conditions have not been found under which this change in structure can be reversed such as to give an active holoenzyme[24, 53, 54]. Thus direct participation of the metal ions as determinants of protein structure in metalloenzymes may be expected to show wide variation in the degree of this participation. Participation of the metal ion in structure ranges from that in a protein like carboxypeptidase A, where little of the protein structure appears to be determined by the metal ion coordination, to alcohol dehydrogenase, where both the tertiary and quaternary structure required for an active enzyme cannot be maintained in the absence of the metal ions. Alkaline phosphatase appears to be an intermediate case.

Apoenzymes have also been prepared from several of the copper oxidases, but have not received extensive physicochemical study. Relatively high concentrations of cyanide can be used to remove the copper ion from ascorbic acid oxidase[55, 56], laccase[57], and tyrosinase[58, 59]. In the case of the blue Cu(II) protein, ascorbic acid oxidase, incubation of the apoenzyme with Cu(II) does not restore activity[55, 56]. However, addition of ascorbate and copper results in activation of the apoenzyme[55]. The metal ion species responsible for restoration of activity is presumably Cu(I)[55]. The intense blue absorption band at 608 mμ typical of Cu(II) in these proteins is also restored to the colorless apoenzyme by this procedure[55]. The exact mechanism requiring prior reduction for the reconstitution of a complex whose final characteristics imply the presence of Cu(II) is not clear at the present time.

Apoascorbate oxidase is not stable indefinitely, but loses activity over a period of days [55]. The freshly prepared apoenzyme contains approximately 10 titrable SH groups which are absent in the native enzyme [55, 60]. Accompanying the aging of the apoenzyme is the disappearance of these free SH groups [55, 60], suggesting that sulfhydryl oxidation may be involved in the structural alteration of the apoenzyme which prevents reconstitution.

The enzyme tyrosinase, in contrast to the blue copper proteins, appears to contain largely Cu(I), at least in the resting state [59, 61], and is colorless. No evidence has been obtained for a reversible Cu(I) ↔ Cu(II) transformation during the oxidation of substrate [61]. The active Cu(I) ion can also be removed by high concentrations of cyanide, and Cu(II) under reducing conditions can also reactivate the enzyme [58, 59]. It appears, however, that addition of Cu(II) alone is sufficient to reactivate the enzyme, the protein inducing the reduction to the Cu(I) state [59]. Excess Cu(II) over that bound at the active center is not reduced [59].

Cyanide treatment of laccase from the lac tree has been reported to result in the formation of an apoenzyme that can be reactivated by Cu(II) ions [57]. Cyanide treatment does result in complete loss of the blue chromophore [62]. Detailed esr studies by Malmström et al. [62] on the interaction of cyanide with fungal laccase show that Cu(II) remains at the original site, but that both oxidase activity and the blue color disappear. An esr signal remains in the cyanide-treated enzyme (see p. 228), the details of which suggest coordination of the Cu(II) to three or four N atoms [62]. All these changes can be reversed by removing the cyanide by dialysis. These findings suggest that a mixed complex forms between cyanide and enzyme-copper [62]. The properties of these copper oxidases are discussed in more detail in Sections 2.2, 2.3, and 3.2.3.

The nonheme iron proteins of the ferredoxin class have been prepared free of iron by treatment with mersalyl [63–65]. Loss of the iron is accompanied by the loss of inorganic sulfide evolved as H_2S [64, 65]. The inorganic sulfide is closely associated with the iron of these proteins and is probably one of the ligands (see Sect. 3.2.4). After treatment with mersalyl, if the iron and sulfide are removed, a material referred to as "apoferredoxin" is produced [64, 65]. This material will not reincorporate iron and sulfide to form an active ferredoxin [63–65]. However, if the mersalyl-treated, decolorized ferredoxin is first treated with excess 2-mercaptoethanol, the original spectrum is restored. This reconstituted ferredoxin is active and appears identical to the native protein [63].

2.2 Absorption Spectra and Optical Activity

In addition to the usual absorption bands present in proteins, metalloproteins may have additional optical absorption bands arising from the *d–d* transitions centered on the metal ion or from charge transfer bands resulting from electronic transitions between the metal ion and the ligand [11, 13, 66]. Use of these special metal-related absorption bands for study of metalloenzymes has the distinct advantage that the chromophore is likely to be situated within the active center of the enzyme and is thus uniquely subject to changes occurring at the active center.

The *d–d* transitions arising in the transition metal ions with unfilled *d*-shells are usually of low intensity, $\epsilon = <1000$, since they are forbidden under the usual selection rules for electronic transitions [11]. The "forbidden" nature of these transitions in actual complexes is presumably lifted to some degree by *d–p* orbital mixing [11, 13]. Mixing of *d*- and *p*-orbitals can occur if the environment of the ion lacks a center of symmetry. Thus complexes of low symmetry are expected to have more intense *d–d* absorption bands. Although the factors contributing to the observed intensities of the *d–d* transitions of transition metal complexes are not completely understood, it is observed that bands for tetrahedral complexes are in general more intense than for octahedral complexes, in keeping with the lack of a center of symmetry in the former geometry. Among the metal ions considered here, this is especially true for Co(II) and Ni(II) complexes [13].

Simplified energy level diagrams for the 3*d*-orbitals in the commonly occurring geometries, tetrahedral, octahedral and square-planar, are shown in Figure 10. The splitting of the five 3*d*-orbitals into two degenerate groups in complexes of regular octahedral or tetrahedral geometries can be deduced from simple geometrical arguments [11, 13]. This degeneracy is lifted by any distortion from a regular geometry. One of the extreme cases, the removal of two vertical ligands, is given in Figure 10.

If the metal ion contains one *d*-electron [Ti(III)] or nine *d*-electrons [Cu(II)], the latter corresponding to one hole in the 3*d*-shell, then the interpretation of the spectra for the regular geometries in terms of the energy level diagram in Figure 10 is rather simple. The observed *d–d* transition represents the transition of an electron from the lower energy set of *d*-orbitals to the higher energy set and has an energy, Δ, often referred to as 10*Dq* (Fig. 10). These energies are relatively low, 3000 to ~30,000 cm^{-1}; hence the *d–d* transitions are observed in the visible and infrared region of the spectrum. If there is more than one electron or less than nine in the five 3*d*-orbitals the precise energy levels giving rise

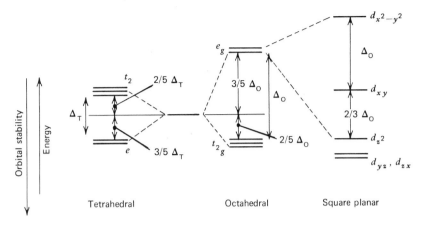

Figure 10   Splitting of $3d$-orbital energies in tetrahedral, octahedral and square-planar complexes of transition metal ions.

to the expected d–d transitions are more complex because the various possible configurations of n electrons must be considered [11–13].

Two types of transitions are seen, depending on the selection rules broken — "spin-forbidden," where the spin direction must change, and "spin-allowed," where the only selection rule broken is the rule against redistribution of electrons in a single quantum shell. The intense d–d transitions, $\epsilon = 10$–100, are the spin-allowed ones, while the much weaker spin-forbidden transitions have ϵ values of 0.01–0.10. The latter are infrequently observed, but can be a prominent part of the spectrum when spin-allowed transitions are not present, in the case of [Mn(H$_2$O)$_6$], for example [13]. In absorption spectra of transition metal complexes the weak bands due to spin-forbidden transitions are usually obscured by the intense spin-allowed transitions. Such transitions might, however, appear in circular dichroism spectra, since their optical activity is not necessarily low.

The charge-transfer bands associated with metal complexes are usually observed to have large intensities ($\epsilon > 10^3$), in keeping with the fact that they are presumably allowed electronic transitions. These transitions are often of high energy and appear in the ultraviolet region of the spectrum. A number of very intense charge-transfer bands occurring in the visible region of the spectrum are known, however, and the complexes in which they occur are frequently used as analytical tools in the quantitative determination of metal ions [67]. There have not been extensive theoretical studies of metal–ligand charge-transfer spectra and their precise nature is often not known with certainty [66]. The factors governing their intensity may not always result in an extinction coefficient of 10^3.

With the above brief introduction it is useful to explore in a qualitative way what has been learned from these metal absorption bands as they appear in metalloenzymes. The low extinction coefficients present an immediate practical problem, since concentrations of protein greater than $10^{-4} M$ must be employed to observe reasonable optical densities. Those metal ions producing the most intense d–d transitions are naturally the most useful. As expected on the basis of model complexes, these have been the Co(II) and Cu(II) metalloenzymes. The band positions and extinction coefficients observed for the visible absorption bands of metalloenzymes are given in Table 3. Extensive data is largely confined to the Co(II) and Cu(II) proteins. One feature which can increase the intensities observed for the d–d transitions of metalloproteins may be the presence of metal-binding sites of low symmetry stabilized by the protein structure [68–72] (see below).

TABLE 3   ABSORPTION SPECTRA OF Co(II) AND Cu(II) METALLOENZYMES

Co(II) Metalloenzymes	Band Position, mμ		Intensity, ϵ	References
Co(II) carboxypeptidase	500			22, 90
	555		160	
	572		160	
	940		~25	
Co(II) carbonic anhydrase	520		205	25, 74, 78, 80
	555		340	
	615		230	
	640		240	
	900		~25	
	1250		95	
Co(II) carbonic anhydrase + CN$^-$		310^a		25, 74, 84
		345		
		450		
	520	545	350	
	570	585	650	
Co(II) carbonic anhydrase + acetazolamide		465^a		25, 74, 84
		515		
	520	550	350	
	570	570	550	
	600	590	500	
Co(II) alkaline phosphatase	640		260	86, 88

TABLE 3 (contd.)

Co(II) Metalloenzymes	Band Position, mμ	Intensity, ϵ	References
	605	220	
	555	378	
	510	335	
Co(II) alkaline phosphatase + HPO$_4^{2-}$	640	120	86, 88
	535	350	
	480	260	
Co(II) alkaline phosphatase + HAsO$_4^{2-}$	500	~240	b
	550	~260	
Cu(II) carboxypeptidase	790	<100	23
Cu(II) carbonic anhydrase	590	50	25, 35, 78
	750	100	
	900	75	
Cu(II) carbonic anhydrase + CN$^-$	700	130	25, 78, 91
	900	80	
Cu(II) alkaline phosphatase^b	~750	~100	

Copper Oxidases			
Laccase [Cu(II)]	730	~500	92
	615	1400	
	532	~300	
Azurin (psuedomonas blue protein) [Cu(II)]	806	~600	92, 93
	621	2800–3500	
	521	~300	
	467	~400	
Ascorbic acid oxidase [Cu(II)]	606	770	94
	412	~500	
Ceruloplasmin [Cu(II)]	605	1200	95
	370	~500	

^{a}Determined from the band positions observed in the circular dichroic spectra.
^{b}M. L. Applebury and J. E. Coleman (unpublished observations).

2.2.1 Co(II) Proteins. Although cobalt metalloproteins (exclusive of cobalamin-containing proteins) do not seem to occur in nature with any frequency, it has been possible to substitute Co(II) at the active centers of several Zn(II) metalloenzymes with retention of catalytic activity (see Sect. 2.6). Figure 11 shows the visible absorption spectra observed for Co(II), Ni(II), and Cu(II) carbonic anhydrases and for several inhibitor

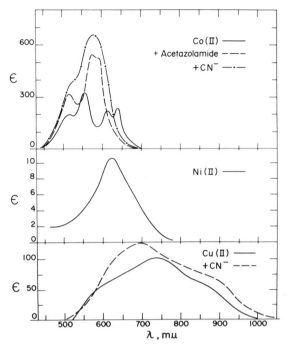

Figure 11　Visible absorption spectra of Co(II), Ni(II), and Cu(II) human carbonic anhydrases B and inhibitor complexes.

complexes of the Co(II) and Cu(II) enzymes. The visible absorption bands for the Co(II) enzyme and its derivatives have ϵ values of several hundred (Table 3). Intensities of this magnitude are more typical of tetrahedral Co(II) complexes rather than octahedral, which are expected to have ϵ values of ~10[13]. The spectrum of the native enzyme at alkaline pH, however, is very widely split, as much as 1800 cm^{-1} separating bands, suggesting that a highly distorted ligand geometry is present [25]. The 2 Å resolution X-ray structure of the active center (see Sect. 3.1.3) indicates the presence of three protein ligands and a fourth site probably occupied by a coordinated water molecule or hydroxide in the native state[37, 73]. Thus the spectrum would appear to arise from a distorted tetrahedron. Cyanide and acetazolamide are both believed to add a ligand to the site occupied by the coordinated water molecule[37, 74–77]. The d–d transitions of the Co(II) ion become more intense and the bandwidth narrows when these inhibitors are added (Fig. 11)[25, 74, 78]. Thus there is a distinct change of geometry at the active center coordination complex when these inhibitors bind. This constitutes some of the strongest evidence in solution that the binding of these inhibitors does involve change within the inner coordination sphere of the metal ion. The shift to a narrower more intense spectrum upon inhibitor binding has been interpreted as a shift to a more regular tetrahedral geometry [79, 80] (see below).

There is also an apparent change in the ligand geometry surrounding the Co(II) ion as the pH is lowered from 9 to 6, as shown in Figure 12 from the data of Lindskog and Nyman [35]. The widely split spectrum is present only in the alkaline form of the enzyme. The change in the spectrum at 640 mμ follows a titration curve with a pK_a very similar to the pH–rate profile observed for carbonic anhydrase (Fig. 12) (see Sect. 3.1.3). Lindskog [81] has determined this relationship in detail and found that both the changes in the Co(II) absorption bands and the pH–rate profile are affected in the same manner by anions (Fig. 13). Thus both processes appear to reflect the same ionization. A good deal of evidence now indicates that this ionization represents that of a coordinated water molecule (see Sect. 3.1.3) and that attack on a substrate carbonyl carbon by a coordinated OH$^-$ is an essential part of the mechanism [37, 82, 83]. The competition between anions and OH$^-$ for the same coordination site would adequately explain both sets of data in Figure 13.

Circular dichroism proves in general to be much more sensitive to small changes in dissymmetry at the coordination site than the absorption spectrum. The ellipticity bands associated with the visible absorption bands of three species and isozyme variants of carbonic anhydrase are

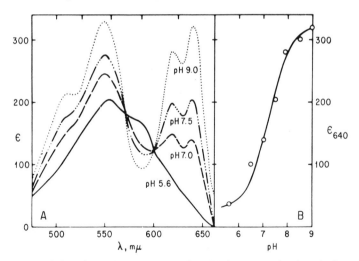

Figure 12   (A) Visible absorption spectrum of Co(II) human carbonic anhydrase B as a function of pH. (B) Molar extinction coefficient at 640 mμ as a function of pH. From Lindskog and Nyman [35], by permission.

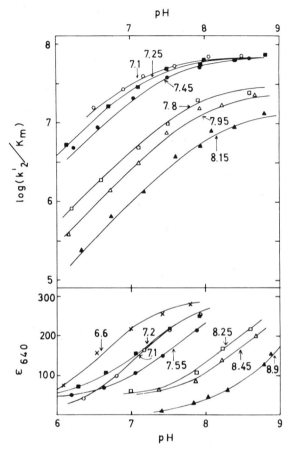

Figure 13 The pH dependence of the activity (*a*) and absorption at 640 mμ (*b*) of Co(II)-carbonic anhydrase. (○) 0.1 *M* F⁻; (■) 0.1 *M* Cl⁻; (●) 0.1 *M* Br⁻; (□) 5×10^{-5} *M* NCO⁻ + 0.1 *M* Cl⁻; (△) 0.1 *M* I⁻; (▲) 0.1 *M* NO₃⁻; (×) 0.0333 *M* SO₄²⁻. Temperature 25°. The curves have been drawn to represent titrations of single groups with the values of pK_{app} indicated in the diagrams. From Lindskog [81], by permission.

shown in Figure 14. While the absorption spectra of the Co(II) derivatives of the human B, human C, and bovine carbonic anhydrases are nearly identical [25, 35, 74], only the human C and bovine enzymes show ellipticity of any magnitude associated with the Co(II) absorption bands. Values of $\epsilon_L - \epsilon_R (\Delta\epsilon)$ vary from 0.5 to 3 [78, 84].

On the other hand, maximum $\Delta\epsilon$ values for the human B isozyme are below 0.2. Thus there are inherent differences in dissymmetry at the active centers of these variants of carbonic anhydrase. Major ellipticity, however, is induced in the Co(II) absorption bands of the human isozyme

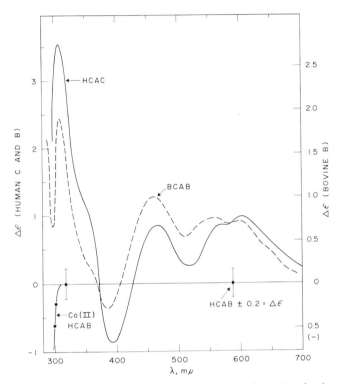

Figure 14   Visible CD of three isozyme and species variants of Co(II) carbonic anhydrase.
(——) human isozyme C; (-----) bovine isozyme B; (●—●) human isozyme B.

B by the addition of sulfonamides (Fig. 15) or cyanide [78, 84], confirming a shift in geometry to a more dissymmetric configuration.

   The mechanism by which the Co(II) chromophore of the bovine and
human C enzymes gain their initial large optical activity is unclear. It
does not appear that the basic coordination geometry is different from
that of the human B isozyme, since the energies of the transitions are
almost the same. A powerful mechanism for inducing optical activity is
the perturbation of the potential field surrounding a chromophore by
the placement of an asymmetric charge or dipole [85]. Such a mechanism
due to an adjacent charged or dipolar amino acid side chain may induce
the ellipticity in the Co(II) bands of the bovine and human C isozymes.

   The data on the human B isozyme suggest that the Co(II) chromophore in the alkaline form of the enzyme has a minimal amount of
dissymmetry. There is some dissymmetry, however, since if the ellipticity
of the human B isozyme is examined at a ten-fold higher concentration
of protein, then small ellipticity bands, $\Delta\epsilon < 0.2$, can be discovered to

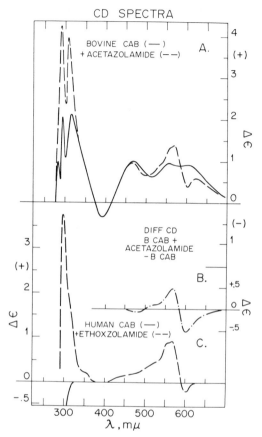

Figure 15   Visible CD spectra of Co(II) bovine carbonic anhydrase B and Co(II) human carbonic anhydrase B plus sulfonamides. From Coleman [84], by permission.

be associated with all the Co(II) absorption bands (Fig. 16). This ellipticity undergoes a two- to threefold increase if the enzyme is titrated to acid pH (Fig. 16).

These findings can best be summarized by supposing that the basic coordination complex in the alkaline form of the enzyme has a geometry that minimizes the dissymmetry present, perhaps not incompatible with a flattened tetrahedron with a hydroxide ion as one ligand. Addition of strong-field ligands, anions or sulfonamides, at the latter position increases the dissymmetry of the complex. This is also not incompatible with a change toward a more regular coordination geometry lacking a center of symmetry. Curve fitting of both the absorption and CD bands of the Co(II) enzyme-inhibitor complexes, however, show that the band

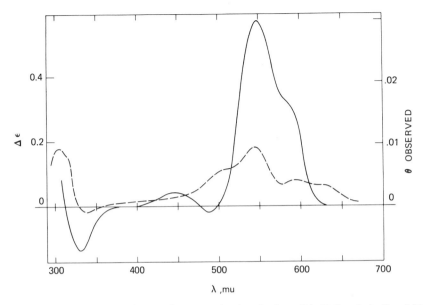

Figure 16   CD spectrum of Co(II) human carbonic anhydrase B in D_2O at (—) pD = 6.81 and (----) pD = 9.21.

multiplicity is not reduced[84], hence a certain degree of distortion must still persist. In the same class as the changes induced by the inhibitors must be the change induced by low pH, presumably an $OH^- \rightarrow H_2O$ transformation[37].

Another interesting feature of the Co(II) chromophore in carbonic anhydrase revealed by the CD spectra is that there are significant near ultraviolet bands associated with the chromophore (Fig. 14). This has proved to be true of a large number of these protein–metal chromophores (see below). Their origin is not clear. Absorption intensities do not appear great enough to significantly add to the protein absorption in this region, although the very large absorption and scattering of the concentrated protein solutions makes examination of the absorption spectrum in this region very difficult. They could represent relatively weak charge-transfer bands. There might also be magnetic dipole forbidden d–d transitions apparent in the CD spectrum that are obscured in the absorption spectrum.

The visible absorption spectrum of Co(II) alkaline phosphatase is remarkably similar to that of carbonic anhydrase, as shown in Figure 17. Another very distorted coordination geometry appears to be present [86–88]. Titration of the development of the spectrum in the apoenzyme as a function of the moles of Co(II) added reveals that two such co-

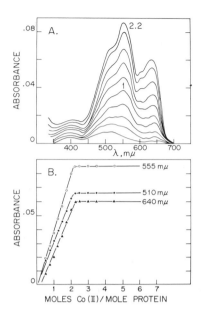

Figure 17 Spectrophotometric titration of apoalkaline phosphatase with Co(II). Addition of one Co(II) ion per dimer is indicated by 1 in part (*A*). Spectral changes are complete on the addition of 2.2 Co(II) ions per dimer. (*B*) Plot of absorbance against equivalents of Co(II) at given wavelengths. From Applebury and Coleman [86], by permission.

ordination sites exist per mole of protein (Fig. 17). The active protein is a dimer of identical monomer units (see Sect. 3.1.4). Circular dichroism of this Co(II) enzyme also reveals a multiplicity of bands [86]. Addition of HPO_4^{2-} causes a major change in these ellipticity bands, as well as in the absorption bands [86, 88]. Phosphate is a powerful inhibitor of the catalysis of the hydrolysis of phosphate esters, but can also be considered the simplest substrate for the enzyme (see Sect. 2.6). Thus, here as well subtle changes in the coordination sphere accompany inhibitor (substrate) binding. Arsenate causes similar changes in the absorption spectrum, but rather different changes in ellipticity. Many of the bands are opposite in sign to those induced by HPO_4^{2-}. Since arsenate and phosphate are isomorphous in their crystalline forms, the fact that they induce rather different symmetry effects on the Co(II) chromophore at the active site of alkaline phosphatase serves as an illustration of the subtle features that must underlie symmetry changes in these metal-protein chromophores.

2.2.2   Cu(II) Proteins. On the basis of the "hole" formalism mentioned above, Cu(II) with nine 3*d*-electrons would be expected to show a single *d–d* absorption band when placed in a ligand field of regular geometry (Fig. 10). In fact, Cu(II) complexes normally show three prominent absorption bands located in the red part of the spectrum. This arises because of the severe Jahn-Teller distortion induced by the d^9 ion if the ligands are free to move[11–13]. The final result of this distortion is that the Cu(II) ion gains considerable stability if it is placed in a square-planar configuration of ligands. This accounts for the observed difficulty of adding fifth and sixth donor atoms to many Cu(II) complexes[12, 13]. X-ray structures of a number of Cu(II) complexes show four close planar ligands, while water molecules or other ligands occupy apical positions at a much longer bond distance[89]. The energy level diagram for the extreme square-planar distortion is shown in Figure 10 and predicts three *d–d* absorption bands, as are observed in many copper coordination compounds[13].

The occurrence of copper-containing enzymes in nature seems to be restricted to oxidases. It has been possible, however, to substitute Cu(II) for the native Zn(II) ion in several enzymes catalyzing hydrolysis and hydration reactions (Table 3). These derivatives are uniformally inactive, perhaps related to the special features of Cu(II) coordination (see Sect. 2.6). The chromophore introduced does give some information about the active center. In the case of the three enzymes for which this substitution has been made (Table 3), all show broad absorption bands in the red region, with molar extinction coefficients $\leqq 100$. In at least one case, Cu(II) carbonic anhydrase, the band can easily be resolved into three absorption maxima at 590, 750, and ~ 900 mμ (Fig. 11). These spectra are not too different from those observed for model complexes, Cu(II) peptide or amino acid complexes[96–98]. There is clearly some degree of distortion present. This may be partly induced by Cu(II), but as seems likely from other data, the protein structure may maintain a distorted coordination geometry.

If the absorption spectra of the naturally occurring Cu(II) oxidases are examined, similar transition energies are observed, but the band intensities are extraordinarily large (Table 3). Many of the bands have intensities expected for allowed electronic transitions, although they appear to be clearly associated with the unfilled *d*-shell of Cu(II). These are intensely blue proteins and absorption bands of this magnitude are unique among Cu(II) complexes. The mechanism by which this intensity is achieved has been a subject of discussion for a number of years. A certain degree of charge-transfer between the ligand and the *d*-shell of Cu(II) has been postulated[69]. A possible interaction between adjacent copper nuclei has also been suggested, since many of these proteins contain more than one copper ion. This seems to be ruled out by the fact that one of the most intensely blue proteins, the *Pseudomonas* blue protein (azurin) is a monomeric protein of MW = 16,400 containing one atom of copper per molecule[99, 100]. Since the intensity of d–d absorption bands depend critically on the symmetry of the complex, irregularities in the geometry increase the intensity. A tetrahedral complex of Cu(II), a rare occurrence in model systems, would be expected to have d–d transitions of greater intensity than other regular geometries because of the absence of a center of symmetry. Tetrahedral Cu(II) has been suggested to be present in these intensely blue copper proteins[69]; however, the multiple transitions observed imply a distorted configuration.

More recently, in an effort to derive possible structures for the copper binding site, at least two laboratories have made extensive calculations of physical parameters describing Cu(II) complexes on the basis of electrostatic crystal field theory. In one model[70], twelve crystal field parameters have been selected from which 12 physical parameters, accessible to measurement in these proteins and related to structure by well developed physical theory, can be calculated. These physical parameters are the three spectroscopic splitting factors of the epr spectrum (see Sect. 2.3), the energies of the three "d–d" optical transitions of highest energy, the oscillator strengths, and the rotatory strengths of these three transitions. By starting with the classical square-planar geometry of Cu(II) and selecting crystal field parameters to describe the covalency of the ligand–metal bonds and a series of even–even and even–odd distortions of the square-planar geometry, Blumberg [70] has calculated the geometry of the electrostatic crystal field required to generate a certain set of physical measurements. On this basis, the physical parameters calculated from theory were made to fit very closely the same set of experimentally observed physical parameters for ceruloplasmin. They included the sign, the magnitude, and the rotatory strengths of the visible bands calculated from visible ORD measurements on the assumption of three transitions.

The geometry necessary to generate the required electrostatic crystal field was characterized by a slight displacement of the copper above the average position of the four strongest ligands, a strong lateral distortion since two ligands were much stronger, and a small amount of "saddle" distortion which moves the four strongest ligands slightly toward tetrahedral positions. The authors caution against considering such an electrostatic field as an actual physical representation of the copper site, but point out that the properties of the copper chromophore in ceruloplasmin, including the intensity, can be explained by sufficiently strong distortions from the more usual square-planar configuration.

Brill and Bryce have set up a 6-parameter model, based on the admixture of 4p- and 4s-character into the primary 3d-orbitals of the cupric ion[71, 72]. By adjusting the six parameters they were able to match the experimental epr parameters of the *Pseudomonas* blue proteins, as well as the oscillator strengths and the signs and magnitudes of the rotational strengths of the visible absorption bands. Their model predicts a 12°7′ out-of-plane distortion toward a tetrahedron for the copper site of the blue proteins from *Pseudomonas*.

The fact that the Cu(II) binding sites in these proteins are highly dissymmetric is shown by the large ellipticity associated with the Cu(II) absorption bands of two of these proteins, *Pseudomonas* blue protein (Fig. 18) and laccase from *Polyporus versicolor* (Fig. 19). In addition to the

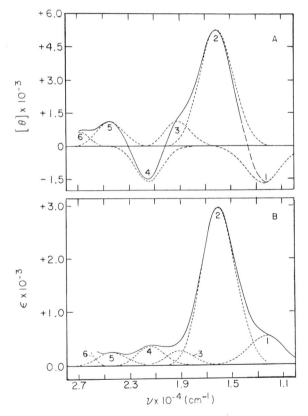

Figure 18   (*A*) Visible CD and (*B*) absorption spectra of the blue protein (Azurin) from *Pseudomonas aeruginosa*. The dashed lines represent the resolution of the spectra into individual Gaussian absorption bands by means of a DuPont curve fitter. From Tang et al.[92], by permission.

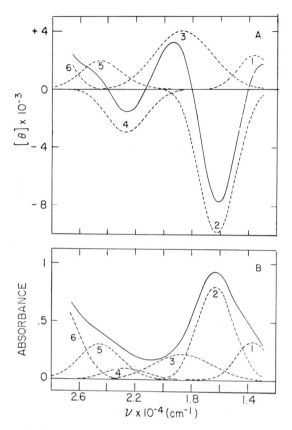

Figure 19 (*A*) Visible CD and (*B*) absorption spectra of laccase from *Polyporus versicolor*. Symbols as in Figure 18. From Tang et al.[92], by permission.

large ellipticity associated with the three major long wavelength bands, there are a number of higher energy bands in the spectrum extending down to 300 mμ. Weak charge-transfer or magnetic dipole forbidden transitions could account for these[92]. It is curious that the two proteins appear to have opposite configurations (chirality) in the region immediately surrounding the copper chromophore[92].

These highly unusual geometries would appear to have something to do with the mechanism of action of these oxidases. All the copper oxidases listed in Table 3 seem to require the reversible reduction of Cu(II) to Cu(I) during the enzymatic reaction (see Sect. 2.3.4). Initial evidence for the participation of the Cu(II)–Cu(I) couple in the actual mechanism of action of the blue oxidases came from the fact that reduction of these enzymes by substrates results in loss of the blue color

[55, 101–104], as expected if the d^{10} Cu(I) ion is produced. This reduction has been amply confirmed by esr studies (see Sect. 2.3.4).

Oxygen is a cosubstrate and the possible metal–oxygen interactions are just beginning to be elucidated[69] (see Sect. 3.2.3). At least one copper oxidase, tyrosinase, appears to contain exclusively Cu(I) in the active form of the enzyme[59, 61]. How this copper complex differs from that of the blue oxidases is unclear at present. Cytochrome oxidase from various sources has also been found to contain copper[105–108]. Electron spin resonance studies of this enzyme-copper have been more extensive than spectral studies, but there is an absorption band near 815 mμ, $\epsilon \sim 1400$, believed to be associated with the copper. This is bleached by reduction of the enzyme[107].

2.2.3 Spectra of Model Co(II) and Cu(II) Complexes.

The hunt for model complexes whose absorption spectra mimic those found in metalloenzymes has been relatively unsuccessful. A great deal of information has been collected on complexes of amino acids and peptides with the metals Co(II), Ni(II), and Cu(II). The X-ray structure is now available for many of them[89]. Cu(II) complexes of amino acids and peptides in the solid state have almost always been found in a distorted octahedral configuration with two very distant axial ligands[89], not incompatible with the square-planar structure postulated from solution chemistry[109–112]. There have been some interesting variations in bond distances and bond angles[89], but none of the distortions observed have been associated with particularly unusual spectra. Most amino acid and peptide Cu(II) complexes have a broad absorption band between 600 and 700 mμ with ϵ values near 100. Circular dichroism of these bands clearly show the presence of three bands which appear to correspond to those theoretically expected[92, 113]. The circular dichroism spectra of 1:1 complexes of Cu(II) with Gly-L-Val and Gly-L-Tyr are shown in Figure 20 as functions of pH. Curve fitting shows that a minimum of three bands are required to fit these spectra[92]. One of the interesting findings from solution chemistry which has been confirmed by X-ray data is that the Cu(II) ion favors coordination with the peptide nitrogen from which it displaces the proton[89, 109–112] giving the structure shown in Figure 21. Thus the peptide N—H must be kept in mind as a possible coordinating group in proteins.

An ionization near pH 9 observed in solutions of Cu(II) peptide complexes has been attributed to the ionization of a coordinated H_2O molecule in the fourth coordination position (Fig. 21). It is of interest that the circular dichroism is markedly affected by this ionization and can be used to titrate the pK_a (Fig. 21). This suggests that a change in

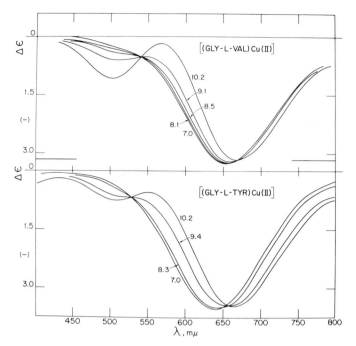

Figure 20　Visible CD of 1:1 Cu(II) dipeptides as a function of pH.

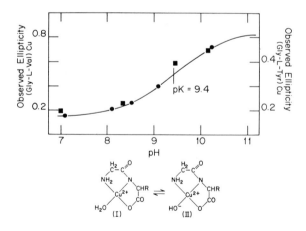

Figure 21　Ellipticity of Cu(II) dipeptides at 500 mμ as a function of pH. (○) [(Glycyl-L-valine) Cu(II)]; (■) [(glycyl-L-tyrosine) Cu(II)].

charge in a dissymmetric environment close to the metal ion can alter the circular dichroism significantly. If several strong-field peptide nitrogens are involved in Cu(II) coordination, the absorption shifts toward 500 mμ [114, 115] and the complexes become reddish in color, but no intense absorption bands have been found in Cu(II) peptide or amino acid complexes. Some tetrahedral Cu(II) complexes have very intense charge-transfer bands in the region of 600 mμ[69]. For example, the complex of Cu(II) with oxalyldihydrazide in the presence of acetaldehyde has an $\epsilon = 14,000$ at 610 mμ[116]. The postulated structure is shown below (**1**) and the color is believed to arise from charge transfer from π-orbitals or σ-orbitals of the ligands to the d-orbitals of the central metal[69]. Frieden et al. has proposed that such a mechanism might operate in the blue Cu(II) proteins[69].

The proteins in which Co(II) has been substituted are probably all high-spin complexes, although magnetic susceptibility data is available only on Co(II) carbonic anhydrase[80] (see p. 210). Models for high-

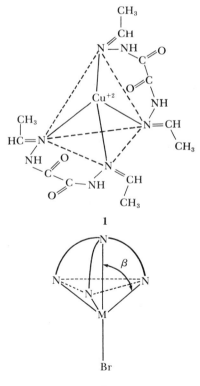

1

2

spin Co(II) complexes are not particularly numerous and none of them have spectra that correspond to those observed in the protein complexes in Table 3. Co(II) complexes are generally either tetrahedral, $[CoCl_4]^{2-}$, or octahedral, $[Co(H_2O)_6]^{2+}$. The former have a single intense d–d transition from the $^4A_2$ ground state to the $^4T_1(P)$ state in the region 600–700 $m\mu$[13]. This accounts for the deep blue color of tetrahedral Co(II). This band often has small splittings which are due to spin–orbital coupling[13]. Additional transitions, $^4A_2 \rightarrow {}^4T_2$ and $^4A_2 \rightarrow {}^4T_1(F)$, are expected in the infrared 3000–8000 cm^{-1}. Infrared bands have been observed in Co(II) carbonic anhydrase and carboxypeptidase (Table 3) [80, 90]. The octahedral complexes are very pale pink, $\epsilon = \sim 10$, and absorb near 500 $m\mu$. Again a single d–d transition is responsible, $^4T_{1g}(F) \rightarrow {}^4T_{1g}(P)$ and a shoulder on the high frequency side is due to spin–orbital coupling[13]. As in tetrahedral complexes another d–d transition, $^4T_{1g}(F) \rightarrow {}^4T_{2g}$, is expected in the infrared at 1200 $m\mu$[13].

Square-planar complexes of Co(II) are found, but all strictly square-planar complexes have been observed to be low spin[13, 117]. Several high-spin Co(II) complexes have recently been reported which appear to have a five-coordinate structure[117–120]. The structure of one of them, ([CoBr(Me$_6$tren)]Br, is a trigonal bipyramid and is indicated by (2). The spectrum is shown in Figure 22.

The splitting of the d–d transitions observed in this spectrum are not unlike those observed in Co(II) carbonic anhydrase or alkaline phosphatase. This suggests that the active center complex in some of the proteins could be five coordinate[79].

Few studies of the optical activity of dissymmetric Co(II) complexes are on record, partly because Co(II) complexes are labile, preventing easy resolution of stable isomers. The CD spectra of [(Histidine)$_2$Co(II)] and its oxygen adduct are shown in Figure 23. This complex has received

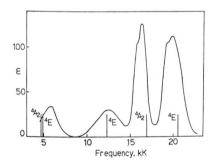

Figure 22 Absorption spectrum of [CoBr(Me$_6$tren)]Br in CH_2Cl_2. From Ciampolini [117], by permission.

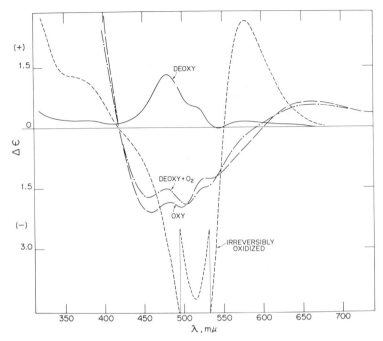

Figure 23   Visible CD changes during the reversible and irreversible oxygenation of [(Histidine)₂ Co(II)].

a good deal of study as a model oxygen carrier, studies which are reviewed in Martell and Calvin[10]. The deoxy complex is high-spin Co(II), and X-ray data show it is to have a very slightly distorted octahedral configuration[89]. Its visible absorption spectrum, $\epsilon = 30$ at ~480 mμ, is typical of octahedral Co(II). The CD contains two ellipticity bands not incompatible with the expected absorption bands (p. 202). Addition of oxygen causes an inversion of the CD bands and the appearance of large ultraviolet ellipticity bands which correspond to large near ultraviolet absorption bands, presumably charge-transfer bands.

The structure of the oxygen complex is not known, but oxygen probably coordinates the cobalt, replacing the carboxylates of the histidine as ligands[10]. The complex is diamagnetic, suggesting that the cobalt has lost an electron; however, the oxygen addition is initially readily reversible by bubbling the solution with N_2. With time a slow irreversible oxidation takes place with further changes in the CD spectrum (Fig. 23). All these complexes are apparently variants of octahedral geometry; hence the circular dichroism may be expected to be very sensitive to microscopic geometry and it may well be difficult to relate to general coordination geometry [121].

2.2.4 Nonheme Fe Spectra. Proteins which contain nonheme iron, such as xanthine and aldehyde oxidases [122], orotate dehydrogenase [122] succinic dehydrogenase [123], and the ferredoxins [63, 65], show absorption bands in the region 300–400 mμ which can be ascribed to the iron chromophore. The spectrum is complicated in the case of the metalloflavoproteins by the presence of the coenzyme, and the nonheme iron chromophore is seen in its purest form in the ferredoxins. Figure 24 shows the absorption spectrum of oxidized clostridial ferredoxin compared to the absorption spectrum of apoferredoxin [65]. Both the absorption band near 400 mμ and the complex band at 300 mμ appear related to the Fe–sulfide chromophore. The absorption spectra of the plant ferredoxins are similar, but have slightly more complex band structure in the visible region. Oxidized plant ferredoxin shows peaks at 463, 420, and 325 mμ [63, 124–126]. The absorption bands above 300 mμ disappear when the protein is reduced [65]. The oxidation and spin state of the iron are discussed in the following section.

Circular dichroic spectra have been recorded for several of the nonheme iron proteins and reveal a complex set of ellipticity bands from 300 to 700 mμ associated with the Fe–protein chromophore (Fig. 25) [127]. One of the most remarkable features of these spectra is the profound change in ellipticity induced by reduction of the enzyme (Fig. 25). The addition of the unpaired electrons to the system appears to have a radical effect on certain features of dissymmetry surrounding the

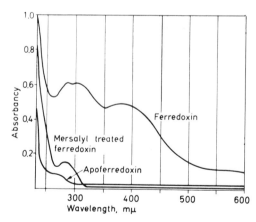

Figure 24 Absorption spectra of native ferredoxin, mersalyl-ferredoxin, and apoferredoxin. From Lovenberg, Buchanan, and Rabinowitz [65], by permission.

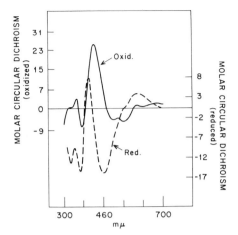

Figure 25 Visible CD of (—) oxidized and (------) reduced ferredoxin. Drawn from the data of Palmer et al. [127].

chromophore. These findings, however, cannot be interpreted in terms of structure at the present time. When CD spectra of ferredoxin and adrenodoxin are compared, the similarity is so great that it suggests very similar coordination sites [127]. Some features, such as the oxidation potentials of these two similar nonheme iron proteins, however, are quite different. The nature of this Fe complex is discussed further in section 2.3.2.

2.2.5 Infrared Spectra. There have been a large number of investigations of the infrared spectra of various coordination compounds [128], but infrared spectroscopy has had only limited application to metalloenzymes. This of course is due in part to the necessity of operating in aqueous solution. In certain suitable cases, however, infrared spectroscopy can be a powerful tool in these systems. The study by Riepe and Wang [82] on the binding of CO_2 and azide to carbonic anhydrase is an example of the successful application of the infrared technique.

Carbon dioxide shows a strong infrared absorption band at 2343.5 cm^{-1} due to the asymmetric stretching of the molecule. Water has relatively low absorbance in this region; hence by carefully balancing by an interference method the path lengths in cells containing ~300 mg of protein per milliliter, Riepe and Wang detected this stretching band due to CO_2 bound at the active site of carbonic anhydrase (Fig. 26). The peak is located at 2341 cm^{-1} and the fact that it is only minimally shifted from that for free CO_2 suggests that CO_2 is bound to the enzyme in an unstrained configuration and is apparently not bound to the metal ion.

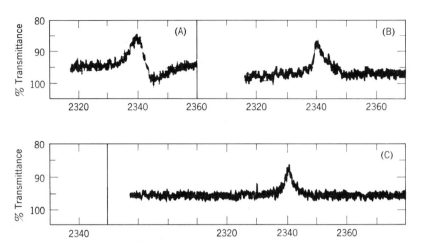

Figure 26 Difference infrared spectra of CO_2 bound to bovine carbonic anhydrase in balanced and unbalanced cells. (*A*) The reference cell has more freely dissolved CO_2 than the sample cell; (*B*) reference cell has less freely dissolved CO_2 than sample cell; Conditions: 0.075 mm sample cell containing 33.3% w/w, dry protein in water at pH 5.5. (*C*) Infrared band of CO_2 bound at the active site of carbonic anhydrase. From Riepe and Wang [82], by permission.

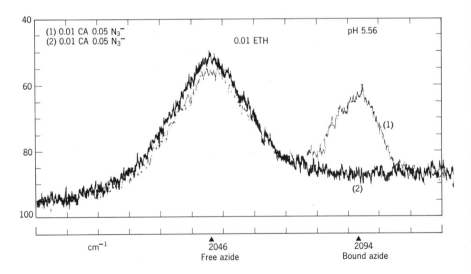

Figure 27 Difference infrared spectrum for bovine carbonic anhydrase plus azide. (Spectrum 1) enzyme plus excess azide. Peak at 2094 cm^{-1} is typical of Zn-complexed azide. (Spectrum 2) enzyme plus azide plus excess ethoxzolamide. From Riepe and Wang [82], by permission.

The azide ion, N_3^-, has a strong band at 2049 cm^{-1} due to the asymmetric stretching of this linear molecule. This band is shifted to higher frequency when azide is bound to metal ions[82]. If azide is reacted with carbonic anhydrase, the shifted azide absorption band appears (Fig. 27), clearly showing that N_3^- is bound to the metal ion. The Co(II) enzyme also shifts the azide band (2081 cm^{-1}), but not as far as the Zn(II) enzyme (2094 cm^{-1}), as expected from studies of model Zn(II) and Co(II) mixed complexes with N_3^- [82]. The sulfonamide ethoxzolamide abolishes the shifted azide absorption band[82]. Azide also displaces the enzyme-bound CO_2 molecule. The latter can, in addition, be displaced by NO_3^- and HCO_3^-. These findings suggest that CO_2 is bound in an unstrained manner to a site adjacent to the metal ion, but not coordinated to it. This is compatible with present postulates of the mechanism of action of carbonic anhydrase (see Sect. 3.1.3).

2.3 Magnetic Susceptibility and Electron Spin Resonance

The unpaired electrons (with unpaired spins) in the d-shell orbitals of the transition metal ions make these ions paramagnetic. Studies of the resultant magnetic suceptibility and electron spin resonance have been of particular value in obtaining chemical information about transition metal complexes because the characteristics of both phenomena depend on the oxidation state of the ion, the spin state of the ion, and the immediate environment (donor atoms) around the metal ion. The magnetic moment of a single electron is given by

$$\mu_s = g\sqrt{s(s+1)} \qquad (6)$$

where s is the spin quantum number and g is the gyromagnetic ratio (g value). The quantity $[s(s+1)]^{1/2}$ is the angular momentum of the electron; hence g can be defined as the ratio of the magnetic moment to the angular momentum of the electron. For a free electron g has the value of 2.00023 which may be assumed to be 2.00 for most purposes. Hence, from equation 6, μ_s (the spin magnetic moment) for a single free electron can be calculated to be 1.73 Bohr magnetons (BM) (one BM = 0.927×10^{-20} erg/gauss).

It is revealing to compare these calculated spin-only magnetic moments for several of the transition metal ions to those calculated from experimental measurements of magnetic susceptibility on complexes of the transition metal ions. Comparisons of these values for the common oxidation states of the metal ions from manganese to copper are given in Table 4. The agreement between the calculated spin-only magnetic

TABLE 4 MAGNETIC MOMENTS FOR COMPLEXES OF THE FIRST TRANSITION METAL IONS

Ion	No. of Unpaired Electrons	Ground State Quantum Numbers (Spin and Orbital Quantum Numbers)		Spectroscopic Symbol	Spin-Only Magnetic Moment μ_S	Spin and Orbital Magnetic Moment (Calculated) μ_{S+L}	Experimental Magnetic Moments BM
		S	L				
Mn(II)[a]	5	$\frac{5}{2}$	0	^6S	5.92	5.92	~5.9
Fe(II)	4	2	2	^5D	4.90	5.48	5.1–5.5
Fe(III)	5	$\frac{5}{2}$	0	^6S	5.92	5.92	~5.9
Co(II)	3	$\frac{3}{2}$	3	^4F	3.87	5.20	4.1–5.2
Ni(II)	2	1	3	^3F	2.83	4.47	2.8–4.0
Cu(II)	1	$\frac{1}{2}$	2	^2D	1.73	3.00	1.7–2.2

[a]High-spin for Mn, Fe, and Co.

moment and the observed magnetic moment is excellent in the case of Mn(II) and the isoelectronic Fe(III). For the others, however, the agreement is not as good, being poorest for Co(II). These deviations arise because there are also orbital contributions to the angular momentum of the electron. Mn(II) and Fe(III) have S ground states, hence no contribution from orbital angular momentum is possible; however, the others all have D or F ground states and do possess orbital angular momentum[13, 129].

The orbital contribution can be calculated and these calculations are given in Table 4 as μ_{S+L}. In those cases where the orbital contribution applies, the observed value is rarely as great as the calculated value of μ_{S+L}. This second discrepancy arises because in most cases ligand orbitals interfere with the orbital motion of the electrons so that the orbital angular momentum is reduced from the theoretical value.

The electrostatic crystal field theory, from which many elegant deductions can be made about the properties of transition metal complexes, assumes that the metal ion and ligand orbitals do not overlap. This is not actually the case even in the most "ionic" of complexes, and in some cases this formalism breaks down completely. Electron spin resonance spectra of the transition metal complexes provide direct evidence for this. A single group of unpaired electrons localized on the metal ion would be expected to give a single esr absorption band. However, esr signals of many transition metal ion complexes, such as $[IrCl_6]^{2-}$ and $[Mo(CN)_8]^{3-}$, show small secondary bands (hyperfine structure) which can be explained by contributions of the nuclear magnetic moments of the ligand atoms[13]. In the case of $[Mo(CN)_8]^{3-}$ this has been demonstrated directly by substituting ^{13}C-enriched CN (^{13}C has a nuclear spin while ^{12}C does not) which markedly enhances the hyperfine structure [130]. Thus even in an "ionic" complex like $[IrCl_6]^{2-}$ the theory must be modified to include some localization (estimated to be 5%) of the electron on the ligand[13]. Thus the observed signal is sensitive to the nature of the donor atoms and may on occasion be used to identify the donor atoms (see below).

Delocalization of the electrons from metal to ligand is also responsible for the shifts observed in nuclear magnetic resonance frequencies of ligand atoms. The effective field at the ligand nucleus is changed by the unpaired spin density arising from the adjacent metal ion. Such shifts in the resonance of certain protons on the porphyrin ring have recently been used to study changes in heme environment when the ligands on the Fe(II) are altered in hemoglobin and myoglobin[131] (see Sect. 2.4). Electron spin resonance lines can also be split by interactions with the nuclear spin of the metal nucleus itself, e.g., Mn(II), where six lines result from the interaction with ^{55}Mn of nuclear spin $\frac{5}{2}$ (see below).

Magnetic susceptibility studies have not been widely applied to metallo-enzymes, with the exception of the heme proteins beginning with the classical experiments of Pauling and Coryell showing that the addition of O_2 to hemoglobin causes the Fe(II) ion to shift from a high- to a low-spin state[132]. Among the studies of the metalloenzymes discussed in this chapter, the study by Lindskog and Ehrenberg of Co(II) carbonic anhydrase and various of its inhibitor complexes is the most extensive [80]. Magnetic susceptibilities of Co(II) carbonic anhydrase and the complexes with several anions and a sulfonamide are shown in Table 5.

TABLE 5 MAGNETIC MOMENTS OF ACTIVE AND INHIBITED Co(II) CARBONIC ANHYDRASE[a]

pH	Inhibitor, mM	Magnetic Moment, μ, BM
7.9	None	4.23 ± 0.10
6.2	Cl^- (100)	4.45 ± 0.08
6.2	Sulfanilamide (0.5)	4.72 ± 0.04
6.2	NCO^- (2)	4.41 ± 0.05
6.3	NO_3^- (100)	4.45 ± 0.03

[a]Data taken from Lindskog and Ehrenberg [80].

All the enzyme complexes contain high-spin Co(II), with magnetic moments from 4.2 to 4.7 BM. They show the expected increase in magnetic moment due to spin–orbit coupling. Extensive susceptibility data on complexes of high-spin Co(II) have shown that octahedral complexes have magnetic moments from 4.7 to 5.2 BM, while tetra-hedral complexes have magnetic moments from 4.1 to 4.9 BM. The square-planar complexes of Co(II) that have been described are all low-spin Co(II) and have magnetic moments from 1.8 to 2.0 BM. This is perhaps not too surprising since the low-spin configuration for the d^7 ion does provide the possibility for stabilization by a Jahn-Teller dis-tortion[13]. The magnetic susceptibilities of the Co(II) carbonic anhydrase complexes are compatible with some variant of tetrahedral geometry. Unfortunately, the interpretation of the fine variations in susceptibility are not possible. Tetrahedral complexes of Co(II) are high spin regardless of the strength of the ligand field because of the distribution of orbital energy levels (Fig. 10). Co(II) carbonic anhydrase does remain high spin in the face of what must be a relatively high ligand field, especially in the sulfonamide complex which interestingly has the highest susceptibility.

A magnetic susceptibility determination is available on clostridial ferredoxin and indicates that the enzyme-iron has an average effective magnetic moment of 2.0 ± 0.2 BM [133]. This moment is compatible with a low-spin Fe(III) complex. However, since the protein contains seven iron atoms other low-spin combinations compatible with an average magnetic moment of 2.0 include, 6Fe(III) + 1Fe(II), 5Fe(III) + 2Fe(II), and 4Fe(III) + 3Fe(II) [133]. Further discussion of the electronic structure of iron in the non-heme iron proteins is given below.

Qualitative magnetic susceptibility measurements have also been made on xanthine oxidase in both the oxidized (resting state) and reduced form [134]. The resting enzyme shows low paramagnetism (−129 scale units). Upon anaerobic reduction with xanthine the susceptibility shows a large increase (to 198 scale units). The increase in susceptibility has been shown to be proportional to the height of the $g = 1.94$ esr signal observed on reduction [134]. This signal is related to the non-heme iron present in the enzyme (see p. 230). Thus the changes in susceptibility have been assigned to changes in the electron distribution on the iron. Unfortunately, it is difficult to quantitate these results. There is some uncertainty about the correction for the diamagnetism of the protein; however, if a calculated correction is applied, the iron in the resting state appears to have only very low paramagnetism or is diamagnetic (compatible with low-spin ferrous). This does not fit too well, however, with the chemical data showing the presence of Fe(III) (see Sect. 3.2.4). If there is some Fe(III) of the low spin form initially present and accounting for the low paramagnetism, then a change to high-spin Fe(II) on reduction would account for the increase in susceptibility [134]. Between four and five molecules of xanthine per enzyme molecule are needed for the reduction; hence 10 electrons must be added to the protein. Not over eight of these should be required to reduce the molybdenum and flavin also present in the enzyme, which suggests that only one to two of the eight iron atoms are associated with the susceptibility change [134].

Electron spin resonance has been more widely applied to the metalloenzymes under discussion in this chapter. It has the great advantage of being extremely sensitive. Under optimum conditions the presence of $\sim 10^{-12} M$ paramagnetic species can be detected. Susceptibility measurements on the other hand require large amounts of material, as much as 100 mg for the best measurements. Electron spin resonance does have the disadvantage that low temperatures (liquid nitrogen or helium) are often required to reduce electron spin relaxation effects sufficiently to produce observable lines (see below). The following sections outline briefly the work done on metalloenzymes containing the transition metals from Mn to Cu, as well as Mo. The section concludes with a discussion of the spin-labelling technique as applied to metalloenzymes.

Detailed discussion of the theory is beyond the scope of this review, and only brief accounts as they pertain to experimental observations in metalloenzymes are presented. Original references should be consulted for details [135, 136].

2.3.1 Manganese.

High-spin Mn(II) in aqueous solution shows an esr signal split into six lines by hyperfine interaction with the ^{55}Mn nucleus of nuclear spin $I = \frac{5}{2}$.* These signals for several manganese complexes including $[Mn(H_2O)_6]^{2+}$, the aqueous species, are shown in Figure 28. The nuclear magnetic quantum numbers, m_I, corresponding to these lines are $-\frac{5}{2}$, $-\frac{3}{2}$, $-\frac{1}{2}$, $+\frac{1}{2}$, $+\frac{3}{2}$, and $+\frac{5}{2}$ from low to high field. While many high-spin Mn(II) complexes are known in solution, only a few give these characteristic esr lines. This is usually explained by the fact that the lines are so broadened by efficient electron spin relaxation mechanisms that they cannot be observed [137, 138]. Theory shows that if the Mn(II) ion occupies a ligand field of perfect cubic symmetry (regular octahedral or tetrahedral geometry) then these efficient electron relaxation mechanisms are not available [137]. Hence only regular octahedral or tetrahedral complexes like those represented in Figure 28 have sufficiently narrow lines to be observed. Even mixing of monodentate ligands to form mixed complexes can distort Mn(II) complexes sufficiently to induce loss of signal [138].

Presumably for the above reasons, most proteins containing incorporated Mn(II) ions do not show detectable esr signals. The aqueous Mn(II) signal disappears as soon as the ion is bound to the protein [139]. Hence in studies such as those of Malmstrom and Vännagård on the binding of Mn(II) to enolase, the signal has served only as an accurate method of titrating the amount of free Mn(II) present [139].

Until recently, the titration of free Mn(II) was the only feature of Mn(II) esr useful in protein chemistry [139–141]. Recently, Reed and Cohn, however, have detected Mn(II) esr signals in solutions of Mn(II) concanavalin A [142]. X-band (9.1 GHz) and K-band (35.0 GHz) esr spectra of this complex are shown in Figure 29. Signals had previously been observed in a powdered solid preparation of this complex [143]. The protein spectra are broadened and show transitions located between the six signals in the isotropic spectrum typical of $[Mn(H_2O)_6]^{2+}$ in solution. Signals of this type are common in solid-state spectra of materials containing Mn(II) and are attributed to forbidden transitions where $\Delta M = \pm 1$ (M = electron spin quantum number) and $\Delta m \neq 0$ (m = nuclear spin quantum number). The broadening of the spectrum in powder

---

*Second-order theory shows that there are actually 30 allowed transitions grouped together into these six hyperfine components [137].

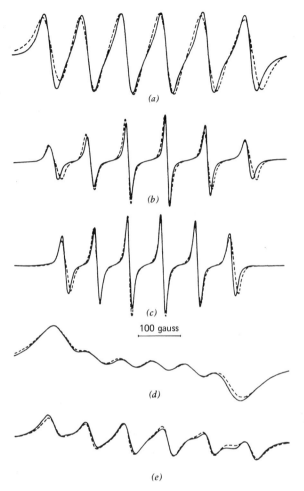

(a)

(b)

(c)

100 gauss

(d)

(e)

Figure 28   A comparison of the computer-simulated spectra of simple Mn(II) complexes with the experimental spectra observed for $[Mn(H_2O)_6]^{2+}$, $[Mn(CH_3CN)_6]^{2+}$, $[MnCl_4]^{2-}$, and $[MnBr_4]^{2-}$ in acetonitrile. The computed spectra are indicated by broken lines. From Chan, Fung, and Lutje [137]. (a) $Mn(H_2O)_6{}^{2+}$; (b) $Mn(CH_3CN)_6{}^{2+}$; (c) $MnCl_4{}^{2-}$; (d) $MnBr_4{}^{2-}$; (e) $5Mn(CH_3CN)_6 + 4MnBr_4{}^{2-}$, $Br^-/Mn(II) = 2.6$.

samples is analogous to that observed in the case of immobilized free radicals (see below).

In the case of the concanavalin complex, if the broadening results entirely from the homogeneous broadening caused by the spin relaxation mechanisms discussed above, then the line width should be temperature dependent. The line widths are not temperature dependent, which suggests that broadening is due to immobilization of the Mn(II) ion in

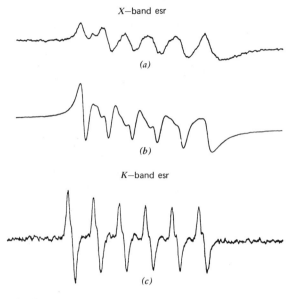

Figure 29 X and K-band esr spectra of Mn(II) concanavalin A. (a) Mn(II) concanavalin A (37 mg/ml) in aqueous solution, 32-mW microwave power, 6.3-gauss modulation amplitude, relative gain 6.2; (b) lyophilized powder, 20-mW microwave power, 5-gauss modulation amplitude, relative gain 1.0; (c) K-band spectrum of Mn(II) concanavalin A recorded at 35.0 GHz, 37 mg/ml protein, 10.5 dB below 50-mW microwave power, 3.5-gauss modulation amplitude, relative gain 3.0. Temperature = 28°. After Reed and Cohn [142].

the protein and results because the rotational motion of Mn(II) is highly restricted [142].

The intensity of forbidden transitions is inversely proportional to the square of the applied magnetic field [144], hence their intensities should be reduced in the K-band spectrum with consequent enhancement of the intensities of the allowed signals. Both phenomena are observed for the K-band spectrum of the Mn(II) concanavalin complex and suggest that the broadening of the X-band spectrum is due to forbidden transitions and not to fast relaxation [142]. Thus the Mn(II) concanavallin spectrum would appear to be characteristic of Mn(II) in the solid state. It has been suggested that such a mechanism may be responsible for signal broadening in other Mn(II) proteins [142]. The general features of the spectrum show that the Mn(II) binding site in concanavalin deviates little from cubic symmetry [142].

2.3.2 Iron. Many states of iron do not demonstrate esr absorption, or such broad bands are present that the signals are not readily detected [145–149]. In many cases satisfactory signals are obtained only at very low temperatures[149, 150]. Electron spin resonance studies of non-heme iron proteins have centered on the study of a type of protein-bound iron giving rise to a very unusual esr absorption. In the reduced state, the iron in these proteins shows a sharp signal at $g = 1.94$, with a minor component at $g = 2.01$[149]. The signal was first observed in iron-containing mitochondrial dehydrogenases, such as NADH dehydrogenase, succinic dehydrogenase, and certain metalloflavoprotein dehydrogenases[145, 146]. The signal was best observed at liquid

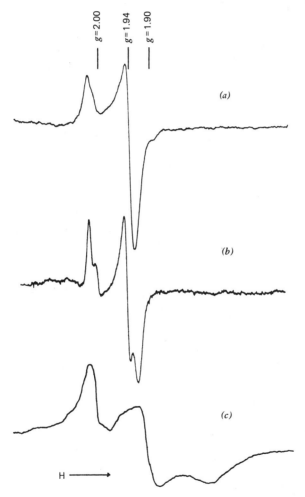

Figure 30 Electron spin resonance signals observed for non-heme iron enzymes of the mitochondrial type. (a) NAD-CoQ reductase; (b) Succinic-CoQ reductase; (c) cytochrome bc_1-iron protein. Temperature $= -178°$. After Beinert[149].

nitrogen temperatures and the signals given by several proteins of the mitochondrial type are shown in Figure 30. A signal of this type had not previously been observed in iron complexes and it was not clear why a g value less than that of a free electron was observed. This signal has been shown unequivocally to be associated with the iron in these proteins by substituting ^{57}Fe in the iron protein isolated from *Azotobacter*[151] (Fig. 31). The resultant shifts in the esr signal due to the change in nuclear spin from 0 to $\frac{1}{2}$ clearly establish the signal as associated with the iron. The line broadening of ~20 gauss observed on substitution of ^{57}Fe for ^{56}Fe indicates that there is unpaired spin density on the central iron atom in the reduced state[149, 151]. A similar broadening of the esr signal is observed in putidaredoxin from *Pseudomonas putida* when the protein is enriched in ^{57}Fe[151a]. Computer simulation of the esr spectrum of the ^{57}Fe-enriched protein shows that the best fit is obtained if it is assumed that the unpaired electron interacts with two iron nuclei[151a].

Both the mitochondrial enzymes and the ferredoxins have certain similar features in their optical absorption spectra in the oxidized state [63, 149, 152]. In addition, "labile sulfide" is associated with the iron in both types of protein and is released as H_2S when the iron is removed [63, 153]. These features have suggested that the iron in the entire group of nonheme iron proteins occupies a rather similar environment.

Initial attempts to observe these unusual esr signals in the ferredoxins failed[149]. Only very broad signals near $g = 2.00$, typical of usual low-spin Fe(III) complexes, could be observed after oxidation of the enzyme with ferricyanide[149]. Subsequently, with the use of relatively mild reducing techniques, either chemical or enzymatic, and very low temperatures (1.9 to 40°K), esr spectra of the reduced ferredoxins revealed the unusual $g = 1.94$ iron signal (Fig. 32)[154, 155]. The measured g values for the spinach enzyme in Figure 32 are $g_x = 1.89$, $g_y = 1.95$, and $g_z = 2.05$[150]. Integration of the signal shows that one-half of the iron

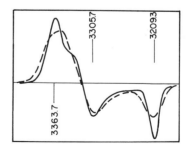

Figure 31   Electron spin resonance spectra of (———) ^{56}Fe and (------) ^{57}Fe iron protein from *Azotobacter vinelandii*. Redrawn from the data of Shethna et al.[151].

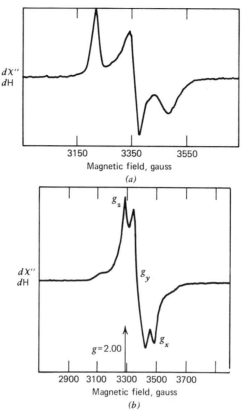

Figure 32 Electron spin resonance signals observed for the reduced forms of the ferre-doxins: (a) spinach ferredoxin; (b) ferredoxin from *C. pasteurianum*. After Palmer et al. [150].

(i.e., one of the two iron atoms per molecule) is responsible for the signal[150]. A large amount of evidence indicates that plant ferredoxin functions as a one electron transfer enzyme between the photo-activated chlorophyl molecule and enzymes reducing NADP[156], although recent evidence suggests that bacterial ferredoxins can be involved in a two-electron transfer[157]. (See Sect. 3.2.4 for discussion.) It would appear that the spinach enzyme in the oxidized state contains two Fe(III) and in the reduced state contains one Fe(III) and one Fe(II) [150, 158] (see below, however, for modification of this model).

The ferredoxins from various plant and bacterial sources contain varying amounts of iron and labile sulfide[63–65]. Ferredoxin from *C. pasteurianum* for example, contains seven iron atoms per molecule[63,

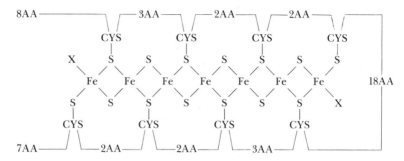

Figure 33   Model proposed for the active center of the ferredoxin from *C. pasteurianum* [63, 159, 160].

65]. Although the number of iron–sulfide units in the ferredoxins may vary, the immediate environment of each iron in the ferredoxins appears rather similar. A model for the active site of *C. pasteurianum* ferredoxin has been proposed on the basis of a large amount of structural information [63, 159, 160] and is shown in Figure 33.

The precise symmetry and spin state of the irons can only be postulated on the basis of models that are compatible with the available data, specifically the esr and Mössbauer spectra. The susceptibility data discussed above, as well as the Mössbauer data (p. 250), are compatible with the presence of low-spin Fe(III) in oxidized ferredoxin, with one of these becoming Fe(II) in the reduced form. Palmer et al., however, have recently suggested that the esr signal could arise from low-spin Fe(III) in a tetrahedral field involving sulfur ligands [150]. In such a model (for

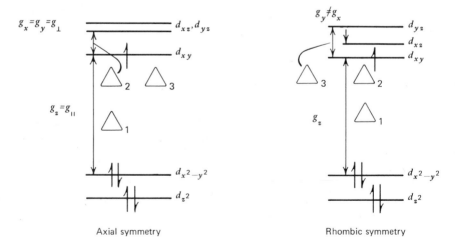

Figure 34   Orbital energy scheme for low spin ferric iron in axial and rhombic symmetry.

which orbital energy level diagrams are given in Figure 34 for both axial and rhombic symmetry) the unpaired electron would be located in the d_{xy}-orbital. Migration of the electron into the empty d_{xz}, d_{yz}-orbitals would create orbital magnetism about the x and y axes leading to shifts in the x and y components of the g value to below 2. The migration of the hole from d_{xy} into the $d_{x^2-y^2}$-orbital would create orbital magnetism along the z axis and give a positive shift in the z component of the g value. If d_{xz}, d_{yz} are degenerate, $g_x = g_y$, and an axial line shape would be observed; if not a rhombic line shape would be observed, $g_y \neq g_x$. The models proposed by Palmer et al.[150] are shown schematically in Figure 35. These authors point out that Fe(III) in an environment of

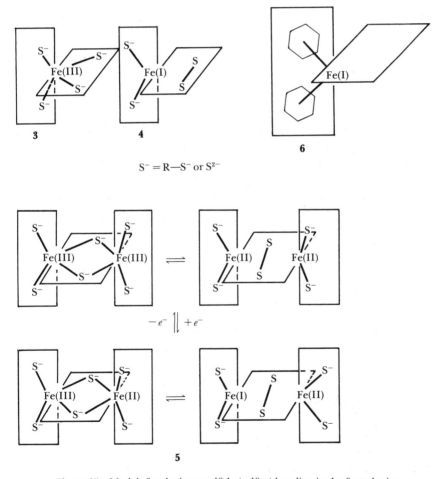

Figure 35 Models for the iron–sulfide (sulfur) bonding in the ferredoxins.

four sulfide ligands (**3**) is formally indistinguishable from Fe(I), with one disulfide and two sulfide ligands (**4**), the Fe(I) and disulfide forming a three-center bonding array[150].

In support of such a model the esr signal of bis-hexamethylbenzene-Fe(I), a complex of similar configuration (**6**), shows the esr signal in Figure 36, which is rather similar to that shown by spinach ferredoxin. The reported g values are: $g_x = 1.86$, $g_y = 1.99$, and $g_z = 2.08$[150]. Extension of this model to the spinach enzyme containing two iron atoms suggests the models of the oxidized and reduced protein shown in Figure 35. Beinert[149] has also proposed an Fe(I) model based on the observed low-temperature esr signal for the pentacyanonitrosylferrate(I) anion, $[Fe(CN)_5NO]^{2-}$, which is also similar to that observed in nonheme iron proteins[149] (Fig. 37). The electron in this complex is partly localized at the NO ligand, and because of this delocalization any of the three possible configurations in equation 7 could describe the complex. The

$$Fe(III)(CN^-)_5NO^- \longleftrightarrow Fe(II)(CN^-)_5\dot{N}O \longleftrightarrow Fe(I)(CN)^-)_5NO^+ \quad (7)$$

unusual signal in Figure 37 is observed only at low temperature, $-172°$, and may be related to certain conformations of the complex which are averaged out by vibrations at higher temperature.

Electron spin resonance has also been used by Hollocher et al.[161]

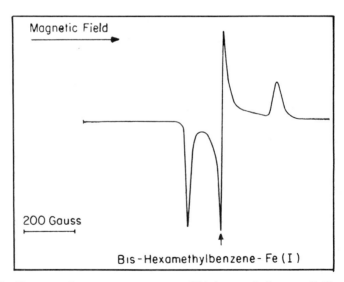

Figure 36   Electron spin resonance spectrum of bis-hexamethylbenzene-Fe(I) at 40°K. The arrow indicates $g = 2$. From Palmer et al.[150], by permission.

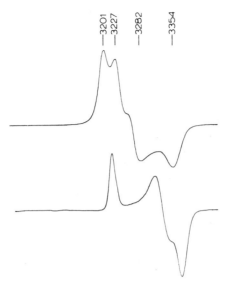

Figure 37 Electron spin resonance spectra of (top curve) pentacyanonitrosylferrate(I) anion, and (bottom curve) *Azotobacter* iron protein. Temperature $= -172°$. From Beinert [149], by permission.

to demonstrate directly the Fe–S interaction in Azotobacter iron protein. If the bacteria are grown on ^{33}S $(I = \frac{3}{2})$ rather than ^{35}S $(I = 0)$, the $g = 1.94$ iron signal is significantly broadened at 52% ^{33}S (Fig. 38). The $g_{\parallel}$ line is broadened considerably more than the $g_{\perp}$ line, thus the Fe–^{33}S superhyperfine interaction appears to be anisotropic. The schematic

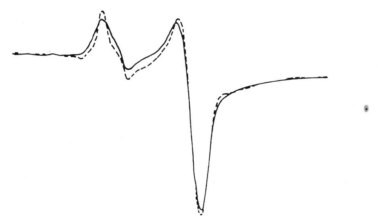

Figure 38 Electron spin resonance signal observed in *Azotobacter* iron protein before and after enrichment with 33S: (—) 52% ^{33}S; (----) ^{32}S. After Hollocher et al. [161].

models in Figure 35 cover most of the features of the electronic configuration of the iron–sulfur site of ferredoxin proposed up to the present time on the basis of esr, Mössbauer, and susceptibility measurements[150, 159]. References 63, 149, 150, and 160 should be consulted for a more adequate discussion of these proposals.

2.3.3 Cobalt and Nickel. Electron spin resonance studies of cobalt and nickel complexes have not been extensive and until recently no metalloprotein spectra have been recorded. Electron spin relaxation mechanisms are very efficient for these ions, and thus sensitivity is poor and low temperatures are required. Spectra can be recorded, however, and it should be possible to observe esr signals for these ions in metalloproteins under the proper conditions. While spectra of model complexes are given as examples, a low-spin dicyanide complex of Co(II) carbonic anhydrase has recently been discovered which gives well-resolved Co(II) esr signals similar to that pictured in Figure 39a (209).

In the case of cobalt, interest has centered on several oxygen-carrying chelates[162–166]. Electron spin resonance spectra of oxygen adducts of Co(acacen) pyridine studied by Hoffman et al.[166] are shown in Figure 39. In position B (7) the complex has a pyridine nitrogen and Y (7) is either open or occupied by oxygen. Figure 39a shows the esr signal of the unoxygenated compound at 77°K. The compound is low-spin (magnetic moment = 2.16 BM) Co(II) in nearly axial symmetry [166]. Seven of the eight lines expected from the hyperfine interaction with ^{59}Co ($I = \frac{7}{2}$) are observed on the high-field side of the signal. Triplets due to coordination to the nitrogen nucleus of a single pyridine molecule are clearly resolved in several of these high-field lines. The spectrum of the oxygen adduct at 77°K is given in Figure 39b. If the temperature is raised to −58° where a fluid toluene solution is obtained, the eight-line spectrum shown in Figure 39c is observed. This is clearly due to the interaction of an unpaired electron with a single ^{59}Co nucleus and proves the monomeric nature of the oxygen adduct[166]. Many of these have been dimeric[10]. The magnetic moment of the oxygen adducts range from

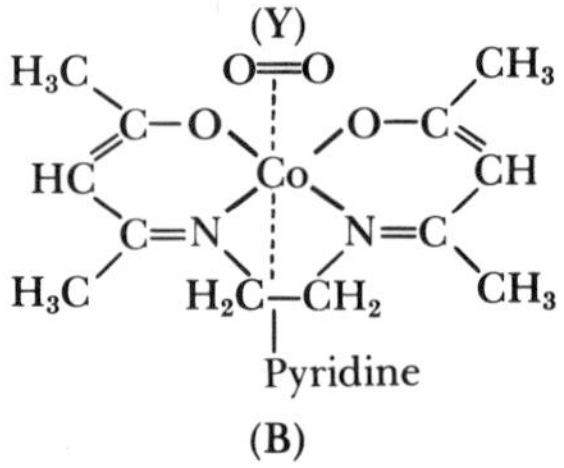

(B)

7

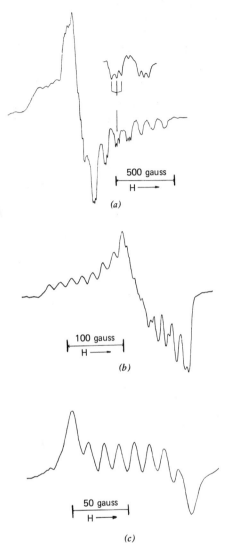

Figure 39   (*A*) Electron spin resonance spectrum of Co(acacen)py in frozen toluene solution at 77°K. (*B*) esr spectrum of Co(acacen)py(O$_2$) in frozen toluene solution at 77°K; (*C*) esr spectrum of Co(acacen)py(O$_2$) in liquid toluene solution at −58°. [acacen = (CH$_3$-C(O$^-$)=CHC(CH$_3$)=NCH$_2$-)$_2$]. After Hoffman, Diemente, and Basolo[166].

1.49 to 1.89 BM, indicating the presence of a single unpaired electron [166]. Analysis by Hoffman et al.[166] suggests that this electron resides primarily on the oxygen molecule and thus the complex can best be thought of as a superoxide adduct of Co(III). Similar spectra have been

observed recently in oxygen adducts of Co(II) phthalocyanine [162] and
vitamin B_{12r} [163, 165]. Electron spin resonance spectra for only a few
simple nickel complexes have been recorded [167, 168].

2.3.4 Copper. Cu(II) complexes give an easily resolved esr signal at
g values only slightly higher than that of a free electron. There are two
major nuclear hyperfine interactions that have been useful in studies of
Cu(II) complexes. The Cu(II) signal is split into four lines by the copper
nuclear hyperfine interaction ($I = \frac{3}{2}$), and if nitrogen is one of the ligand
donor atoms, hyperfine lines appear from the ^{14}N nuclei ($I = 1$). A
valuable aspect of the latter interaction is that the number of nitrogen
hyperfine lines observed is a function of the number of nitrogen donors
involved [169]. These features of the Cu(II) esr signal are illustrated in
Figure 40 by the room temperature esr spectrum of Cu(II) glycylglcyl-
glycine which has square-planar coordination to three nitrogens and an
oxygen (8) [169]. There are seven hyperfine lines observed on the high-
field side of the spectrum for the three nitrogen nuclei in addition to the
copper hyperfine splitting.

In enzymology Cu(II) signals were initially used to confirm in an
elegant manner the postulate that copper underwent reduction

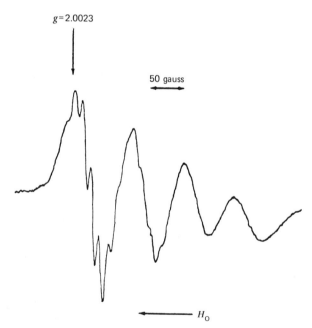

Figure 40   Room temperature spectrum of [Cu(II)(glycylglycylglycine)]. After Wiersema
and Windle [169].

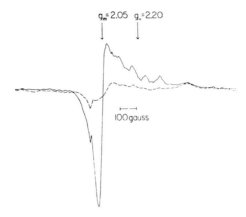

Figure 41   Electron spin resonance signal of mushroom laccase (———) before and (------) after treatment with 10 mM catechol. Temperature = 77°K, field modulation = 6.6 gauss. From Nakamura and Ogura[170], by permission.

during the action of the blue copper oxidases. Since Cu(I) lacks a signal the disappearance of the signal offered a particularly convincing argument, as illustrated for mushroom laccase in Figure 41 from the work of

8

Nakamura and his colleagues[170]. Perhaps more Cu(II) proteins have been examined by esr than any other type of metalloprotein and a great number of studies are now available. These are briefly summarized in Table 6, which gives the characteristics of the signals observed. Nitrogen hyperfine interactions have been observed in a number of Cu(II) complexes[171–178], including proteins like the Cu(II) conalbumins[176, 177], Cu(II) transferrins[176, 177], Cu(II) carboxypeptidase[179], and Cu(II) insulin[180]. An example of the details that can be observed in the protein spectra is given in Figure 42, which shows the spectrum of Cu(II) in fungal laccase and its mixed complex with cyanide as studied by Malmstrom et al.[186]. A detailed analysis of the signal in the untreated enzyme shows that it can be best accounted for by assuming the

TABLE 6 ELECTRON SPIN RESONANCE SIGNALS OBSERVED IN COPPER OXIDASES AND CERTAIN COPPER PROTEINS

Protein	g Values	Remarks	References
Laccase	$g_m = 2.048, g_{\parallel} = 2.197, A = 0.009 \ \mathrm{cm}^{-1}$	Observed in oxidized protein; 2 types of signal observed; detailed hyperfine structure in CN^- complex;	170, 171
Tyrosinase		Cu(II) signal very low, detected only in aged preparations, substrates have no effect	181–183
Ceruloplasmin	$g_m = 2.056, g_{\parallel} = 2.209, A = 0.008 \ \mathrm{cm}^{-1}$	Observed in oxidized protein; 2 types of copper are present, perhaps related to preparative procedure	171, 184, 185
Cytochrome oxidase		Varies somewhat with preparation; hyperfine structure has been observed	105, 106, 187
Ascorbate oxidase		Two superimposed signals for Cu(II) are observed in the oxidized state	55
Cu(II) carboxypeptidase	$2.060 = g_m, g_{\parallel} = 2.24, A = 0.019 \ \mathrm{cm}^{-1}$	N,N,O coordination, Nitrogen hyperfine structure has been observed	171, 179
Cu(II) transferrins and conalbumins	$g_{\parallel} = 2.30$ g_m cannot be determined because of hyperfine splitting	Nitrogen hyperfine structure has been observed	176, 177
Cu(II) Carbonic Anhydrase	$g_{\parallel} = 2.32 g_{\perp} = 2.08 A = 0.0156 \ \mathrm{cm}^{-1}$	Nitrogen hyperfine structure indicating coordination to 3 (209)	
Cu(II) Carbonic Anhydrase · CN	$g_{\parallel} = 2.24 g_{\perp} = 2.09 A = 0.013 \ \mathrm{cm}^{-1}$	equivalent nitrogens.	

Stellacyanin (*Rhus vernicifera* blue protein)	$g_z = 2.30$ $g_x = 2.03$ $g_y = 2.06$	190, 191
Dopamine β-hydroxylase	$g_z = 2.26$ $g_x = 2.08$ $g_y = 2.05$	193
Pseudomonas blue protein	$g_m = 2.055, g_{\parallel} = 2.157, A = 0.006 \text{ cm}^{-1}$	100

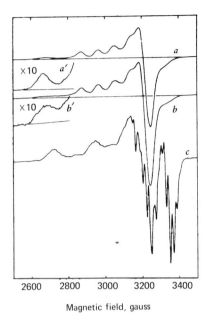

2600   2800   3000   3200   3400

Magnetic field, gauss

Figure 42   Electron spin resonance spectra of (*a*) fungal laccase, plus (*b*) KCN, 3 m*M* for 70 min; plus (*c*) KCN, 40 m*M* for 175 min. Temperature = 90°K. After Malmstrom, Rein-hammar, and Vanngard[186].

superposition of signals from Cu(II) in two slightly different environments[186]. When cyanide is added in relatively high concentration, a new signal appears which has a large amount of nitrogen hyperfine structure. The number of nitrogen hyperfine lines indicates coordination to three or four nitrogen atoms[186].

Rapid-flow esr experiments have also provided some interesting results on the copper oxidases. A rapid flow experiment on the reduction of laccase with catechol demonstrates the appearance of a catechol semiquinone[170]. This implies that the electrons can be transferred one at a time from catechol to the enzyme and that the semiquinone is not the product of an equilibrium reaction between catechol and *o*-quinone produced by the enzymatic oxidation. It has similarly been concluded on the basis of esr results that a free radical derived from ascorbate is involved in the mechanism of ascorbate oxidase[188]. On the other hand, during the oxidation of catechol by tyrosinase, no semiquinone could be detected, indicating that this Cu(I) enzyme apparently requires a two-electron transfer[170]. Implications of the two-electron transfer to the mechanism of action of tyrosinase are discussed in Section 3.2.3. In the case of all the intensely blue copper proteins (Table 6), if the spectroscopic splitting factors, g_z, g_x, and g_y, of the esr spectrum and the hyperfine splitting constants, A, are carefully analyzed (or determined by the constants needed to generate the spectrum by computer program [191]) a very unusual ligand field must be invoked to explain the observed values [70–72, 191].

2.3.5 Molybdenum and Xanthine Oxidase.
Xanthine and aldehyde oxidases contain Mo, flavin adenine dinucleotide, (FAD), and non-heme iron [194–198]. Thus, potentially these enzymes contain three paramagnetic species, including a flavin semiquinone. In an elegant series of studies Beinert and co-workers have demonstrated all three paramagnetic species by esr and provided information on the kinetics of their appearance during reduction of the enzyme by substrate [147, 148]. Xanthine oxidase contains 2Mo, 2FAD, and 8Fe per unit of molecular weight 300,000 [194–197]. The oxidized enzyme shows practically no esr signal except for a small absorption at $g = 2.00$. The progressive development of a complex group of esr signals during reduction of the enzyme by xanthine is shown in Figure 43 from the work of Palmer et al. [148]. The signal at $g = 2.00$ has been identified with the flavin free radical. The complex signal with greatest amplitude near $g = 1.97$, labeled, α, β, γ, δ, has been associated with Mo, while the signal at 1.93 appears similar to that observed for Fe in non-heme iron proteins [148]. Mo(V) coordination compounds show signals of major amplitude at $g = 1.95–1.99$ [148]. The hyperfine structure is complex because of several nuclear interactions. An intense central signal is due to molybdenum isotopes without a magnetic moment. Six minor components on either side of the central line are due to the hyperfine interaction with ^{95}Mo and ^{97}Mo (combined abundance 25%) which have nuclear spins of $\frac{5}{2}$ [148]. There is often some fine structure at the central line due to donor hyperfine interaction, in $K_3Mo(CN)_8$, e.g., due to the natural abundance of ^{13}C [148].

In xanthine oxidase it has been possible to separate the α,β signals from the γ,δ signals and they appear to reflect slightly different states of molybdenum. The expected hyperfine lines have been resolved (Fig. 44) [148]. The assignment of these spectra to Mo(V) seems to be fairly certain on the basis of their similarity to model Mo(V) compounds [148].

Bray et al. [199] have examined these two signals in considerable detail in an enzyme enriched in ^{95}Mo and also in an enzyme dissolved in D_2O. The γ,δ signal is relatively simple and corresponds to Mo in nearly axial symmetry. Enrichment with ^{95}Mo results in splitting of each of the main signals into six hyperfine lines as expected [199]. The α,β signal is more complex. Many of the lines are observed to be doublets, suggesting that

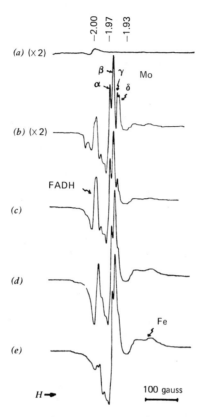

Figure 43   Electron paramagnetic resonance signals obtained from xanthine oxidase (a) before and (b, c) during reaction with xanthine. The solutions, in 0.05 M pyrophosphate buffer, pH 8.3, were mixed in the rapid flow apparatus at 22°C. The concentrations after mixing were: xanthine oxidase, 0.12 mM; xanthine, 1.25 mM; and oxygen, 0.76 mM. The mixture was frozen at the indicated times after mixing: (b) 26 msec; (c) 77 msec; (d) 860 msec; (e) 1410 msec. Curves a and b were recorded at twice the amplification of curves c through e. The microwave power was 25 mW, the modulation amplitude 5 gauss, and the temperature − 174°. From Palmer, Bray and Beinert[148].

this species of paramagnetic molybdenum is interacting with protons ($I = \frac{1}{2}$)[199]. This has been confirmed by dissolving the enzyme in D_2O, which collapses the doublet structure of the α,β signal, but does not affect the γ,δ signal[199].

The origin of the two different environments for molybdenum in the reduced enzyme is not clear. The proportion of the two signals induced varies with substrate, for example, methylxanthine induces only the α,β signal. The γ,δ peak appears unique to xanthine[199]. In addition, the kinetics of their appearance is somewhat different for the two signals

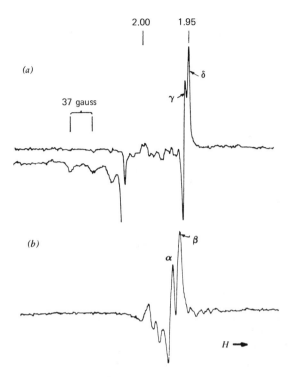

Figure 44   Electron paramagnetic resonance signals obtained from xanthine oxidase during reaction with xanthine in 0.05 *M* pyrophosphate, pH 6.0 and 9.6. The conditions were as follows. (*a*) (pH 9.6): xanthine oxidase 0.63 m*M*; xanthine, 2.9 m*M*; oxygen, 1.0 m*M*; reaction time, 8 msec at 1°; microwave power, 2.5 mW; modulation amplitude, 2.5 gauss (5-mW power and 15-gauss modulation for inset in lower left). (*b*) (pH 6.0): xanthine oxidase, 0.11 m*M*; xanthine, 0.10 m*M*; oxygen, 0.76 m*M*; reaction time 71 msec at 22°; microwave power, 25 mW; modulation amplitude, 3 gauss. All concentrations refer to the state after mixing. The observation temperature = − 173°. After Palmer, Bray, and Beinert[148].

(see below). The nature of the proton interaction which affects the α,β signal is not clear. It does not appear to be a proton from the substrate, since it exchanges very rapidly with D_2O and must not be carbon bound [199]. It might be a proton on the protein, perhaps a thiol group, since the iodacetamide-inactivated enzyme does not show the proton interaction [199].

Bray, Palmer, and Beinert determined the kinetics of the development of the various esr signals during "single turnover" experiments [147]. The results of an experiment with xanthine as substrate are shown in Figure 45. The Mo-δ signal rises most rapidly, followed by FADH,

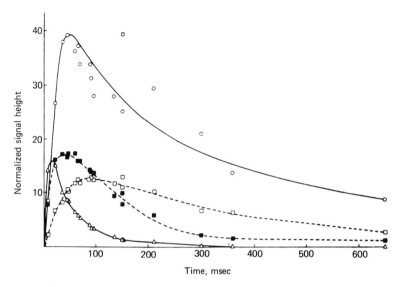

Figure 45　Time course for the development of the esr signals in xanthine oxidase during a "single turnover" experiment. The enzyme was reduced with xanthine, incubated at 22° for the times indicated and rapidly frozen. (○) FADH; (■) molybdenum β; (△) molybdenum δ; (□) iron. After Bray, Palmer, and Beinert[147].

then Mo-β, and then the Fe signal. On the basis of these results the authors postulated that the electron transfer sequence in the enzyme was Mo → FAD → Fe, the molybdenum accepting electrons from the substrate, the iron donating them to molecular oxygen[147].

The detailed study of the molybdenum signals suggests that several sequences between molybdenum and flavin may be possible, depending on substrate[199]. These are listed below as given by Bray et al.[199]:

1. Substrate → FADH → Mo(α,β)
2. Substrate → Mo(α,β) → FADH
3. Substrate → Mo(γ,δ) → FADH

The molybdenum signals at no time account for more than 37% of the Mo present in the enzyme[199]. Hence, Mo(V) never appears to be the predominant valence state, or if it is, there is interaction with other paramagnetic species, perhaps a Mo(V)–Mo(V) interaction[199]. Since the oxidation–reduction potential of the Mo(VI)–Mo(V) couple is of the right value for the operation of xanthine oxidase[200], it has been attractive to assume that Mo(VI) is present in the resting enzyme and that reduction gives rise to the paramagnetic Mo(V)[199, 200]. This appears to be a sound assumption, but some type of change in the interaction of paramagnetic species already present induced by substrate reduction cannot be ruled out.

2.3.6 Spin Labels. The introduction of a paramagnetic species into a protein by attaching a stable free radical to the protein is potentially a valuable spectroscopic probe [201, 202]. The resultant esr signal from the incorporated free radical can be used to monitor changes in the protein or to deduce something about the environment of the free radical. McConnell and co-workers have introduced this technique and presented a number of applications [203–205]. The stable nitroxide free radical [206–208] has been the most satisfactory because it is inherently rather unreactive and can be attached to a variety of compounds, containing any one of a number of reactive groups which can be directed toward specific protein side chains or active sites [202–205].

Studies of the nitroxide radical suggest that the free electron is largely confined to a $2p\pi$ atomic orbital on the nitrogen. The signal from this free radical is split into three lines by the hyperfine interaction with the nitrogen nucleus [201, 202] (see Fig. 46). This hyperfine splitting depends on the orientation of the nitroxide relative to the applied field. Thus three different splittings are observed along the three crystal axes of a nitroxide-doped crystal [202]. However, in a dilute solution of a small molecule containing the free radical, the tumbling of the molecule is much more rapid than the time for spin reversal, hence the various positions of the free radical relative to the external field are averaged out. On the other hand, if the radical is attached to a macromolecule with a much slower tumbling rate, such that spin reversal occurs before the molecule has assumed all possible orientations, then the spectrum will be a composite of the spectra for various orientations of the free radical, for example, a composite of the spectra observed along the three crystal axes of a nitroxide-doped crystal. Such a composite can be produced by grinding up a crystal such that the resultant powder is a composite of all crystal orientations [204]. Hence spectra of such immobilized free radicals are referred to as powder-type spectra [201, 202].

Since the paramagnetic species in the free radical represents essentially a free electron confined to the $^+N\!\!-\!\!O^-$ group, it does not in general interact with the environment (surrounding electron cloud or nuclei) in the way the free electrons on a transition metal ion do. The changes in the esr spectra represent primarily degrees of immobilization. The technique is only beginning to be applied to metalloenzymes; however, a couple of examples are now available.

Sulfonamides bind to the active site of carbonic anhydrase by coordination to the metal ion (see Sect. 3.1.3). Their binding affinity is not

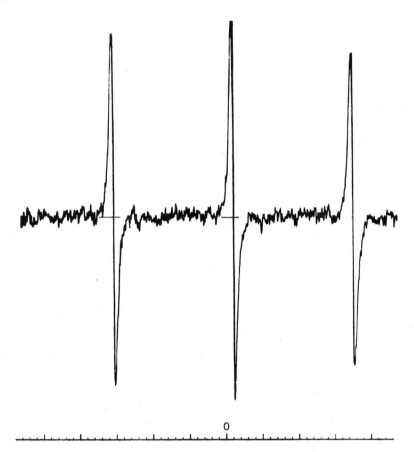

Figure 46   Electron spin resonance spectrum of compound 9 in solution. Field set = 3401 gauss. Modulation amplitude = 0.5 gauss. Data from Taylor et al. [209].

affected in major fashion by the nature of the ring structure attached to the sulfonamide. Hence, it has proved relatively easy to incorporate a nitroxide into carbonic anhydrase by attaching it to a sulfonamide. The nitroxide-labeled sulfonamide in **9** has been synthesized and introduced into carbonic anhydrase by Mushak and Coleman [209].

     The esr spectrum of the compound in solution is shown in Figure 46, while the esr spectra of the 1:1 complexes of this nitroxide with

9

Zn(II) bovine B, Zn(II) human B, and Co(II) human B carbonic anhydrases are shown in Figure 47. Binding to all three enzymes immobilizes the free radical. Immobilization appears slightly greater in the case of the bovine enzyme, in agreement with the finding that the sulfonamides are more tightly bound[210]. On the basis of the comparison of this spectrum to those calculated by Itzkowitz[201, 211] for immobilized free radicals, the correlation time, τ, for the radical bound to the bovine enzyme is 2×10^{-8} to 4×10^{-8} sec. It is of interest that the rotational relaxation time of bovine carbonic anhydrase calculated by Chen and Kernohan from fluorescence data on the complex with a fluorescent sulfonamide is 2.89×10^{-8} sec [212].

Co(II) and Zn(II) both induce almost identical immobilization of the nitroxide. Since the binding is metal induced[75], the sites must be nearly identical in the two active metallocarbonic anhydrases. Neither Hg(II) nor Cd(II) carbonic anhydrase immobilizes the nitroxide sulfonamide, in agreement with binding data showing that sulfonamides do not bind to these derivatives[75].

Another significant feature is that there appears to be practically no spin–spin interaction between the paramagnetic Co(II) and the nitroxide. This is perhaps not surprising, since according to present models of the mode of binding of the sulfonamide, 15–20 Å must separate the sulfonamide-coordinated Co(II) from the free radical.

Taylor et al.[213] have recently reported a striking spin–spin interaction between a spin label bound to creatine kinase and the paramagnetic metal ions in the metal adenosine diphosphate complexes bound at the active sites of the spin-labeled enzyme. Spin-labeled creatine kinase was prepared by reacting the enzyme with N-(1-oxyl-2,2,5,5-tetramethyl-3-pyrrolidinyl) iodoacetamide, which reacts with the essential sulfhydryl groups of the enzyme (one on each monomer)[214]. Although the derivative is inactive it retains the ability to combine with metal nucleotide substrates[213]. Addition of the nucleotide complexes of the alkaline earths Ca(II) and Mg(II) causes minor shifts and changes in amplitude of the esr signal of the immobilized spin label[213]. These are interpreted as indicating a conformational change at the active site when the complexes with the activating metals are bound[213]. On the other hand, when the spin-labeled enzyme combines with the nucleotide complexes of the paramagnetic ions Mn(II), Co(II), and Ni(II) (all inactive with the native enzyme), the amplitude of the esr signal shows a radical decrease (Fig. 48). The Mn(II) ADP complex reduces the signal to 9% of the original amplitude. This indicates a strong spin–spin interaction between the two paramagnetic species[213]. The lack of line broadening is apparently due to the fact that both species are buried in

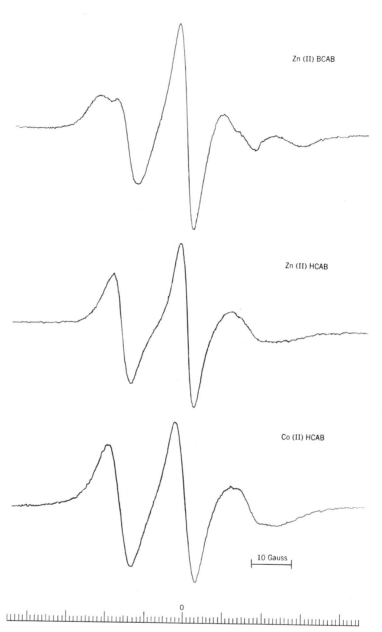

Figure 47 Electron spin resonance spectra of bovine Zn(II) carbonic anhydrase B (BCAB), Zn(II) human carbonic anhydrase B (HCAB), and Co(II) HCAB spin-labeled with compound 9. Field set = 3401 gauss; modulation amplitude = 6.3 gauss. Data from Taylor et al. [209].

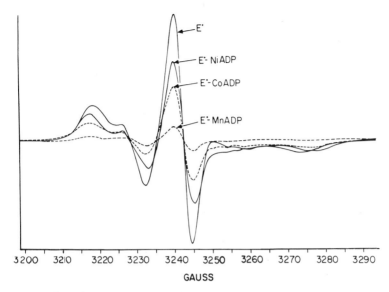

Figure 48   Effect of paramagnetic metal nucleotide complexes on epr spectrum of spin-labeled creatine kinase. Temperature = 22°. From Taylor, Leigh and Cohn[213], by permission.

a rigid protein matrix[213]. Calculations show that the two paramagnetic centers must be 7–10 Å apart, indicating that the nucleotide complexes bind near the active sulfhydryl groups[213].

A spin-labeled analog of NAD [adenosine diphosphate 4-(2,2,6,6-tetramethylpiperidine-1-oxyl)] has recently been used to study the binding of the coenzyme to liver alcohol dehydrongenase[215, 216]. Binding immobilizes the free radical. Two classes of sites were detected, two sites of weak binding affinity and two sites of strong binding affinity per molecule. Removal of the zinc atoms does not apparently destroy the tight binding sites[215, 216].

2.4 Nuclear Magnetic Resonance Spectra

The application of high resolution nuclear magnetic resonance to proteins in general is just beginning to produce exciting results[131, 217]. One specific aspect of proton magnetic resonance, however, namely the effect of a paramagnetic metal ion on the resonance of protons in its vicinity, has been used in the study of metalloenzymes for some time. The magnetic moment associated with an unpaired electron is $\sim 10^3$ times that associated with a nucleus, hence it is not suprising that the randomly fluctuating magnetic field due to the paramagnetic species often dominates nuclear magnetic phenomenon associated with nuclei in the immediate vicinity of the metal ion. These effects have been most often detected by the enhancement of proton relaxation observed for protons on various ligands caused by the unpaired spin(s) on an adjacent or coordinated metal ion. The theory relating observed nuclear relaxation times to physico-chemical features of the system, such as interatomic distances, number of coordinated water molecules, and coordination geometry, is complex and is covered in reviews by Mildvan and Cohn [218] and Jardetzky [219].

Only a very elementary attempt will be made here to point out some of the striking structural information that can be deduced from the application of these techniques. If one pictures a nuclear spin placed in an applied magnetic field, H_0 (Fig. 49), it will line up either with or against the field. If it is displaced from this orientation by another force, e.g., an applied field at right angles to H_0, H_1 (Fig. 49), then it will precess around the z axis with the Larmor precession frequency; ν (as defined in equation 8 where μ is the magnetic moment and H is the field

$$\omega = 2\pi\nu \quad \text{and} \quad 2\mu H = h\nu \tag{8}$$

at the nucleus). The usual technique for nmr absorption is to drive the precession by making H_1 an oscillating field of frequency ν. However, if the perturbing field is removed, the nucleus will gradually realign itself with the field, as indicated by the spiral in Figure 49. The time con-

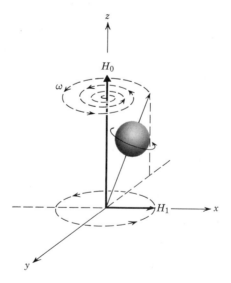

Figure 49   Nuclear precession in an applied magnetic field.

stant characterizing this realignment is defined as T_1, or the longitudinal relaxation time. The cessation of the precession is due to what can figuratively be called "frictional forces" provided by interaction with the environment. If these are large the relaxation time is short; if they are small, the relaxation time is long. These frictional forces result from the fluctuating magnetic field provided by the random motion of magnetic moments in neighboring molecules. The relaxation rate, $1/T_1$, of the nuclei will be proportional to the intensity of the component of this field at the Larmor frequency. Since a paramagnetic species in solution can generate a fluctuating local field 10^3 as great as those due to solvent nuclei with a corresponding increase in intensity of the fluctuating field component at the Larmor frequency, they can greatly enhance the relaxation rate, $1/T_1$. Likewise, the rotational motion of the nuclei can influence T_1. In liquids and solutions, the frequencies of thermal motion are about 10^{11} Hz, while the frequencies of nuclear magnetic resonance are on the order of 10^7–10^8 Hz. Hence the frequency spectrum from molecular motion which can induce T_1 relaxation is small and the T_1 relaxation process is slow. As much as 3 sec are required for the relaxation of water protons. As molecular motion becomes slower, for example, by incorporation of the nuclei into a macromolecule, then the component of the randomly fluctuating magnetic field at the Larmor frequency becomes more intense and the T_1 relaxation process becomes faster.

In the case of the precessing nucleus there is also a component of the magnetization in the xy plane (Fig. 49). This gradually goes to 0 as the nucleus realigns itself. The time constant associated with this process is termed T_2. T_2 is not independent of T_1 since the longitudinal relaxation also reduces the magnetic component in the xy plane. A certain amount of dephasing of the spins in the xy plane may reduce the xy component before T_1 is complete, hence T_2 is shorter than or equal to T_1.

There are a number of experimental methods available for measuring T_1 and T_2 for a given nucleus or set of nuclei under observation[218]. In the case of paramagnetic ions, there are two types of electron–nuclear interactions which will affect T_1 and T_2, a dipolar interaction which depends on the ion-nuclear distance and a scalar or spin–spin interaction which depends on the electron spin density at the nucleus under examination. The correlation time, τ_c, which characterizes the rate process which modulates the dipolar interaction is given by equation 9, where τ_r is the correlation time for the rotational motion of the ion-nucleus radius vector, r; τ_s is the electron spin relaxation time; and τ_M is

$$1/\tau_c = 1/\tau_r + 1/\tau_s + 1/\tau_M \qquad (9)$$

the residence time of the nuclear species in the first coordination sphere of the paramagnetic ion ($1/\tau_M$ is therefore the ligand exchange rate between the bound and unbound form). τ_c is determined by the fastest of the three rate processes in equation 9. In the aquo-complexes of the paramagnetic ions of the first transition series elements τ_r is $\sim 10^{-11}$ sec, while τ_M is several orders of magnitude larger. τ_s is longer than τ_r for Cu(II), Mn(II), and Cr(III), while τ_s is on the order of τ_r for most other paramagnetic ions [218].

Thus in the case of the relaxation rate of protons in [Mn(H$_2$O)$_6$], $1/\tau_r$ dominates the right-hand side of equation 9. Hence if a Mn—H$_2$O is bound to a macromolecule in a mixed complex which rotates more slowly, τ_c increases. These correlation times are related to T_1 and T_2 by a set of complex equations known as the Bloembergen-Solomon equations [218, 220–222]. They are given in simplified form by equations 10–13 to illustrate how certain molecular parameters like the ion–nuclear distance and the number of coordinated water molecules can be derived from measurements of relaxation times.

$$\frac{1}{NT_1} = \frac{6}{15} Dq\tau_c \tag{10}$$

$$\frac{1}{NT_2} = \frac{7}{15} Dq\tau_c - Cq\tau_\epsilon \tag{11}$$

where

$$D = \frac{S(S+1)\, g^2\beta^2\gamma_I^2}{r^6\,[\mathrm{H_2O}]} \tag{12}$$

is the dipolar interaction between ion and proton at a distance r and

$$C = \frac{1}{3}\frac{S(S+1)A^2}{h^2[\mathrm{H_2O}]}. \tag{13}$$

N is the molar concentration of the metal ion, q is the number of water molecules in the first coordination sphere, τ_c is the correlation time for the dipolar interaction (Eq. 9), and τ_ϵ is the correlation time for the scalar or spin–spin interaction. S is the electron spin quantum number, γ_I is the nuclear gyromagnetic ratio, g is the gyromagnetic ratio for the electron, β is the Bohr magneton, A is the hyperfine coupling constant, and h is Planck's constant. This very brief and sketchy theoretical background will serve only to illustrate some of the experimental results. Readers should consult references 218–222 for a satisfactory development of theory behind nuclear relaxation processes.

From equation 10 it is clear that if τ_c increases, T_1 will decrease or the longitudinal relaxation will be enhanced. For example, studies by

Shulman et al.[223] on Mn(II) carboxypeptidase have given the T_1 values listed in Table. 7. Coordination of Mn(II) to the active site of carboxypeptidase reduces the relaxation time of the coordinated water protons by almost 20-fold. This is often referred to as relaxation enhancement defined by the ratio ϵ_1, given in equation 14. The results also show that at least some water remains in the coordination sphere of Mn(II) when it is coordinated to the enzyme, and by the use of equation 11 the only

$$\epsilon_1 = \frac{(1/T_1)_{[(CPD)Mn]}}{(1/T_1)_{(H_2O)Mn}} \tag{14}$$

consistent value for q appears to be 1 [223]. In the presence of the competitive inhibitor β-phenylpropionate, the relaxation enhancement is mostly abolished (Table 7), which suggests that the inhibitor coordinates the metal ion and displaces the water molecule [223], in agreement with the earlier model based on inhibitor binding studies [44].

Another aspect of the paramagnetic ion–nuclear interaction has been used by Navon et al.[224] to examine carboxypeptidase inhibitor complexes. These investigators examined the resonance of protons located on the inhibitor. The effect on the resonance signals of the methyl and methylene protons of the inhibitor methoxyacetic acid, CH_3—O—CH_2—COO^-, when bound to Mn(II) and Zn(II) carboxypeptidases is shown in Figure 50. Both peaks are broadened in the complex with the Mn(II) enzyme compared to those observed in the Zn(II) enzyme, but not chemically shifted. The relaxation of both sets of protons on the inhibitor has been significantly enhanced by the Mn(II) ion. Calculations of r from equation 12 shows that the ion–nuclear distance for the methylene protons is 4.3 ± 0.8 Å compared to a value of 4.8 Å obtained from a model in which the carboxyl is coordinated to the metal ion. The difference is expected because $1/r^6$ in equation 12 is averaged over all instantaneous configurations of all the methylene protons and in such an average the configurations with smaller distances have a larger contribution.

TABLE 7 RELAXATION ENHANCEMENT OF WATER PROTONS BY Mn(II) CARBOXYPEPTIDASE [223]

Species Present	Concentrations, M^a			
[Mn(II)]	1.0×10^{-4}	0.9×10^{-4}	1.0×10^{-4}	0.90×10^{-4}
[(CPD)Mn]	0	0	5.3×10^{-4}	4.8×10^{-4}
[β-Phenylpropionate]	0	9.0×10^{-3}	0	9.0×10^{-3}
Relaxation time	T_1	T_1	T_1	T_1
	0.78	0.85	0.049	0.23

^{a}Conditions: 1.0 M NaCl–0.05 M Tris, pH 7.5.

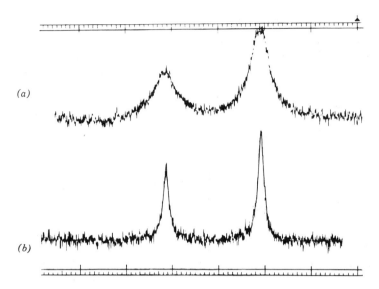

Figure 50   100-Mc nmr spectra of methoxyacetic acid, 0.4 M in 1 M NaCl, pH 7.5, dissolved in D_2O. (*a*) 2 mg/ml Mn(II) carboxypeptidase A; (*b*) 2 mg/ml Zn(II) carboxypeptidase A. Reproduced from Navon et al. [224], by permission.

The effect of Co(II) carbonic anhydrase on the relaxation rate of solvent water protons has also recently been used to demonstrate the presence of a coordinated OH^- or H_2O at the active site of this enzyme [228]. Enhancement of the relaxation rate of the solvent protons is observed at high pH and is decreased by inhibitors such as azide and ethoxzolamide. The inhibitable part of T_1 is pH dependent, with a pK_a of 7.0 ± 0.2 for the bovine Co(II) enzyme and a pK_a of 8.2 ± 0.2 for the human B cobalt enzyme. These titration curves are shown in Figure 51 from the data of Fabry, Koenig, and Schillinger [228]. Very similar pK_a's are reported by Coleman [37] using a difference titration method: pK_a 7.5 for the bovine Zn(II) enzyme and 8.1 for both the Zn(II) and Co(II) human B enzymes. Fabry et al. [228] calculated a distance of 2.2–2.5 Å between the paramagnetic center and the proton, compatible with an OH^- or H_2O as the proton-containing species. The nmr data also suggest that the less active human B enzymes exchanges protons more slowly than the more active bovine B isozyme. This is of interest in light of the possibility that proton transfer may be the rate-limiting step in the hydration–dehydration of CO_2 by this enzyme (see Sect. 3.1.3).

These examples illustrate the remarkable feature of these methods — the ability to get direct information on molecular structure. This approach has now been applied to a couple of dozen enzymatic systems

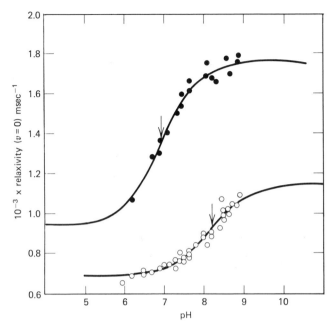

Figure 51 pH dependence of the relaxivity in the limit of low frequency for (●) Co(II) bovine and (○) Co(II) human B carbonic anhydrases in 0.1 M Tris, 0.1 M SO_4^{2-} at 25°C, ~ 10 wt%. The solid lines are a least squares fit of the data to a titration curve with pK_a values indicated by the arrows. After Fabry, Koenig, and Schillinger [228].

in which a paramagnetic metal ion is either part of the active site or has been introduced in one of the substrate complexes, such as in the case of the kinases requiring metal ion ATP complexes. Space prevents a detailed report of these studies, and a brief indication of the results on a number of these systems is given in Table 8. In most of these enzyme systems, relaxation enhancement of water protons is observed, clearly indicating that at least one water molecule remains in the coordination sphere of the enzyme-bound metal ion. This enhancement is modified by the addition of substrates or cofactors. Relaxation enhancement has also been observed for protons on the ligands. In some cases this has indicated the addition of a donor atom of the ligand (inhibitor or cofactor) to the inner coordination sphere of the metal ion as in carboxypeptidase A. The precise nature of the coordination, of course, can only be deduced from models based on the approximate distances indicated by the nmr data. Some of the "ligands" listed may be very close to the metal, but might not have a donor within the inner coordination sphere. Ligands in this general sense are indicated in Table 8. In the latter cases approximate distances from the metal ion to the ligand protons can be calculated.

TABLE 8 METALLOENZYME – LIGAND (H₂O SUBSTRATE, INHIBITOR, COFACTOR) INTERACTIONS DETERMINED BY NUCLEAR RELAXATION STUDIES

Enzyme	Paramagnetic Center	Ligands	Ion–Nuclei Radius Me–Proton Radius, Å	Type of Complex	References
Arginine kinase	Mn(II)	H_2O ADP		En—Mn—ADP	225
Adenylate kinase	Mn(II)	H_2O		Mn—ATP—E Mn—2—dATP—E	226
Alcohol dehydrogenase (liver)	ADP—Ṙ	H_2O Ethanol, acetaldehyde, isobutyramide		Substrates decrease the accessibility of water to the free radical (region corresponding to the pyridine N—C₁ ribose bond of NAD)	215, 216
Aldolase (yeast)	Mn(II) Co(II)	H_2O Fructose diphosphate		Evidence for both Me—E—S and E—Me—S	216, 227
Carbonic anhydrase	Co(II)	H_2O		Active Co—O̅H or Co—O^H^H complex	228
Carboxypeptidase	Mn(II)	H_2O Indoleacetate, t-butylacetate, bromoacetate, methoxyacetate, methoxyacetate	Mn → —CH₂ 8.3 Mn → —CH₂ 6.9 Mn → —CH₂ 5.7 Mn → —CH₂ 4.3 Mn → —CH₃ 4.7	E—Mn—OOC E—Mn—OOC E—Mn—OOC E—Mn—OOC E—Mn—OOC	224
Citrate lyase	Mn(II)	H_2O		Mn—E—S	229
Creatine kinase (muscle)	Mn(II)	H_2O		MnATP—E	141,230,231

Enzyme	Metal	Substrate	Interaction	Structure	Ref.
Creatine kinase, spin-labeled	Ṅ—O	Creatine	Mn → —CH₂ 7.5–7.8; Mn → —CH₂ 5.3–8.9		213, 218
Creatine kinase, spin-labeled	Ṅ—O	Creatine / ADP	Ṙ → —CH₂ 10.6; Ṙ → —CH₃ 10.5; Ṙ → —H₂ 8.4; Ṙ → —H₈ 7.8; Ṙ → —H₁ 8.5		213, 218
Enolase	Mn(II)	H₂O		E—Mn—S	218, 231
Histidine deaminase	Mn(II)	H₂O / Urocanate / Imidazole	Mn → HC₂ 3.5; Mn → HC₅ 3.6; Mn → HC₂ 2.9–4.5; Mn → HC₅ 3.2–4.8	E—Mn—S	218
Phosphoenolpyruvate carboxykinase	Mn(II)	H₂O		E—Mn—S	218, 232
Phosphoglucomutase	Mn(II)	H₂O			218
Pyruvate carboxylase	Mn(II)	H₂O / Pyruvate / α-Ketobutyrate / Oxalacetate / D-Malate	Mn → —CH₃ 3.5–6.6; Mn → —CH₃ 3.5–6.6; Mn → —CH₂ 3.5–6.6; Mn → —CH₂ 3.5–6.6	E—Mn—S	218, 233
Pyruvate kinase (muscle)	Mn(II)	H₂O / FPO_3^{2-}		E—Mn—PEP; E—Mn structure (see below)	218

$$\text{E—Mn—PEP} \qquad \begin{array}{c} \text{E—Mn} \overset{\displaystyle ^-O}{\underset{\displaystyle O^-}{<}} \overset{O}{P} \text{—F} \quad\text{or}\quad \text{E—Mn—}\bar{O}\text{—}\overset{O}{P}\text{—F} \\ O^- \end{array}$$

245

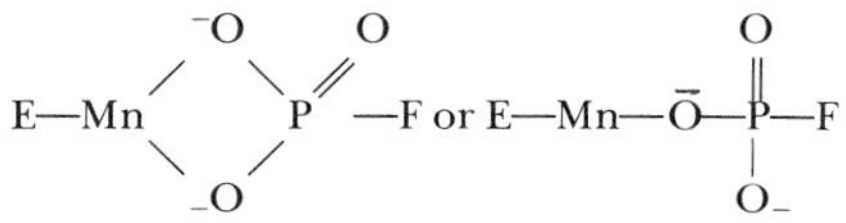

These distances and the type of complex believed to be formed among enzyme, metal, and ligand [enzyme–metal–substrate (E–Me–S), enzyme–substrate–metal (E–S–Me), or substrate–enzyme–metal (S–E–Me)] at the active site on the basis of the nmr data is also given in Table 8. The proton relaxation data provide some of the strongest evidence that complexes of substrates, inhibitors, and cofactors with metalloenzymes can involve direct mixed-complex formation with the enzyme-bound metal ion. The one case where significant relaxation enhancement of water protons has not been observed is in the ferredoxins [234]. This is in keeping with present models of the active sites of these enzymes, which suggests that the iron is coordinated to four sulfur ligands without a water molecule in the coordination sphere. While oxidized ferredoxin does enhance the relaxation of water protons slightly, it is not as effective as the ferric ion in water. This indicates that there is some paramagnetism in the oxidized protein, but the magnitude of the effect on T_1 and the equality of T_1 and T_2, as well as the temperature dependence, have all been interpreted to indicate relaxation of water protons in an outer sphere rather than in the first coordination sphere [234].

The findings on pyruvate carboxylase are of particular interest because this is the first Mn(II) metalloenzyme to be found in nature. The enzyme contains four Mn(II) and 4 biotin molecules per mole and is a tetramer [235]. Pyruvate carboxylase catalyzes the reactions shown by equations 15. Reaction 15c occurs at only 3% of the rate of reaction 15b [218]. The

$$\text{Enz-biotin} + \text{ATP} + \text{HCO}_3^{-} \overset{\text{Mg(II), acetyl CoA}}{\rightleftharpoons} \text{ADP} + \text{P}_i + \text{Enz-biotin-CO}_2 \tag{15a}$$

$$\text{Enz-biotin-CO}_2 + \text{pyruvate} \rightleftharpoons \text{Enz-biotin} + \text{oxalacetate} \tag{15b}$$

$$\text{Enz-biotin-CO}_2 + \text{ketobutyrate} \overset{\text{slow}}{\rightleftharpoons} \text{Enz-biotin} + \beta\text{-methyloxalacetate} \tag{15c}$$

Mn(II) incorporated into the native enzyme does enhance the relaxation of the water protons [218, 233] (Table 8). The substrates of reactions 15b and 15c (pyruvate, α-ketobutyrate, oxalacetate, and β-methyloxalacetate) all appear to interact with the enzyme-bound Mn(II). This interaction can be demonstrated by the effect of the Mn(II) on the relaxation of substrate protons (Table 8). The proposed mechanism for the formation of the enzyme–substrate complex [218, 236] is shown in equation 16. As calculated from the ϵ values of the water protons, the relatively inactive substrate α-ketobutyrate appears to be a bidentate ligand, while pyruvate appears to be monodentate [233, 237]. The range for the distances between Mn(II) and various substrate protons is consistent with direct coordination to Mn(II). These distances don't dis-

$$\text{Enz-Mn-(OH}_2\text{)} + \text{(OH}_2\text{)S} \xrightleftharpoons{\text{fast}} \text{Enz-Mn (OH}_2\text{)S} + \text{H}_2\text{O}$$

$$k_1 \left\| k_{-1} \right. \qquad\qquad\qquad k_3 \left\| k_{\text{off}} \right. \qquad\qquad (16)$$

$$\text{Enz-Mn} + \text{H}_2\text{O} \qquad\qquad \text{Enz-Mn-S} + \text{H}_2\text{O}$$

tinguish, however, between carboxyl or carbonyl coordination. If Mn(II) coordinated the carbonyl, the electron withdrawal would facilitate proton departure from the methyl group of pyruvate or carboxyl departure from the methylene group of oxalacetate. Features of a possible active complex are shown in Figure 52, drawn from the work of Mildvan and co-workers[233, 238].

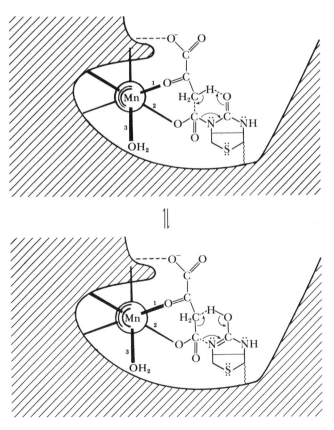

Figure 52   Features of the active site of pyruvate carboxylase as proposed by Mildvan, Scrutton, and Utter[233].

2.5 Mössbauer Spectra

The only Mössbauer isotope that has had wide biochemical application is ^{57}Fe. Mössbauer isotopes are restricted in number, since recoil-free absorption of nuclear gamma rays must be possible[158, 239]. The decay scheme for ^{57}Co, the parent isotope of ^{57}Fe, is shown in Figure 53. Following K-shell electron capture, 95% of the subsequent transitions of ^{57}Fe occur between the 136-keV isomeric level and the metastable 14.4-keV isomeric level, followed by a transition to the ground state. The 14.4-keV metastable state has a half-life of 1×10^{-7} sec and an energy width according to the Heisenberg uncertainty principle of 4.8×10^{-9}

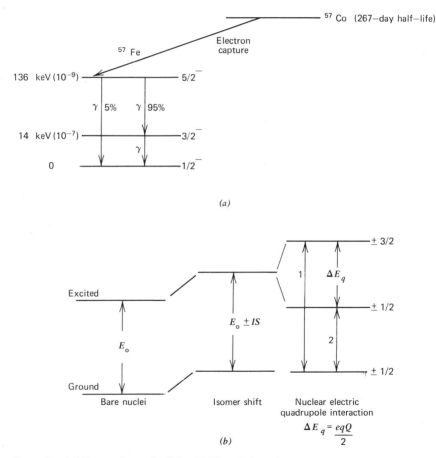

(a)

(b)

Figure 53   (A) Decay scheme for ^{57}Co. (B) The relation of E_0, gamma ray transition energy for the bare nucleus to the observed Mossbauer transition energies under the influence of isomer shift and quadrupole splitting.

eV. Hence absorption of this gamma ray from a ^{57}Co source by a ^{57}Fe nucleus in a given environment in the absorber can potentially detect energy differences of the order of 5×10^{-9} eV caused by the interaction of the ^{57}Fe nuclear state with the surrounding electronic charge distribution. These small changes in energy are detected by utilizing the first-order relativistic Doppler effect induced by moving the source relative to the absorber. The altered energy of the gamma emission can then be made to correspond to the resonant energy of the absorbing nucleus in a slightly different environment, for example, a protein, by changing the relative velocity of source and absorber. The movement requirement for most applications is on the order of millimeters per second (1 mm sec^{-1} = 4.9×10^{-8} eV). Absorption (or transmission) of gamma radiation is thus expressed as a function of this movement in millimeters per second.

The alterations in Mössbauer nuclear absorption energies reflecting the state of the surrounding electron cloud are the features of Mössbauer spectroscopy that are of interest to the biochemist. This isomer or chemical shift is dependent both on the electron charge and the nuclear charge distributions in the ground and excited states. The d-shell population of electrons formally have no density at the nucleus, however, changes in the d-shell configuration do produce changes in the s-wave functions which do have a finite contribution at the nucleus; hence changes are observed in the Mössbauer energy levels due to changes in the d-shell electronic configuration. For example, ionic Fe(II) compounds show isomer shifts about an order of magnitude greater ($+1.0$–2.0 mm sec^{-1}) than those shown by ionic Fe(III) compounds (0.1–0.4 mm sec^{-1}). A further energy-level splitting is caused by the nuclear electric quadrupole moment. The interaction of the electric field gradient with a nuclear quadrupole moment removes some of the degeneracies in the nuclear energy states, hence splitting the observed Mössbauer lines. In the case of the ^{57m}Fe (14.41 keV) $\rightarrow$ ^{57}Fe (ground state) transition, $I_e = \frac{3}{2}$ (for the excited state) and $I_g = \frac{1}{2}$ (for the ground state). These states are split according to the formula $(2I + 1)/2$, hence the excited state is split into two levels and the ground state remains unsplit as shown in Figure 53. The observed spectrum will thus show two lines shifted from the zero velocity position by an amount equal to the isomer shift. In general, Fe(II) compounds show significantly larger quadrupole splittings (2–3 mm sec^{-1}) than Fe(III) compounds (0–0.5 mm sec^{-1}). Generalizations, however, are difficult since the symmetry of the complex and the spin state influence the electric field gradient and the quadrupole splittings. Both the quadrupole and the isomer shifts are also functions of temperature [158, 239]. Original references should be consulted for details

[133, 158, 239]. Of the proteins discussed in this chapter, the ferredoxins are the only ones that have received extensive study by means of Mössbauer spectroscopy[133, 158]. The work on spinach ferredoxin is given here as an example of the type of important information that can be obtained.

The Mössbauer spectrum at 4.6°K of oxidized spinach ferredoxin enriched in ^{57}Fe consists of a narrowly split quadrupole pair (Fig. 54) and indicates that the iron is present as Fe(III). It also shows that all the Fe nuclei must occupy equivalent environments. Unfortunately, the spin state cannot be unambiguously assigned. If the enzyme is reduced with sodium dithionite a pair of lines with much greater quadrupole splittings appears and shows that reduction does result in reduction to Fe(II) and that only one of the two iron atoms is reduced[158]. The quadrupole splitting observed suggests high-spin Fe(II).

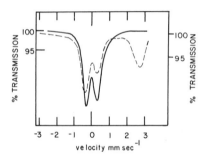

Figure 54   Mössbauer spectra of (———) oxidized and (-------) reduced spinach ferredoxin. Drawn from the data of Bearden and Moss[158].

Mössbauer studies of non-heme iron proteins which contain more iron atoms, such as *Chromatium* ferredoxin, show a slightly different picture [158]. In the *Chromatium* enzyme, which is believed to contain three atoms of iron per mole, there appear to be two iron environments present in the oxidized protein[158]. Additional studies are reviewed by Bearden and Moss[158], while Johnson et al.[240] have reported a Mössbauer study of iron in xanthine oxidase.

2.6   Metal Ion Specificity in Catalysis of Enzymatic Reactions

It has been a hope in studies of the immensely complicated biological systems that comparisons of their chemical properties to those of simple molecular analogs or model systems in which variable functions can be severely limited might yield solid information on the fundamental chemistry involved in the function of the more complex molecule. While this approach has provided pieces of solid and necessary information for the understanding of biological function, it cannot be said that it has been an unqualified success. Many facets such as the remarkable speed of enzyme catalysis and the remarkable specificity of enzymes, have not received any satisfactory explanation on the basis of model system chemistry. It is, in fact, at the very point where the actual biological molecule deviates from the model system that perhaps the most interesting chemical features are to be found. Several of these have been mentioned in connection with the spectra discussed in Section 2.2.

In the area of metalloenzymes, the models have been the great variety of coordination compounds that have been prepared since the tremendous stimulation of this branch of chemistry by Werner in the late 19th century [241, 242]. The chemistry of these complexes has been systematically studied [15] and classified on the basis of some fairly well understood theoretical principals. A number of attempts have been made to compare these properties to those observed in metalloenzymes, and the purpose of this section is to cover these comparisons in terms of their successes and failures.

As discussed in Section 2.1, determination of the stability constants of a given organic (or inorganic) ligand with a series of metal ions has been a convenient means of semiempirically classifying these complexes. The largest amount of data exists for complexes with the metals of the second half of the first transition series. Comparison of the somewhat limited stability data available on several metalloenzymes to the Irving-Williams series has been given in Section 2.1. In the case of N—O or N—N ligands, there exist some solid theoretical reasons for this series.

Much of the favorable energy driving complex formation between small cations and their negatively charged or dipolar ligands resides in the relative electronegativity of the metal atom: "the power of an atom in a molecule to attract electrons to itself" [Pauling (243)]. It has also been discovered that the transition metal ion complexes are additionally stabilized by the so-called crystal or ligand field. The reasons for this stabilization have been formalized in the crystal or ligand field theories. Space precludes a detailed discussion of this aspect and discussions in references 11–13 should be consulted. Briefly, the basis of these theories is that the five $3d$-orbitals, all of the same energy in the free ion, acquire different energies in the complex. The nature of the split in the energies of these d-orbitals is determined by the geometry of the complex and arises because the electron density of the metal avoids that of the ligand, thereby producing a more stable complex of lower energy. The geometry of these *d*-orbitals was briefly outlined in connection with the spectra in Section 2.2.

The reflections of this ligand field stabilization energy as seen in certain thermodynamic parameters measured for complexes of the divalent metal ions from Ca(II) to Zn(II) are shown in Figure 55. These are the lattice energies of the dichlorides, the hydration energies, and the dissociation energies of the hexamines of each divalent metal ion for which data are available. Electronegativity for these ions rises in a smooth curve from Ca(II) to Zn(II). The energies of complex formation, while showing a general increase along the series, rise and fall in two distinct periods. A straight line can be used to connect Ca(II), Mn(II), and Zn(II) ions, which because of their spherical symmetry even in the ligand field are not expected to gain crystal field stabilization energy [13]. Energies of the absorption bands of the aquo complexes corresponding to the *d-d* transitions of the central metal ion can be used to independently calculate the ligand field stabilization energy present in the complexes of the transition metal ions other than Mn(II). If these Δ (or 10 Dq) values (see p. 184) are subtracted from the observed hydration energies, the open symbols in Figure 54 are obtained which indicate the expected energies of complex formation in the absence of ligand field stabilization.

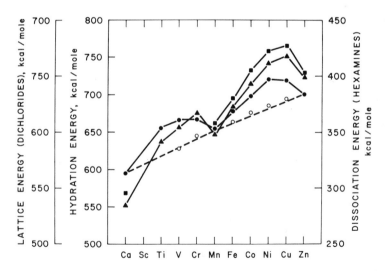

Figure 55 Thermodynamic parameters for complexes of the divalent metal ions from Ca(II) to Zn(II): (▲) lattice energies of the dichlorides; (●) hydration energies; (■) dissociation energies of the hexamines; (○) hydration energies minus ligand field stabilization energies. Data from refs. 13 and 43.

They do fall along the line connecting Ca(II), Mn(II), and Zn(II). The additional energy gained from CFSE is only 25–50 kcal/mole out of a total bond energy of several hundred kilocalories per mole. However, this additional energy is very important in explaining some of the observed physicochemical properties of these complexes. For example, the additional stability caused by the CFSE suggests why the observed stabilities of the complexes of divalent Me(II) from Mn(II) to Zn(II) with small organic ligands follow the Irving-Williams series (Sect. 2.1).

Such data, however, would be expected to follow such a regular series only if similar geometries, ligands, and oxidation states (or even spin states) are being compared. The most reliable data has been collected for ligands containing mixtures of oxygen and nitrogen as donor atoms in aqueous solution, and it is possible that another type of donor or other environments in a protein molecule might cause severe deviations from the observations expected on the basis of model complexes. The highly polarizable sulfur atom acting as a donor atom is known to produce significant deviations in the relative stabilities of Co(II), Ni(II), Cu(II), and Zn(II) complexes, stabilizing the IIB ions Zn(II) and Cd(II) relative to the transition ions Co(II), Ni(II), and Cu(II)[15, 28].

Examination of the two metalloproteins for which complete stability series are available, carboxypeptidase A and carbonic anhydrase (Table 1), reveals that the Zn(II) complex appears particularly stable in comparison to the complexes with the first transition metal ions. Both carbonic anhydrase and carboxypeptidase achieve this stability without the participation of a sulfur donor atom with a combination of three ligands. Apparently two nitrogens and an oxygen in the case of carboxypeptidase and three nitrogens in the case of carbonic anhydrase (see Sect. 3.1) bind to the metal ion.

The great stability of the Zn(II) complex of carboxypeptidase A was initially interpreted as indicating the presence of a sulfur donor because of the excellent correlation of the observed stability constants with N—S bidentate ligands[27, 28]. This data was collected on carboxypeptidase Aδ[244]. The 2 Å X-ray data on crystals of carboxypeptidase Aα show the two cysteinyl residues to be present as a disulfide some distance from the Zn(II) ion[245]. The presence of a disulfide has recently been confirmed by peptide sequence determination[246]. Since the structures at the N-terminal end of the two molecules (carboxypeptidase Aα and Aδ) are different[244], it must always be kept in mind that the other features of structure might be different as well. However, at the present time it appears as if carboxypeptidase also stabilizes the Zn(II) complex through two nitrogens and one oxygen as donor atoms[245]. The nature and precise geometry of the complexes at the active site of carboxypeptidase

A, however, must be quite different from that at the active site of carbonic anhydrase, as has been discussed in Sections 2.1.2 and 2.2.1. If model complexes, like those with histidine, involving N, N, O coordination are examined, the stabilities follow the Irving-Williams series[15].

Catalysis by metal ions of reactions of small organic molecules have been of particular interest, since such reactions might reveal character-istics of similar reactions catalyzed by metalloenzymes. It is a fair assumption that if metal ions participate directly in catalysis they do so in their capacity as Lewis acids. Catalysis utilizing this function would also be expected to bear a relationship to the affinity of the particular metal ion for the donor atom that is attacked. The rate constants for the decarboxylation of acetone dicarboxylate and oxaloacetate catalyzed by first transition and IIB metals from Mn(II) to Cd(II) are plotted in Figure 56[247–249]. The rate constant k_1 rises to Cu(II) and falls to Zn(II) and Cd(II), mimicking the stability series previously discussed (Figs. 5 and 55). The mechanism of this reaction involves enolate coordination, as shown in equation 17. The oxygens are relatively good donors and, hence, the catalytic efficiency of the metal ions follows the stability of the complexes formed. On the other hand, in model systems involving poor donor atoms, for example, phosphate ester hydrolyses, the divalent

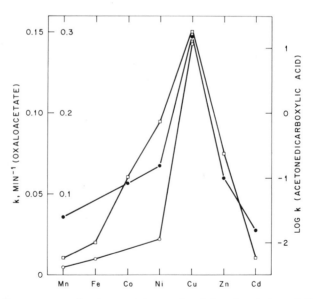

Figure 56 Rate constants for the metal ion catalyzed decarboxylation of (○) [outer left hand ordinate[247] and (●) [inner left hand ordinate[249] oxaloacetate and (□) acetone-dicarboxylic acid[248].

$$O=C-C-CR_2-C \overset{O}{\underset{O_-}{\diagdown}} \rightleftharpoons O=C-C-CR_2-C \overset{O}{\underset{O_-}{\diagdown}} \rightleftharpoons$$

$$O=C-C=CR_2 + CO_2 \rightleftharpoons O=C-C=CR_2 + H^+ \rightleftharpoons \quad (17)$$

$$O=C-C-CHR_2 \rightleftharpoons O=C-C-CHR_2 + Me^{2+}$$

metal ions all show similar catalytic efficiency. While some variation in catalytic efficiency between divalent metal ions is observed in model systems, the general observation is that if one of the first transition of IIB metal ions catalyzes the reaction, the others will also unless they inhibit by forming a stable nonreacting complex [23].

There are some plant and bacterial enzymes which catalyze the decarboxylation of oxaloacetate and are activated by metal ions. The metal ion activation of the parsley root enzyme has been studied in detail by Speck [250], and a summary of the data is shown in Figure 57. The metal ion activation of the enzyme shows a very different pattern from that shown by the model system. The least effective metal ion in the model system, Mn(II), is the most effective in the enzyme, while the most effective metal in the model, Cu(II), is least effective in the enzyme. One

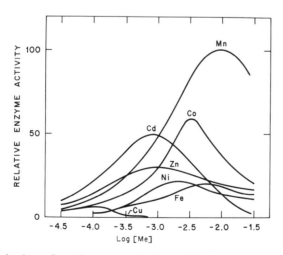

Figure 57 Activation of parsley root oxaloacetate decarboxylase by metal ions as a function of metal ion concentration. Data plotted from Speck [250].

complication in the enzyme system is immediately apparent: the metal ions can also inhibit as well as activate, especially at the higher concentrations. Cu(II) is the most effective inhibitor, which may account for its failure to activate. Inhibition probably occurs by binding of these ions to a variety of protein side chains besides those at the active site which indirectly interferes with the process at the active center. Aside from the complications of metal ion inhibition, it appears clear that the catalytic activity of the enzyme cannot be due solely to the stability of the complexes formed between metal ion and substrate.

For several metalloenzymes it is apparent that the same metal binding site has affinity for a number of metal ions [22, 33, 52, 251]. The three for which data is available for the metals from Mn(II) to Zn(II) plus Cd(II) and Hg(II) are carboxypeptidase A, carbonic anhydrase, and alkaline phosphatase of *E. coli*. The stability data were presented in Section 2.1 and catalytic data are presented in Tables 9–11. The striking finding is that no consistent or predictable metal ion specificity applies for the catalysis of the various substrate reactions. In the case of both carbonic anhydrase and alkaline phosphatase, significant catalytic activity is limited to the Co(II) and Zn(II) derivatives. For carbonic anhydrase this specificity applies to both hydration and hydrolysis reactions (Table 9).

Carboxypeptidase A, on the other hand, shows activity with all the metal ions except Cu(II) (Table 10). There are, however, some striking distinctions between the catalysis of peptide and ester hydrolysis. Zn(II), Mn(II), Co(II), and Ni(II) catalyze both reactions, while Cd(II) and

TABLE 9   ENZYMATIC ACTIVITIES AND BINDING OF ACETAZOLAMIDE TO HUMAN METALLOCARBONIC ANHYDRASES [75]

Metal[a]	Hydration of CO_2^b (U)	Hydrolysis of p-NO_2-Phenylacetate[c]	Moles of Acetazolamide Bound per Mole Enzyme[d]
Apoenzyme	400	0.09	0.04
Mn(II) B	400	0.38	0.40
Co(II) B	5,700	8.70	1.00
Ni(II) B	500	0.32	0.02
Cu(II) B	127	0.50	0.14
Zn(II) B	10,200	2.70	1.00
Zn(II) C	30,000	7.30	1.00
Cd(II) B	430	0.20	0.06
Hg(II) B	5	0.09	0.05

[a]B and C refer to the isozymes of human erythrocyte carbonic anhydrase.
[b]Determined by the method of Wilbur and Anderson [252].
[c]Conditions: 0.025 M Tris, 5% acetonitrile, pH 7.5, 23°.
[d]Determined in the presence of 1×10^{-6} M free acetazolamide.

TABLE 10 ENZYMATIC ACTIVITIES OF METALLOCARBOXYPEPTIDASES

Metal	Hydrolysis of Carbobenzoxy-glycyl-L-phenylalanine [308] C (Proteolytic Coefficient)	Hydrolysis of Hippuryl-β-phenyllactate [5] $K_m, M \times 10^5$	$V_{max}, min^{-1} \times 10^{-3}$
Mn(II)	9.5	32.0	56.8
Fe(II)	[a]	—	—
Co(II)	73	9.8	37.7
Ni(II)	17	21.0	27.6
Cu(II)	0[b]	[b]	0
Zn(II)	34	7.6	28.6
Cd(II)	0	55.0	61.5
Hg(II)	0		[c]

[a]Fe(II) carboxypeptidase appears to be active as a peptidase [30], however, the transformation Fe(II) $\rightarrow$ Fe(III) hydroxide in aqueous solution makes study of this enzyme difficult.
[b]Cu(II) carboxypeptidase appears inactive in both peptide and ester hydrolysis [28]. Substrate binding studies show, however, that both peptides and esters bind to the Cu(II) enzyme [46].
[c]Hg(II) carboxypeptidase under standard conditions of assay has esterase activity slightly greater than the Zn(II) enzyme, but no peptidase activity [28].

TABLE 11 PHOSPHATE BINDING AND ACTIVITY OF APO AND METALLOALKALINE PHOSPHATASES AT pH 8.0[a]

Enzyme	Moles of $^{32}P_i$ Bound[b] Mole Dimer	No. of Determinations	Phosphatase Activity[c] μmoles/hr/mg protein
ZnAP[b]	1.36 ± 0.19	20	3000
ApoAP	0.22 ± 0.15	10	100
MnAP	1.23 ± 0.06	4	200
CoAP	1.25 ± 0.19	4	600
NiAP	0.20 ± 0.12	4	100
CuAP	0.5–0.8	6	100
CdAP	1.31 ± 0.07	6	100
HgAP	0.08 ± 0.04	4	100

[a]Data from reference 52.
[b]Phosphate binding determined at $1 \times 10^{-5} M$ free phosphate.
[c]Assay in 1 M Tris, pH 8.0.

Hg(II) catalyze only the esterase reaction, the latter with more efficiency than the native Zn(II) ion (Table 10). Clearly something more is involved than a simple relationship to stability of the complex.

"Complex" in the metalloenzyme case implies an apoenzyme–metal–substrate complex, and stringent requirements forced on such a complex by enzyme–substrate and substrate–metal contacts, as well as apoenzyme–metal bonds, could radically affect the specific chemistry required for catalysis. Considerable information is available on mixed enzyme–substrate or enzyme–inhibitor complexes in the case of carbonic anhydrase and carboxypeptidase. Spectral [25, 37, 74], titration [37], infrared [82], and nmr [228] data indicate that one coordination site of the active metal complex in carbonic anhydrase is filled with a water molecule. It also appears that anions, sulfonamides, and intermediates on the catalytic pathway may also occupy this site [37, 73, 74, 77] (see Sect. 3.1.3). The water ligand is replaced by the sulfonamide group when any one of a large number of these powerful inhibitor molecules is bound [37]. The X-ray diffraction data on the complex of the human C isozyme with a modified sulfanilamide shows the center of gravity of the sulfonamide group to be within bonding distance of the Zn(II) ion [77] (See Sect. 3.1.3). It is thus not surprising that the binding of sulfonamides to carbonic anhydrase is metal-ion dependent [75], as can be directly demonstrated using tritiated acetazolamide (Diamox) (Table 9).

The most interesting feature of this binding is that only the metal ions that induce activity, Zn(II) and Co(II), induce the effective binding of the sulfonamide (Table 9). This finding implies that only the geometry around these two ions at the active center of carbonic anhydrase favor the addition of the additional monodentate ligand. Such a finding suggests that the ease of addition or exchange of the monodentate ligand in this coordination site may be related to activity [37]. Very small differences in preferred coordination geometry could influence this, especially since the geometry favoring catalysis appears to be one of low symmetry [25, 79] (see Sect. 2.2.1). The precise mechanisms proposed for carbonic anhydrase on the basis of these findings are presented in Section 3.1.3

In carboxypeptidase A the metal ions effective in catalysis are less limited than for carbonic anhydrase (Table 10). The spectrum of the Co(II) enzyme also suggests a geometry of low symmetry (lower than a regular tetrahedron), but apparently somewhat different from that in carbonic anhydrase (see p. 186). As in carbonic anhydrase, both nmr [223] and X-ray data [245] indicate that at least one coordination position on the metal ion is occupied by a water molecule. When a peptide substrate is bound, the water is apparently replaced by the carbonyl oxygen of the susceptible peptide bond [245] (see Sect. 3.1.1).

The findings at the active center, however, give no obvious clue to the reasons for the metal ion specificity. The complexes of the active center with the first transition metal ions [except Cu(II)] and with Zn(II) all have the properties necessary for catalysis of both peptide and ester hydrolysis (Table 10). The activity of Cd(II) and Hg(II) in ester, but not peptide, hydrolysis is not readily explained. It is tempting, however, to relate this to some feature of ionic or covalent radius[245, 253], since the two largest metal ions are involved. There is some preliminary X-ray data on the Hg(II) enzyme suggesting that a distortion due to size might be responsible[245] (see further discussion in Sect. 3.1). Loss of activity in the case of Cu(II), Cd(II), and Hg(II) carboxypeptidases A due to gross loss of substrate binding can be ruled out[45, 46], although a non-productive mode of binding may well be induced.

Alkaline phosphatase of E. coli is a more complex system than the two monomeric enzymes discussed above. The active form of this enzyme is a dimer of identical subunits containing two Zn(II) ions[26, 254, 255]. All evidence indicates that the dimer containing at least two metal ions is the minimal requirement for activity[26, 256–259], and only a single active site is operating on the dimer at any one moment[52, 260–267]. Whether a single active site is present or some sort of negative cooperativity is responsible for effective catalysis at only one of two identical sites is as yet not clear[52, 267]. In any event, the series of first transition and IIB metal ions discussed above can occupy the metal binding sites of the dimer[26, 51, 52, 86, 251]. As in the case of carbonic anhydrase effectual catalysis is limited to Co(II) and Zn(II) (Table 11). Two bound metal ions occupy sites of unique and probably highly distorted geometry (see Sect. 2.2)[86]. Phosphate can be considered the simplest substrate for this enzyme, since the enzyme catalyzes ^{18}O exchange from water into phosphate[52, 268, 269]. Binding of a single phosphate ion to the dimer does alter geometry at the metal coordination sites[86, 88].

Phosphate binding does prove to be metal-ion dependent (Fig. 58). In contrast to the finding for the binding of sulfonamides to carbonic anhydrase, the ability of the metal ion to induce phosphate binding is not limited to the two active metal ions, Co(II) and Zn(II). Mn(II), Co(II), Zn(II), and Cd(II) all induce the binding of one phosphate ion (Table 11). Cu(II) is only partially effective, while Ni(II) and Hg(II) are ineffective. The reason for this rather unique specificity is not suggested by known properties of coordination complexes of these ions.

The number of intermediates involved in the alkaline phosphatase reaction may be greater than in the case of the carboxypeptidase and carbonic anhydrase reactions. Phosphate and phosphate monoesters have been shown to form covalent intermediates during reaction with the enzyme. This covalent intermediate has been isolated as an extraor-

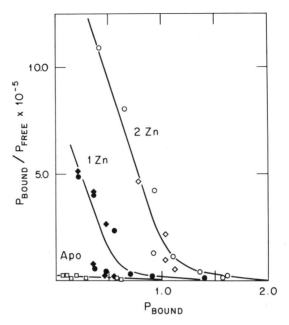

Figure 58   Scatchard plot of (□) ³²phosphate binding to apoalkaline phosphatase, (●, ◆) apoalkaline phosphatase + 1 Zn(II), and (○, ◊) apoalkaline phosphatase +2 Zn(II). From Applebury, Johnson, and Coleman [52], by permission.

dinarily stable phosphoserine [260–262, 269, 270]. The native Zn(II) enzyme forms this intermediate only at low pH, but evidence has been obtained that the phosphoryl enzyme may be an intermediate on the normal reaction pathway at alkaline pH [271–273].

Additional features of the metal-assisted catalysis of phosphate ester hydrolysis are apparent from comparing the pH stability of this phosphoryl enzyme intermediate formed by the several metalloalkaline phosphatases (Fig. 59). The Zn(II) enzyme forms no significant equilibrium concentration of this intermediate at alkaline pH. Very small equilibrium concentrations of this intermediate must be a requirement for rapid hydrolysis, since the enzyme is maximally active at alkaline pH. The other active metallophosphatase, the Co(II) enzyme, also forms very small equilibrium concentrations of this intermediate at alkaline pH. At acid pH the equilibrium concentration of the phosphoryl enzyme begins to increase, but falls rapidly when structural changes begin in the enzyme at pH 5 and below (see Sect. 3.1.4). On the other hand, two of the inactive enzymes, Cd(II) and Mn(II), both form significant equilibrium concentrations of the phosphoryl enzyme at alkaline pH.

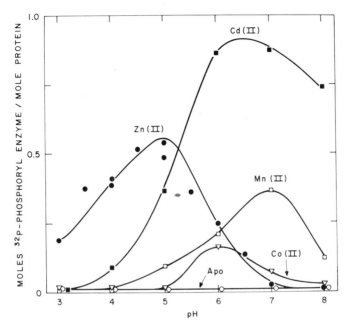

Figure 59   pH stability of phosphoryl alkaline phosphatase as formed by Mn(II), Co(II), Zn(II), Cd(II), and apo-alkaline phosphatases. The data represent moles of phosphoserine formed per mole of enzyme dimer. Isolation of the ^{32}P-labeled peptide from the Cd(II) enzyme shows the same amino acid composition as for the peptide isolated from the Zn(II) enzyme. Data from Applebury et al. [52].

The observed equilibrium concentration of the phosphoryl enzyme must reflect the rate at which its formation is catalyzed relative to the rate of its breakdown [273]. The data in Figure 59 suggests that the metal ion is involved in catalysis of the breakdown of the phosphoryl enzyme and perhaps also in its formation. Very slow catalysis of this breakdown by Cd(II) could, of course, explain the complete lack of phosphatase activity shown by this derivative.

Once again the chemical features of Cd(II) and Mn(II) coordination which radically change this function as compared to Zn(II) and Co(II) coordination are not apparent. As a confession of ignorance it can be attributed to small changes in preferred coordination geometry. Such detailed information on the nature of the coordination site unfortunately is not available as it is in the case of carboxypeptidase A and carbonic anhydrase. If, as seems possible, the active site contains structural features contributed by both monomers [274], small shifts in geometry could have major effects on the enzyme–substrate interactions required for catalysis.

Enzymes differ from the model systems studied thus far in that there are mutliple interactions leading to a relatively rigid, highly specific, enzyme–substrate complex. Small changes in preferred geometry induced by one metal ion compared to another, while not evident in gross coordination properties (stability for example) may exceed the tolerated limits for variation in the enzyme–substrate complex which can still lead to the proper function of all steps in the mechanism.

Another aspect of metal ion specificity in the catalysis of enzymatic reactions is represented by the findings on arginase shown in Figure 60. For this metal–enzyme complex, several of the first transition period metal ions are effective in catalyzing the hydrolysis of arginine, but the order of efficiency varies with pH[275, 276]. In all likelihood this represents the pH stability (or pH–formation curves) for each of the protein complexes with the activating metal ions. In this case the behavior is probably quite analogous to model systems. Such pH stability must be kept in mind in interpreting pH–rate profiles observed for metal–enzyme complexes.

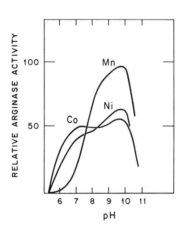

Figure 60 Activation of arginase by divalent metal ions, pH dependence. Drawn from the data of Greenberg [276].

In the case of the oxidative enzymes that contain metal ions the metal ion specificity appears to be more stringent than that discussed above for enzymes catalyzing hydrolysis or hydration reactions. Three metal ions, Fe, Cu, and Mo, are found in this group. For the copper oxidases from which stable apoenzymes have been prepared (see p. 182), only Cu of the first transition and IIB metal ions have restored activity [58, 59].

If it is assumed that one of the functions of the metal ion in oxidative enzymes will be to undergo a reversible transition between two oxidation states, then examination of the relevant oxidation potentials for the metals from Mn to Zn reveals a plausible reason for the restriction noted above [11]. The first two electrons are removed from all the first transition and IIB metals with relative ease, accounting for the frequent occurrence of the divalent state. With the exception of Fe, the third ionization potential is high, >30 eV [11]. The divalent species is thus stabilized and the trivalent state is relatively infrequently encountered. Copper has the highest second ionization potential, 20.29 eV, hence Me^+ is stabilized relative to Me^{2+} more than in any other case, accounting for the relative ease of reversible oxidation–reduction of the Cu^+–Cu^{2+} couple. Likewise, among the third ionization potentials, that for Fe is lowest, 30.64 eV; thus the third electron is removed with relative ease compared to the other metals of this series, accounting for the frequency of the Fe^{2+}–Fe^{3+} couple. The energies involved in these two reversible metal ion oxidations are comparable to the energies involved in the oxidation of a number of biologically important substrates.

The energy change on going from Co(II) to Co(III) is considerably greater than in the case of Fe, but some ligand environments do stabilize Co(III). Only a few instances of cobalt oxidation state changes related to function are known in biological systems, and they all involve cobalt incorporated into the unique environment of the corin ring system in the cobalamins.

Molybdenum has a number of relatively stable oxidation states, Mo(I) to Mo(VI) [200]. A comparative study of the stabilization of the various oxidation states of molybdenum by various ligands and their relative oxidation potentials [200] suggests that the Mo(V)–Mo(VI) couple is likely to be the one found in metalloflavoproteins, the only instance where biologically active molybdenum has been found. The oxidation–reduction potential of the Mo(V)–Mo(VI) couple is -0.2 to -0.4 V, rather similar to the -0.25 V observed for flavins [200]. The proof that reversible oxidation–reduction of molybdenum is a feature of xanthine and aldehyde oxidases has been discussed in Section 2.3.5.

Studies of the precise mechanisms involved in the oxidative metalloenzymes are not as advanced as in the case of those catalyzing hydration and hydrolysis reactions. The reasons for this include the complex and poorly understood chemistry involved in the interaction of oxygen with metal ions and the experimental difficulties in studying the oxygen interaction with many of the oxidative enzymes. There have been a number of studies of model oxidative reactions catalyzed by metal ions, and this field will probably become more active. Udenfriend et al. [277] reported in 1954 that aromatic compounds are hydroxylated by O_2 or H_2O_2 when Fe(II), ascorbic acid, and ethylene diaminetetraacetic acid are present in buffered solution at neutral pH. A number of investigations of this system have been made[278]. Different hydroxylated products are obtained with the two oxidants[279, 280] and there is evidence that the hydroxyl radical (HO·) is the hydroxylating agent when H_2O_2 is the oxidant[279, 281]. Hamilton and co-workers[280–283] have carried out a number of studies on possible model systems for mixed function oxidases. In one of these anisole is hydroxylated by H_2O_2 when catalytic amounts of Fe(III) and an enediol catalyst, such as catechol, are present [281]. Cu(II) is the only other active catalyst, although Cr(III), Co(III), Zn(II), Mn(II), Al(III), and Mg(II) were all tried. The hydroxylating agent appears to be a complex between the metal ion, the enediol, and H_2O_2[281]. These systems do appear to have limited metal ion specificity and it is of interest that the hydroxylation reaction catalyzed by tyrosinase does require catalytic amounts of the enediol catechol (see Sect. 3.2.3). The precise relationship of these model systems to the mechanisms of the copper- or iron-containing oxidases remains for future investigation.

3 INDIVIDUAL ENZYMATIC SYSTEMS – CATALYTIC MECHANISMS

3.1 Hydrolysis and Hydration Reactions

3.1.1 Carboxypeptidase A. Carboxypeptidase A isolated from bovine pancreas is a globular protein of molecular weight 34,600 which catalyzes the hydrolysis of the C-terminal peptide bond of proteins and peptides [284]. Substrate length may vary from a dipeptide to a large protein. The enzyme was crystallized by Anson in 1937[285] and is one of the most thoroughly studied pancreatic enzymes. The enzyme is synthesized in the bovine pancreas[286] as a zymogen of molecular weight 87,000 which can be separated into three subunits, one of which is carboxypeptidase A[287, 288]. Neurath and co-workers have studied the physicochemical properties and activation of the zymogen in detail[287–291]. The zymogen contains significant amounts of iron and nickel in addition to Zn(II)[5]. The total metal content appears to be 1g-atom per mole of proenzyme[5]. A complexometric titration performed on the metal-free zymogen indicates that one of the Zn(II) ligands present in the active enzyme is missing in the zymogen[292].

Numerous studies have established its specificity in great detail [293–297]. Inhibition by chelating agents suggested its metalloenzyme nature[298], and the definitive demonstration that it was a Zn metalloenzyme was carried out in 1954 by Vallee and Neurath[299]. The detailed studies of the interaction of the enzyme with Zn(II) and other first transition and IIB metals[22, 28, 300], the studies of the chemistry and structure of the organic portion by chemical modification[5] and amino acid sequence determination [301], and the elegant X-ray diffraction studies at 2.0 Å resolution [245] make this the most extensively studied metalloenzyme. The wealth of information that can be brought to bear on its mechanism of action makes it worth while to discuss this enzyme in considerable detail.

Two general types of synthetic substrates are hydrolyzed, peptides (**10**) and the analogous esters (**11**). Two of the best substrates, benzoyl-gly-L-phenylalanine and its analog benzoyl-gly-L-phenyllactic acid, have been chosen for illustration (**10** and **11**). The enzyme requires a free

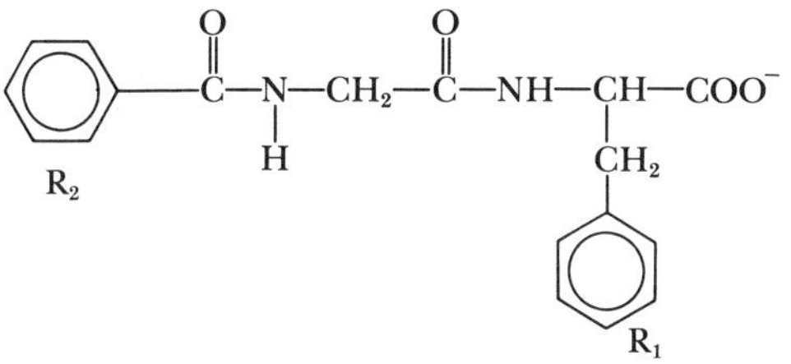

Benzoyl-gly-L-phenylalanine
(BGP)

10

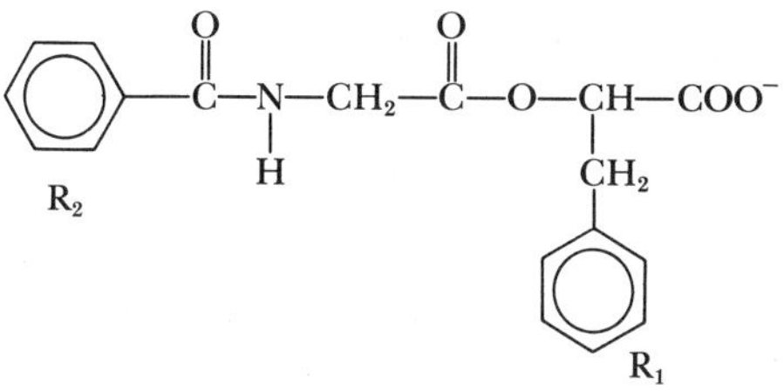

Hippuryl-β-phenyllactate
(HPLA)

11

C-terminal carboxyl on a residue of the L-configuration containing an aromatic or branched aliphatic side chain. The penultimate residue (glycine in the example) does not have strict stereospecificity. Substrates with residues of D- or L-configuration at this position are both hydrolyzed [302]. Dipeptide substrates are much more rapidly cleaved (1000:1) if the terminal amino group is acylated [295]. A metal ion [Zn(II) in the native state] is absolutely required for hydrolysis [299]. The metal ion specificity has been discussed in Section 2.6.

The features of chemical structure at the active center leading to the participation of the metal ion in catalysis can now be described rather precisely with the data available from investigations of the solution

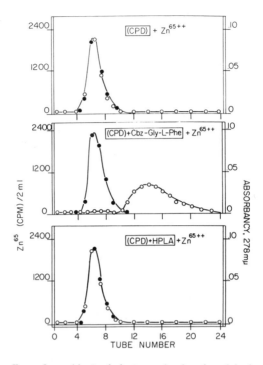

Figure 61   The effect of peptide (carbobenzoxyglycyl-L-phenylalanine) and ester (hip-puryl-β-phenyllactate) substrates on the binding of ^{65}Zn to apocarboxypeptidase A; (upper curve) apocarboxypeptidase + ^{65}Zn passed over a G-25 Sephadex column; (middle curve) apocarboxypeptidase + CGP + ^{65}Zn passed over a G-25 Sephadex column; (lower curve) apocarboxypeptidase + HPLA + ^{65}Zn passed over a G-25 Sephadex column. From Coleman and Vallee [45, 46], by permission.

chemistry and the X-ray studies in the solid state. Peptide substrates bind to the enzyme in the absence of the metal ion [45]. Part of the experimental evidence for this is shown in Figure 61 which demonstrates that peptide substrates bind to the apoenzyme and prevent the binding of ^{65}Zn. Thus the peptide binding site would appear to be in close proximity to the metal coordination site, since the bound substrate prevents access of the free metal ion to this site. Restoration of activity to the apoenzyme is also prevented by the prior addition of peptide substrates [303]. Peptide substrates also prevent the exchange of metal ions when an inactive metal ion like Cd(II) is present at the active center [45, 46].

Five general variations of carboxypeptidase substrates are shown in Table 12. Of this group only 1 and 2 show the phenomenon of binding to the apoenzyme. Substrates 3, 4, and 5 require the presence of the metal ion for binding. The latter three, however, can prevent the exchange of an inactive for an active metal ion [29, 46]. These findings suggest a type of cooperative binding to the apoenzyme involving contacts of three groups on the peptide substrate, the C-terminal side chain in the L-configuration, the N—H group of the C-terminal residue, and the N—H group of the penultimate residue (indicated by square brackets in Table 12). The latter must be in the L-configuration. If any one of these contacts is missing, binding does not occur to the apoenzyme, but requires the participation of the metal ion. If the bulky C-terminal side chain of the L-configuration is missing, the small synthetic peptides are

TABLE 12   BINDING OF SUBSTRATES TO APOCARBOXYPEPTIDASE A

Substrate	Binding to Apoenzyme
Acylated dipeptide	
1. 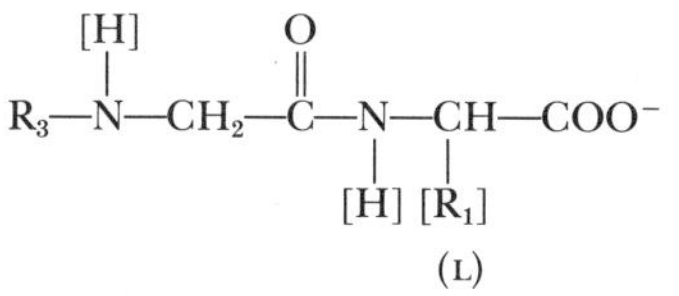 R_3—N—CH$_2$—C—N—CH—COO$^-$ with [H] on first N, O double bond on C, [H] and [R$_1$] below, (L)	Bound
Dipeptide	
2. 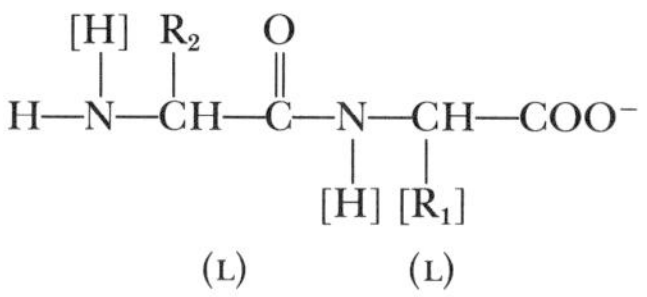 H—N—CH—C—N—CH—COO$^-$ with [H] R$_2$ on first part, O double bond, [H] [R$_1$] below, (L) (L)	Bound
3. Same as 2 except configuration is (D)—(L)	Not bound
N-acyl amino acid	
4. 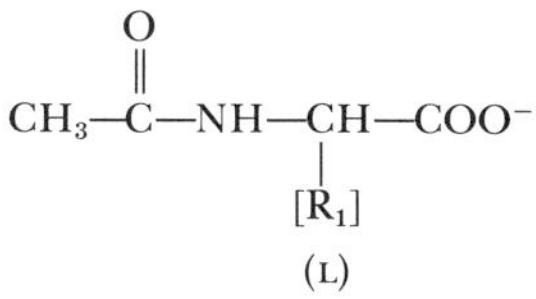 CH$_3$—C—NH—CH—COO$^-$ with O double bond on C, [R$_1$] below, (L)	Not bound
Ester analog of acylated dipeptide	
5. 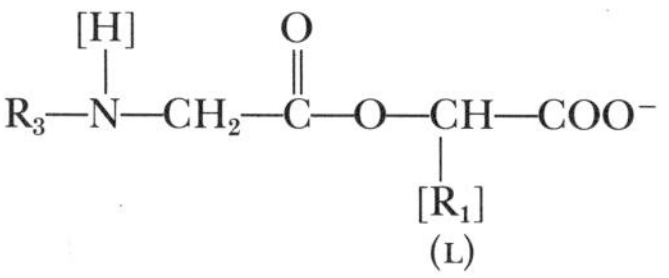 R_3—N—CH$_2$—C—O—CH—COO$^-$ with [H] on N, O double bond on C, [R$_1$] below, (L)	Not bound

not substrates. Evidence does indicate, however, that the C-terminal residue may be glycine in large polypeptides [304], in which case the interaction of the long chain with the protein must provide sufficient binding. This agrees with the observation that carboxypeptidase A has been known to liberate glycine from the C-terminal end of proteins.

The binding phenomena described above apparently do not distinguish between the active and inactive metal ions listed in Table 10, since peptide substrates are bound in the presence of metal ions like Cd(II) and Hg(II), which are inactive in peptide hydrolysis [45, 46]; and Cu(II) induces the binding of the ester substrate [45, 46], although this ion does not catalyze either ester or peptide hydrolysis [28]. Hence the subtle influences of metal ion chemistry on function evident in Table 10 do not appear to reside in the gross binding function.

In addition to the coordinated metal ion, organic groups of the protein are also involved in the catalytic mechanism, as has been demonstrated by numerous chemical modification studies on carboxypeptidase A. The classical experiment is shown in Figure 62 from the work of Simpson, Riordan, and Vallee [305]. Acetylation of the enzyme with N-acetyl-imidazole results in abolition of peptidase activity and increase of esterase activity by 600% [5, 253, 305, 306]. Spectral studies show that two tyrosyl residues of the enzyme have formed O-acetyl tryosine [305, 306].

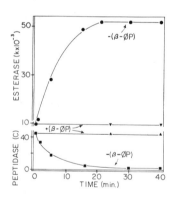

Figure 62   Progression of changes in (●) esterase and (■) peptidase activities during acetylation of carboxypeptidase A with a 60-fold molar excess of acetylimidazole in the (▼, ▲) presence and absence of β-phenylpropionate. From Simpson et al. [305], by permission.

The reaction is completely prevented by the competitive inhibitor [β-phenylpropionate (Fig. 62)]. The residues can be deacylated by the addition of hydroxylamine with regeneration of a completely normal enzyme [253, 305, 306]. In addition to acetylation, a number of other chemical modifications can be performed that apparently involve these tyrosyl residues and produce qualitatively similar results. Iodination of the enzyme produces results almost identical to acetylation [307, 308], while acylation with succinyl, isobutyryl, *n*-valeryl, *n*-butyryl, and propionyl groups produce smaller changes, apparently related in part to the rapidity of deacylation of the *O*-acyl tyrosyl residues [309].

Reaction of one of these two tyrosyl residues can be achieved by selective nitration with tetranitromethane [310], while the second tyrosyl can be modified selectively with a small excess of 5-diazonium-1 *H*-tetrazole [311]. The nitration reaction destroys peptidase activity like acetylation and iodination, while the reaction with a small excess of 5-diazonium-1 *H*-tetrazole leaves peptidase activity intact. The spectrum of the nitro-enzyme has been useful as a spectroscopic probe of the active center since the visible aborption band is shifted by the binding of β-phenyl-propionate [310].

The molecular structure of crystalline carboxypeptidase A has now been determined by Lipscomb and co-workers to a resolution of 2.0 Å by X-ray diffraction techniques [245]. By diffusing the very slowly hydrolyzed substrate Gly-L-Tyr into crystals of glutaraldehyde cross-linked crystals of carboxypeptidase [312, 313], an enzyme–substrate complex was also obtained and a structure determined to 2 Å resolution. Crystals of carboxypeptidase A crosslinked with glutaraldehyde have less than 1% of the activity of a comparable amount of enzyme in solution [312]. If glutaraldehyde crosslinked crystals are placed in flow cells and solutions of metal ions and substrates or inhibitors flowed over them, metal ion exchange proceeds much as in solution and small substrates and inhibitors can be diffused into the crystal [312, 313]. The crystalline complex used for the X-ray data contained only one molecule of Gly-L-Tyr per molecule of enzyme [245].

The general results of the X-ray data as concerns the folding of the polypeptide chain are shown in Figure 63 from Lipscomb et al. [245]. Three hundred and seven amino acid residues can be fitted into the electron density map. This figure is in close agreement with the results of amino acid analysis, 306 residues [301]. About 20% of the total residues are contained in eight extended chains which form the twisted pleated sheet structure shown in Figure 63*b*. The uppermost of these chains is twisted 120° with respect to the lowermost chain, apparently chiefly because of the packing of hydrophobic side chains [245]. The sheet is made up of four pairs of parallel and three pairs of antiparallel chains and forms the lining of one side of a pocket which extends into the interior of the molecule [245].

Helical regions of the molecule represent about 30% of the residues and the helix is mostly α-helix. Most of the helical sections are on the surface of the molecule on one side of the pleated sheet and approximately parallel to the adjacent extended chains of the pleated sheet

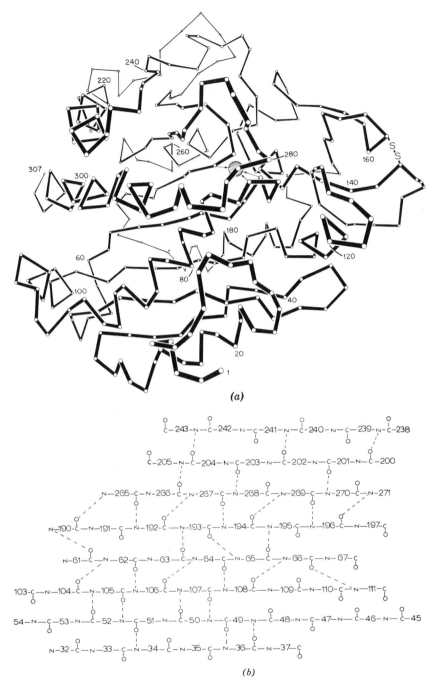

Figure 63 (a) Folding of the polypeptide chain in carboxypeptidase A (b) β Structure in the "core" of carboxypeptidase A. From Lipscomb et al. [245], by permission.

(see upper left of Fig. 63*a*). The major helical regions involve residues 14–29, 72–88, 215–233, and 288–305. There are also a few turns of helix of various degrees of perfection involving residues 94–103, 115–123, 174–184, and 254–262. Sections of α_{II} helix [314] are also present at the *C*-terminal ends of three of the helical sections, including residues 26–29, 100–103, and 260–262. The combination of the long helical sections and the extensive pleated sheet structure should confer great stability, and carboxypeptidase is known to be an extremely stable protein [284].

The portion of the electron density map including the Zn(II) ion is shown in Figure 64. There are three ligands from the protein, His 69, Glu 72, and His 196. The identity of these residues is as indicated by the amino acid sequence of the protein determined by Neurath and co-workers [245, 245a, 301]. Some of the chemical data, such as a pK_a of 9.3 for one of the ligand groups [28] (p. 165), and the stability of the complex would seem to indicate that one of the nitrogen ligands has a high pK_a. Data on acetylation of the apoenzyme had implicated an

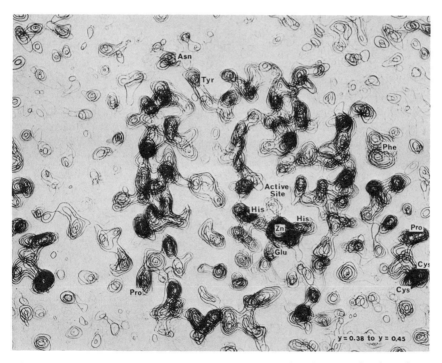

Figure 64 Electron density map of the region surrounding the Zn(II) ion in carboxypeptidase A. From Lipscomb et al. [245], by permission.

ε-amino group of lysine as a possible coordinating group[5, 292]. However, the amino acid sequence coupled with the X-ray data seems to rule this out. A possible interpretation is that one of the histidyl residues has a very high pK_a if the data from the difference titration of the active site with Zn(II) reflect the pK_a's of the donor groups (see p. 165). It is possible that dissociation of the proton from one of them makes it more difficult to dissociate the proton from the other. It is curious that a similar situation seems to apply to carbonic anhydrase where the donors are tentatively identified as three histidine nitrogens (see p. 288). In this case as well the stability constant rises sharply with pH until pH 10, suggesting that H^+ continues to interact with the binding site until relatively high pH. Associated with zinc coordination site in carboxypeptidase is the long groove and a pocket to contain the side chain of the substrate demonstrated in the glycyl-L-tyrosine complex (see below).

The determination of the structure of the active center with a glycyl-L-tyrosine molecule in place makes possible a detailed discussion of structural features of the molecule which may be involved in the catalytic mechanism. The X-ray findings on this complex are summarized in schematic form in Figure 65 and will be discussed as they relate to the observations made in solution. There are several interactions between glycyl-L-tyrosine and the crosslinked crystalline enzyme that are probably characteristic of productive enzyme–substrate complexes. The C-terminal aromatic side chain is inserted into a pocket, large enough for a tryptophan side chain, but containing no specific binding groups[245]. This finding is compatible with the known specificity of the enzyme, and a pocket for binding the side chain has been a feature of

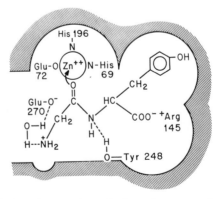

Figure 65   Model of Gly-L-Tyr complex with the active center of carboxypeptidase A as drawn from the data of Lipscomb et al.[245].

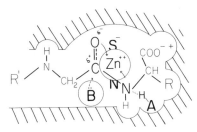

Figure 66   Postulated features of the active center of carboxypeptidase involved in peptide hydrolysis as proposed by Vallee, Riordan, and Coleman [253] on the basis of chemical data in solution.

mechanisms such as that proposed by Vallee et al.[253] on the basis of the solution chemistry (Fig. 66). The C-terminal carboxylate group interacts with the guanidinium group of Arg 145, providing the positive center for the binding of this group (Figs. 65 and 66). Upon binding of the substrate, Arg 145 moves 2 Å toward the carboxylate group [245]. The electron density difference maps do not allow a precise visualization of the peptide bond, since the binding apparently displaces some bound water molecules, with the result that there is little change in density in the region of the peptide bond [245]. Placement of the other parts of the substrate, as well as the requirement for the L-configuration, however, limits the location of the peptide bond pretty well to the vicinity of the zinc ion. This had been predicted on the basis of the solution chemistry and from the observation that zinc does not have access to the binding site when peptide substrates are bound to the apoenzyme (Fig. 66). The mechanism postulated by Vallee et al.[253] visualized a possible electron withdrawal by Zn(II) from both the carbonyl oxygen and the peptide nitrogen (Fig. 65). A single interaction with the carbonyl oxygen has recently been favored [245], although interaction with the peptide nitrogen at some stage of the mechanism cannot be ruled out.

In the case of the dipeptide Gly-L-Tyr, the X-ray data show the α-amino group to bind through a water molecule to Glu 270 (Fig. 65). In addition, the phenolic hydroxyl group of Tyr 248 moves about 12 Å when the dipeptide is bound such that this OH is in the vicinity of the peptide bond of the substrate, as indicated schematically in Figure 67 which is drawn from the data of Lipscomb et al. [245].

The interaction of the peptide nitrogen with Tyr 248, the binding of the side chain in the pocket, and the interaction of the α-amino group with Glu 270 would account for the three groups of the dipeptide required for binding of this substrate to the apoenzyme (see p. 267). For binding to occur to the apoenzyme the α-amino group of the pen-

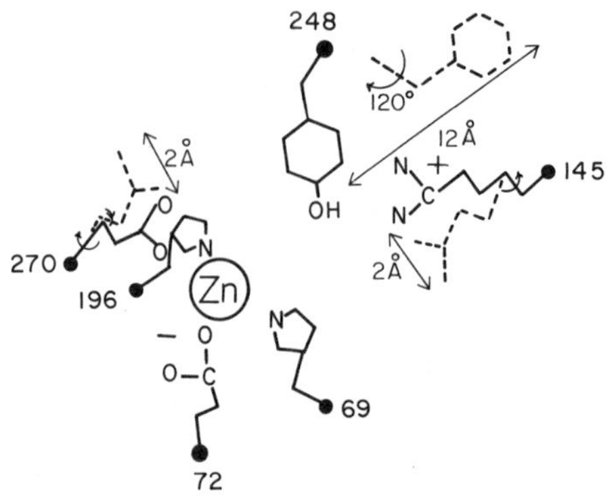

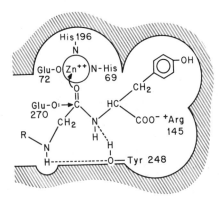

Figure 67   Schematic representation of the motions of tyrosine 248, arginine 145, and glutamate 270 induced by dipeptide substrate binding to the active center of carboxypeptidase A.

ultimate residue must be in the L-configuration; however, this group (either NH_2 or N—H) is not required for hydrolysis, since acylamino acids are hydrolyzed[46, 295]. The N—H on a substituted dipeptide works equally well for binding to the apoenzyme (p. 267). In this case the specific interaction with Glu 270 seems less likely, although X-ray data is not available on the more efficiently hydrolyzed substituted dipeptides. Lipscomb et al.[245] have postulated that in the case of the substituted dipeptides there is an additional hydrogen bond between the penultimate N—H and the oxygen of Tyr 248 (Fig. 68). The specific interaction of the α-amino group of the dipeptide with Glu 270 forming an essentially nonproductive complex might account for the observed very slow hydrolysis of dipeptides with a free amino group[245, 295].

Figure 68   Postulated features of the interaction of acylated dipeptide substrates with the active center of carboxypeptidase A suggested by the X-ray data of Lipscomb et al.[245].

The conformational changes induced in carboxypeptidase structure by substrate binding are particularly interesting and are summarized below before a discussion of possible mechanisms of action based on the present structural information and the known solution chemistry. These conformational changes around the active center induced by the binding of Gly-L-Tyr are indicated in Figure 67. The most striking change in enzyme conformation on substrate binding is the 12 Å shift in the position of the OH group of Tyr 248. This motion involves a 120° rotation of the side chain about the C_α—C_β bond, as well as some motion of the peptide backbone [245].

Spectral data in solution have suggested a change in environment for tyrosine or tryptophan residues, since spectral changes at 298 mμ have been observed when the enzyme binds β-phenylpropionate [315]. In addition, β-phenylpropionate causes a shift in the absorption band at ~ 425 mμ observed in nitrocarboxypeptidase [310]. This spectrum is due to the presence of a single nitrotyrosyl residue, the nitration of which reduces peptidase activity to < 15% of the control values [310].

As mentioned above the guanidinium group of Arg 145 moves about 2 Å when the substrate is bound. This is made possible by a rotation about the C_β—C_γ bond. The motions of Tyr 248 and Arg 145 are coordinated in interesting fashion, as described by Lipscomb et al. [245]. "In the native state there is a system of four hydrogen bonds which link Arg 145 and Tyr 248. The guanidinium group of Arg 145 is bonded to the backbone carbonyl group of residue 155. The adjacent carbonyl 154 is bonded to Glx 249, which interacts via a water molecule with the hydroxyl of Tyr 248. When Gly-L-Tyr is bound, all these groups move. The interaction between Glx 249 and Tyr 248 is destroyed. This change results in a motion of about 1 Å of the backbone of residues 248 and 249. Stability in this portion of the molecule is maintained by the presence of the disulphide bond and the persistence of certain H bond interactions."

Mechanism of Action. The static structural data can only be used to infer a mechanism of action. This structural data in conjunction with the enormous amount of physicochemical and kinetic data available on carboxypeptidase A makes a far larger number of pieces of the mechanism puzzle available than is the case with most other metalloenzymes.

Referring to Figure 68, the OH of Tyr 248 is near enough to the peptide nitrogen to donate a proton; thus this could be one of its functions in peptide hydrolysis. Donation of a proton to the peptide nitrogen by an enzyme group has appeared to be a likely feature of the mechanism [253] (Fig. 66). Acetylation demonstrates the presence of two reactive tyrosines, one or both of which are necessary for peptidase action [305, 306]. Acetylation or iodination both abolish the binding of the peptide substrates to the apoenzyme as measured by the prevention of ^{65}Zn binding or metal ion exchange [308]. Thus it seems likely that tyrosine 248 is one of these reactive tyrosines. Since nitration of one of these tyrosyls reduces peptidase activity to 15%, while reaction of the other with 5-diazo-1H-tetrazole leaves peptidase activity essentially unchanged [5, 310, 311], tyrosine 248 would seem to be the tyrosyl residue selectively nitrated. This interaction of the peptide nitrogen with Tyr 248 could also be the binding function associated with the C-terminal N—H (see p. 267). Following iodination of carboxypeptidase A with ^{131}I-labeled hypoiodite, Roholt and Pressman [305a] have isolated a peptide containing an iodinated tyrosyl residue whose iodination is prevented by the presence of β-phenylpropionate. The sequence of this peptide is isoleucyl-tyrosyl-glutam(in)ylalanyl. The tyrosyl residue is apparently residue 248. These features of the interaction of Tyr 248 and the substrate are entirely in keeping with the chemical modification of the active tyrosyl residues which is prevented by β-phenylpropionate or substrates [5, 305]. This interaction of Tyr 248 is apparently not required for ester (hippuryl-β-phenyllactate) hydrolysis, but may actually inhibit it, since O-acetylation of the active tyrosyls actually increases ester hydrolysis by 600% when HPLA is used as substrate [5, 253, 305]. It is also clear that modification of the active tyrosyl residues does not interfere with the metal-dependent ester binding, while it does seriously interfere with peptide binding [308]. It does, however, alter K_m for the ester substrate [5]. In addition to the alteration of K_m for the ester substrate, acetylation also removes the pronounced substrate inhibition by the ester substrate observed at low concentrations of the ester when the native enzyme is employed [5, 308a]. The apparent removal of this substrate inhibition by acetylation results because this chemical modification shifts the region of substrate inhibition to much higher concentrations of ester. Significant substrate inhibition of the acetylated enzyme by HPLA does not occur until $5 \times 10^{-2} M$ HPLA, rather than $10^{-3} M$ HPLA as observed in the case of the native enzyme [5]. It is not clear whether ester binding is exactly analogous to peptide binding [5, 316]. The specificity requirements do appear similar and esters are observed to be competitive inhibitors of peptide hydrolysis and vice versa, suggesting that the sites do at least overlap [5, 317]. Additional evidence indicating that at least some chemical features of the active center are common to both the peptidase and esterase sites has been obtained by Kaiser and Carson [317a], who showed that 2-phenylacetate, 3-phenylpropionate, and 4-phenylbutyrate are competitive inhibitors of both peptidase and esterase activity. The K_I values calculated for these inhibitors using esterase kinetics are very similar to the values determined using peptidase kinetics, suggesting that the same enzyme–inhibitor complex is responsible for both inhibitions.

The N—H interaction of peptides at Tyr 248 coupled with the side chain binding in the pocket and the salt bridge between the free carboxyl and Arg 145 limits the terminal residue to the L-configuration if the carbonyl of the peptide bond is going to be placed in the vicinity of the Zn(II) ion. The likely function for the Zn(II) ion would be to withdraw electrons from the carbonyl, hence potentiating attack of a nucleophile on the carbonyl carbon. The presence of such a nucleophile seems a likely requirement if one attempts to postulate a mechanism for peptide hydrolysis [5, 245, 253]. Lipscomb and co-workers have concluded that Glu 270 is in the best position to serve as this nucleophile [245]. These authors have proposed two possibilities. Oxygen of Glu 270 could attack the carbonyl carbon directly and form an anhydride intermediate which is rapidly hydrolyzed. Alternatively, Glu 270 could promote a general base mechanism via an intervening water molecule as the nucleophile. Steric factors tend to favor the anhydride mechanism, but a definitive choice cannot be made on the basis of present data [245].

The participation of the penultimate N—H group in some sort of binding interaction, probably the donor or acceptor of a hydrogen bond, has seemed likely from both kinetic data [295] and the apoenzyme binding data [45, 46] (p. 267). Since the X-ray data are available only on the substrate with a free amino group in this position, the total picture concerning this interaction is not entirely clear. The free α-amino of the dipeptide interacts with Glu-270 via a water molecule (Fig. 65). Model building shows, however, that if longer peptides are placed with all other positions identical to Gly-L-Tyr, their N-terminal amino groups cannot interact with Glu 270 via a water molecule, and since no data is available on the interaction of a substrate with a penultimate N—H, its interaction remains speculative. An NH_2 or N—H at this position on the penultimate residue is equally effective in inducing binding of the peptide to the apoenzyme (see p. 267). Lipscomb et al. [245] suggest that this N—H might hydrogen bond with the oxygen of the phenolic hydroxyl of Tyr 248. These workers also point out that this might assist hydrolysis in that it would facilitate proton transfer from Tyr 248 to the N—H of the susceptible peptide bond. Such a hydrogen bond might also produce distortion at the carbonyl carbon of the susceptible bond. Lastly, no other group which could form a hydrogen bond with the penultimate N—H is present in the X-ray structure.

Model building generated by extending the number of peptide units of the substrate suggests some interesting interactions between longer peptides and side groups of the protein in the groove of the enzyme more distal to the zinc atom. The region containing Tyr 198 and His (or Phe) 279 is in a position to interact with the R group of an N-acylated dipeptide (the best synthetic substrates). Tyr 198 may possibly be the second reactive tyrosine indicated to be present by chemical modification and may be the one specifically modified by 5-diazo-1H-tetrazole [5, 311]. Details of further interactions are covered in reference 245. A detailed study of the hydrolysis of longer peptides has been carried out by Abramowitz, Schecter, and Berger [304, 318]. One of the most interesting findings is that if the peptide is sufficiently long, the C-terminal side chain of the substrate can be removed and hydrolysis of the C-terminal peptide bond is still observed. This certainly implies additional binding interactions further down the peptide chain of the substrate.

In the case of hydrolysis or hydration reactions catalyzed by metal-containing enzymes, the participation of a metal-coordinated water molecule or hydroxide ion has had considerable appeal. Nuclear magnetic resonance studies of the relaxation times of solvent water in solutions of Mn(II) carboxypeptidase have demonstrated the presence of a coordinated H_2O [223]. These nmr data show that this water is displaced when β-phenylpropionate is bound to be Mn(II) enzyme [223] (see p. 241). The difference electron maps of the Gly-L-Tyr complex versus the non-complexed enzyme also suggest some displacement of water upon substrate binding [245], although the precision is not good enough to determine whether this includes coordinated water. Model building shows that it is difficult to accommodate a Zn(II)-coordinated H_2O or OH^- as attacking the carbonyl carbon of the susceptible peptide bond and still preserve acceptable interactions with Tyr 248 [245].

The various interactions of substrate molecules of the L-configuration at the C-terminal residue postulated from the solution studies and found in the X-ray model allow some speculation on the nature of the binding of the powerful carboxylic inhibitors such as β-phenylpro-pionate. If it is assumed that the aromatic side chains of such inhibitors occupy the same pocket in the enzyme as the side chain of the substrate, then the carboxyl of β-phenylpropionate can either point toward Arg 145 or toward the Zn(II) ion. However, if the inhibitor is asymmetric, e.g., phenylalanine, it is the D-series that contains the powerful inhibi-tors [44, 284, 295], suggesting that the carboxyl must coordinate the metal ion (Fig. 69). Nuclear magnetic resonance data on the binding of a series of these inhibitors suggests the same thing [224] (see Sect. 2.4).

The wealth of physicochemical data discussed above has allowed some detailed postulates to be made on the mechanism of action of carboxy-

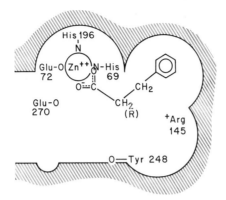

Figure 69   Model of the binding of aromatic carboxylate inhibitors to the active center of carboxypeptidase A based on chemical [44], X-ray [245], and nmr [224] data.

peptidase A. However, if one applies the dictum that the physico-chemistry must ultimately explain the observed kinetics of the reaction, then a number of features of carboxypeptidase A action still appear somewhat difficult to explain. The observed pH– rate profile for the hydrolysis of carbobenzoxy-Gly-L-Phe is approximately bell shaped, with inflection points at pH 6.7 and 8.5. The pH– rate profile for the hydrolysis of the ester benzoyl-Gly-L-phenyllactate, however, demonstrates neither of these inflections, but shows a twofold rise from pH 5.5 to 10.5. The rise, however, is steeper at the two extremes of pH, a relatively constant rate being observed between pH 7.0 and 9.0 [5, 253, 306].

If an attempt is made to explain the pH-rate profile for peptidase activity on the basis of the groups observed in the X-ray structure, it would be logical to assign the upper pK_a of 8.5 to the hydrogen donor, Tyr 248, and the lower one, 6.7, to the nucleophile, Glu 270 [245]. Both pK_a's would then have to be considered very abnormal, perhaps made so by the immediate protein environment. The lack of obvious participation in HPLA hydrolysis of the groups with these pK_a's is not entirely clear. Since modification of the tyrosine actually enhances ester hydrolysis [253, 305, 306] and does not interfere with ester binding [308], it does appear that the tyrosyl interactions are not required for HPLA hydrolysis. If the pK_a of 6.7 is assigned to the nucleophile, Glu 270, it is not clear why this pK_a does not appear in the esterase (HPLA) pH-rate profile unless this particular nucleophile is not required for HPLA hydrolysis. It was initially proposed that OH⁻ might assist in catalyzing ester hydrolysis [253], however, this has been objected to on the grounds that the rise in esterase activity (two-to threefold over 3 pH units) is too small to be compatible with such a mechanism [319].

There is a group of small ester substrates including O-acetylphenyl-lactate and O-acetylmandelate [320] which appear to require a mechanism similar to peptide hydrolysis in that the pH–rate profiles for hydrolysis are bell-shaped [320]. They are hydrolyzed much less efficiently than HPLA [320]. The pK_a's involved in the hydrolysis of O-acetylmandelate have been derived as 6.9 and 7.9 by fitting the bell-shaped pH–rate profile observed. The precise differences in mechanism implied by these findings are not clear as yet. It may be that the requirements for the activation of certain substrates (HPLA, e.g.) to a suitable transition state do not require all the interactions necessary for peptide activation.

Another general feature of carboxypeptidase A action which complicates a straightforward interpretation of mechanistic features is the anomalous kinetic behavior observed with practically all synthetic substrates for this enzyme. HPLA hydrolysis, for example, is under severe substrate inhibition, even at the lowest concentrations employed with the native enzyme [5, 321a]. Acetylation removes some of the inhibition, but it still occurs at higher HPLA concentrations, 10^{-2} or greater [5, 321]. The increase in hydrolysis rate observed on acetylation is a combination of a tenfold increase in the substrate concentration where inhibition is observed, a rise in K_m from 7.6 to $250 \times 10^{-5} M$, and an increase of V_{max} from 28.6 to 42.0×10^3 min^{-1} [5]. Substrate inhibition by peptide substrates is also great at higher substrate concentrations [321a]. Substrate inhibition would appear to occur because of the interaction of several substrate molecules with the enzyme in nonproductive modes of binding. Vallee and Riordan [5, 317] have proposed models for the modes of binding of both peptide and ester substrates to the active center. Some larger polypeptide substrates do not show these kinetic anomalies, suggesting that if the substrate is large enough it binds to the protein in such a way as to prevent the nonproductive modes of binding of additional substrate molecules [5, 318].

Another remarkable feature of carboxypeptidase A chemistry that the X-ray data has not clarified much as yet is the reason for the metal ion specificity for peptidase and ester hydrolysis outlined in Section 2.6. Some X-ray data are available on the Hg(II) enzyme which catalyzes only ester hydrolysis [245]. The Hg(II) ion displaces the native Zn(II) ion, but the final coordination of Hg(II) places its coordinates about 1.2 Å from the position occupied by Zn(II). Lipscomb et al. [245] point out that this must be related to the longer metal–ligand bonds (the covalent radii are 1.48 Å for Hg and 1.31 Å for Zn) and van der Waals contact between the metal ion and the carbon skeleton of His 196 [ionic radii are 0.74 Å for Zn(II) and 1.10 Å for Hg(II)]. Since the van der Waals radii are 1.55 Å for C and 1.2 Å for H, the authors point out that if it is assumed that the carbonyl remains bound to the metal ion, then the susceptible peptide bond of the substrate is displaced about 1 Å further away from the nucleophile, Glu 270. Such changes could radically interfere with the nucleophilic attack on the carbonyl. It could be that the metal–carbonyl interaction is all that is needed in HPLA hydrolysis. The totally inactive Cu(II) enzyme could well have a distinctive geometry because of the marked Jahn-Teller distortion preferred by this metal ion. It is not clear yet whether this distortion does in fact occur at the carboxypeptidase binding site.

3.1.2 Carboxypeptidase B.

Carboxypeptidase B as isolated from porcine or bovine pancreas has physicochemical properties that are very similar to those of carboxypeptidase A. The enzyme has a molecular weight of ~ 34,000 and contains 1g-atom at of zinc per mole, which is essential for activity[284, 321–323]. This protein is also an exopeptidase with substrate requirements like those of carboxypeptidase A, except that the specificity is directed toward arginine and lysine side chains on the C-terminal peptide residue of the substrate rather than aromatic or branched aliphatic side chains[284]. The best synthetic substrate is hippuryl-L-lysine[284]. Carboxypeptidase A apparently has no specificity for carboxypeptidase B substrates[284] and they do not appear to bind to the enzyme[45]. Carboxypeptidase B on the other hand does hydrolyze some of the substrates for the A enzyme[284]. Like the A enzyme, the B enzyme has esterase activity, hydrolyzing esters analogous to the peptide substrates, e.g. O-hippuryl-L-argininic acid[284].

Procarboxypeptidase B has a molecular weight of 57,000 and thus differs rather substantially from the much larger procarboxypeptidase A[322]. The active B enzyme also contains more cysteine, seven residues per mole[322]. There is considerable evidence that some of the sulfhydryl groups are involved in zinc binding[323].

There is not enough evidence available as yet to predict whether any of the features of the mechanism in the B enzyme are similar to those in the A enzyme. It is of significance, however, that metal substitutions analogous to those in carboxypeptidase A can be performed[321]. The substitution of Co(II) in carboxypeptidase B causes marked enhancement of the peptidase activity with hippuryl-L-lysine as the substrate[321]. Likewise, Cd(II) carboxypeptidase B has esterase activity, but is inactive in the hydrolysis of peptides[321]. This striking similarity between the A and B enzymes in the effects of metal ion substitution on the hydrolysis of their analogous substrates certainly suggests that the metal ion function may be very similar.

3.1.3 Carbonic Anhydrase. Carbonic anhydrase was the first Zn(II) metalloenzyme to be discovered[324, 325]. Keilin and Mann described many of the basic properties of the erythrocyte enzyme in 1940, including the presence of zinc, the inhibition of metal-complexing agents, and the powerful inhibition of the enzyme by sulfonamides[324, 325]. Carbonic anhydrase was initially believed to be an enzyme of great specificity, catalyzing only one known reaction, the reversible hydration of CO_2 (eq. 18). The reaction may be formulated with either water or

$$CO_2 + H_2O \underset{k_{-1}}{\overset{k_1}{\rightleftharpoons}} H_2CO_3 \rightleftharpoons HCO_3^- + H^+ \qquad (18a)$$

$$CO_2 + \bar{O}H \underset{k_{-2}}{\overset{k_2}{\rightleftharpoons}} HCO_3^- \qquad (18b)$$

hydroxide as the active species. Which of these applies in the enzyme-catalyzed reaction is not certain at present (see below). In the last few years a number of other reactions have been found to be catalyzed by this enzyme. These include the hydrolysis of p-nitrophenyl acetate [75, 326, 327] (eq. 19), the hydrolysis of 2-hydroxy-5-nitro-α-toluene-sulfonic acid sultone[83, 328] (eq. 20), and the hydration of acetaldehyde [329–331] (eq. 21), as well as a number of other aldehydes[332, 333].

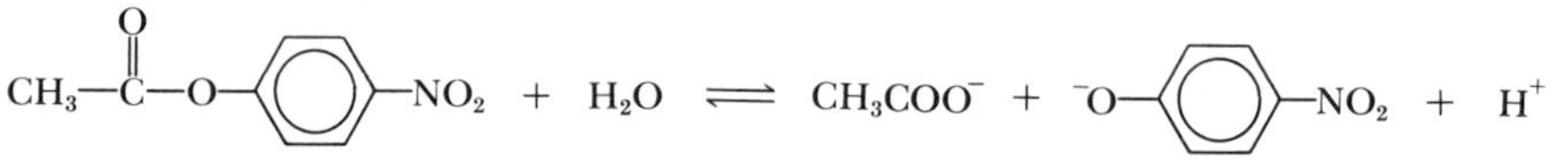

Equation (19): $CH_3{-}\overset{O}{\overset{\|}{C}}{-}O{-}\langle\!\langle\bigcirc\rangle\!\rangle{-}NO_2 + H_2O \rightleftharpoons CH_3COO^- + {}^-O{-}\langle\!\langle\bigcirc\rangle\!\rangle{-}NO_2 + H^+$ (19)

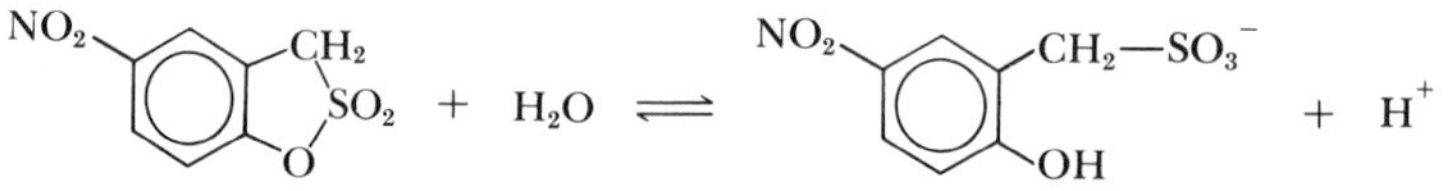

Equation (20) (20)

Equation (21): $CH_3{-}\overset{O}{\overset{\|}{C}}{-}H + H_2O \rightleftharpoons CH_3{-}\overset{OH}{\underset{OH}{\overset{|}{\underset{|}{C}}}}{-}H$ (21)

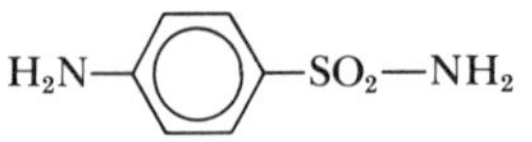

$H_2N{-}\langle\!\langle\bigcirc\rangle\!\rangle{-}SO_2{-}NH_2$

Sulfanilamide

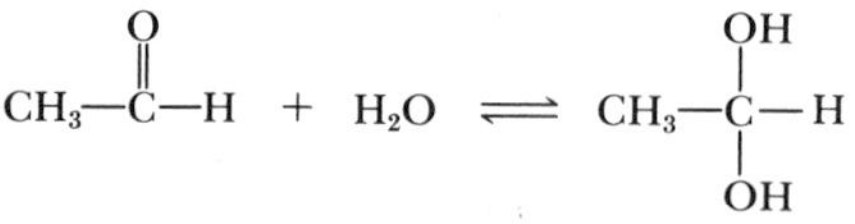

12

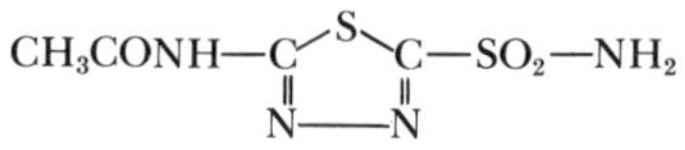

$CH_3CONH{-}C\overset{S}{\underset{N{-}\!{-}N}{\diagdown}}C{-}SO_2{-}NH_2$

Acetazolamide

13

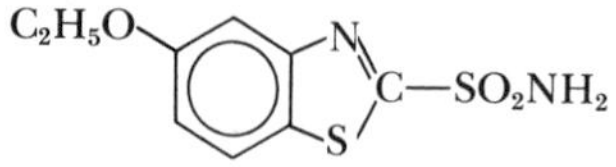

Ethoxzolamide

14

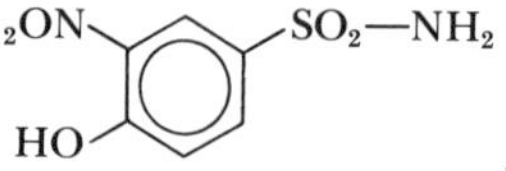

4-Hydroxy-3-nitrobenzene-sulfonamide

15

The hydrolysis of the sultone was initially described by Lo and Kaiser [328] and its hydrolysis is catalyzed extremely efficiently by the enzyme. Turnover numbers with the sultone as substrate vary from 600 to 20,000 moles/min per mole of enzyme, depending on the isozyme and species variant of carbonic anhydrase used, the pH, and the presence or absence of inhibiting anions and organic solvents[334, 335]. The rapid catalysis of the sultone hydrolysis compared to the relatively slow hydrolysis of p-nitrophenylacetate (turnover numbers from 6 to 20 moles/min per mole of enzyme) perhaps resides in certain special interactions with the active center, since the structure of the sultone is similar to that of the potent sulfonamide inhibitors shown above (**12–15**).

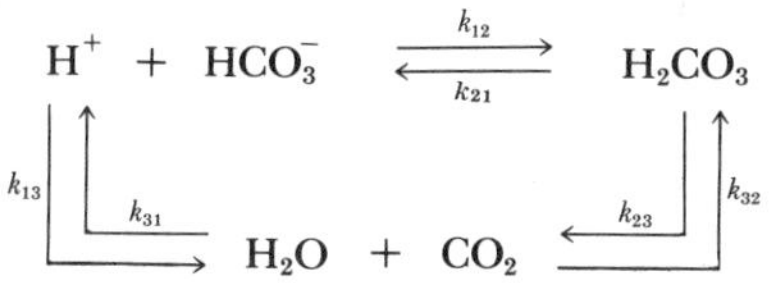

$$
\begin{array}{ccc}
H^+ + HCO_3^- & \underset{k_{21}}{\overset{k_{12}}{\rightleftharpoons}} & H_2CO_3 \\
\end{array}
\tag{22}
$$

with rate constants k_{13}, k_{31}, k_{23}, k_{32} and

$$H_2O + CO_2$$

Whether the reactions of equations 19–21 are of any physiological significance is unknown at present. The reversible hydration of CO_2 is of obvious physiological importance. The reaction proceeds relatively rapidly in aqueous solution without benefit of enzyme catalysis. The formulation of this equilibrium at pH values below 8 as proposed by Eigen et al.[336] is given in equation 22, which has the advantage of picturing both HCO_3^- and H_2CO_3 in direct equilibrium with CO_2. The kinetic constants in equation 18a defined in terms of this scheme are:

$$k_1 = k_{31} + k_{32} \quad \text{and} \quad k_{-1} = k_{13}K_a + k_{23} \tag{23}$$

where K_a is the true ionization constant for carbonic acid, $pK = 3.6$. The first-order hydration rate constant,

$$k_1 = \frac{-d[CO_2]}{dt} \Big/ [CO_2] \tag{24}$$

has been determined by several studies, and values of 0.03 sec^{-1}[337], 0.0358 sec^{-1}[338], and 0.0375 sec^{-1}[339] have been reported for 25° and 0.0434 sec^{-1}[340] for 37°. The best value for the dehydration rate constant, k, appears to be 15 sec^{-1}[338, 339] at 25°. The enzyme catalyzes the hydration of CO_2 at approximately 10^7 times the rate of the uncatalyzed reaction (see below). At pH > 10, however, the reaction between OH^- and CO_2 predominates and this second-order rate constant,

$$k_2 = \frac{d[CO_2]}{dt} \Big/ [OH^-][CO_2] \tag{25}$$

is near $8500 \sec^{-1} M^{-1}$ and k_{-2} is $2 \times 10^{-4} \sec^{-1}$[337]. This may have implications as to the mechanism of action of the enzyme (see below).

Calculations based on physiological considerations, such as the circulation time of the blood through the capillary beds of the lung alveoli, originally suggested that uncatalyzed equilibrium between HCO_3^- and CO_2 would not be achieved nearly rapidly enough to allow the required amount of CO_2 to leave the red cell during one pass through the capillaries[341, 342]. The original calculations predicted "speedy death" from complete inhibition of carbonic anhydrase[343]. More recent data suggest that the consequences might not be immediately so catastrophic[344], but carbonic anhydrase clearly plays a most significant role in the rapid adjustment of the $HCO_3^- \leftrightarrow CO_2$ equilibrium [see Maren[345] for review].

Carbonic anhydrase also plays a major role in other tissues, including the kidney, eye, pancreas, and stomach, where the enzyme is associated indirectly with ion transport. Fundamentally its role arises from the fact that it catalyzes a reaction that allows for the uptake or release of H^+ and OH^-. These ions are then involved in various transport processes characteristic of the cell which may result in bicarbonate secretion (pancreas) or acid secretion (stomach).

The physiological, pharmacological, and chemical aspects of carbonic anhydrase have been thoroughly and elegantly reviewed recently by Maren[345], and the intention of this section is to summarize the physicochemical data on the pure enzyme which bear directly on the mechanism of action.

In the early 1960's several laboratories developed techniques for the preparation of highly purified homogeneous erythrocyte carbonic anhydrase[346–352]. Several of these have now been crystallized [91, 357]. The pure enzyme has been prepared from human[346, 347], monkey[91], bovine[349, 350], equine[353], canine[354], and elasmobranch[335] erythrocytes. The initial purification of the human enzyme was developed by research groups at Harvard[346], Marseilles[347] and Uppsala[349, 350], and these groups made the interesting discovery that the human red cell enzyme exists as two isozymes, designated B and C, which differ rather significantly in amino acid composition[346–350]. Both are globular proteins of molecular weight $\sim$30,000 and contain 1 g-atom of Zn(II) per mole.

The purification of the bovine enzyme by the groups in Uppsala and Göteborg also revealed a protein of molecular weight $\sim$ 30,000. On chromatography and electrophoresis the bovine enzyme can be resolved into several species, but these do not appear to differ in amino acid composition or activity[349, 350]. The two major species have been designated A and B. Thus far only primates appear to possess the two isozymes differing in amino acid composition, designated B and C [351, 352]. Isozymes almost identical to the two human isozymes have been crystallized from the erythrocytes of the *rhesus monkey* [91].

The C isozyme is present in both human and monkey red cells in much smaller quantity than the B isozyme, but the C enzyme has a much higher specific activity [355]. Specific activities expressed in terms of units equal to the quantity of enzyme required to catalyze the hydration of 1 μmole of CO_2 per minute at 25° are 44,000 units per milligram of protein for the B isozyme and 1,300,000 units per milligram of protein for the C isozyme [355]. The latter figure corresponds to a turnover number of 6×10^5/sec or 36×10^6/min. This makes the hydration of CO_2 by carbonic anhydrase the most rapid enzyme reaction known and must be remembered in considering possible mechanisms for carbonic anhydrase (see below).

Zn(II) in carbonic anhydrase is not in rapid equilibrium with ionic zinc in the environment as documented in Section 2.1. Treatment of the enzyme with 1,10-phenanthroline at pH 5.5 over a period of 15 days, however, is successful in removing the Zn(II) [33, 356]. The resultant apoenzyme is completely inactive. The apoenzyme is stable and can be completely reactivated by readdition of metal ions. Some of the properties of the metallocarbonic anhydrases prepared from Mn(II), Co(II), Ni(II), Cu(II), Cd(II), and Hg(II) have been described in Section 2. Although all the above metal ions occupy the metal-binding site, only the Zn(II) and Co(II) complexes have the characteristics required for catalysis [35, 74, 75] (see Sect. 2.6). Possible reasons for this involving a postulated mechanism of action and relating to the coordination chemistry and ease of exchange of monodentate ligands have been discussed in Section 2.6.

The X-ray studies on the structure of human carbonic anhydrase C in the crystalline state by Strandberg and co-workers [48, 73, 77, 357] have now progressed to the stage of 2 Å resolution. Some of the 2 Å as well as the earlier 5 Å, data now make it possible to discuss the solution data bearing on the mechanism of action in terms of known structural features of the enzyme. The data on the structure at 2 Å resolution were most graciously supplied by Dr. Bror Strandberg and Dr. Anders Liljas prior to publication [73].

The general structure of the molecule at 5 Å resolution is shown in Figure 70, based on a polystyrene model of the electron density maps [77]. The Zn(II) ion is located near the center of the molecule in a deep crevice. *N*-Terminal and *C*-terminal regions of the polypeptide chain were tentatively identified at positions marked 1 and 33 respectively.

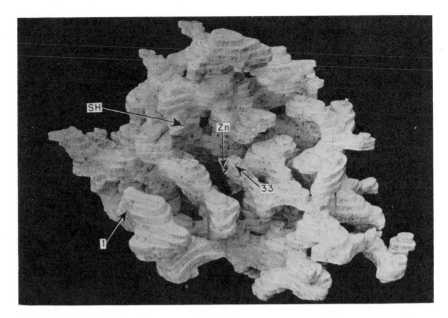

Figure 70   Polystyrene model of human carbonic anhydrase C based on the electron density map at 5.5 Å resolution. From Fridborg et al. [77], by permission.

The identification of the N-terminal has been found to be correct in the 2 Å map [73]. The position of the single free SH group of the C-isozyme has been identified from the difference between the enzyme electron density maps of the native enzyme and the enzyme reacted with acetoxy-mercurisulfanilamide. Two molecules of acetoxymercurisulfanilamide react with carbonic anhydrase, one at the active center and one at the free —SH group as an organic mercurial [77]. A model of this complex is shown in Figure 71, which was built by placing atomic models of the inhibitor on the electron density model of the enzyme according to the positions indicated by the densities in the difference map [77]. The —SH group is 14 Å away from the Zn(II) ion.

The 2 Å map shows the center of gravity of the sulfonamide to be 3.2 Å away from the zinc ion. This is within binding distance and is certainly compatible with coordination of the *nitrogen* of the sulfon-amide group to the Zn(II) ion. This is not unexpected in terms of the absorption and CD spectra of the Co(II) enzyme (see Sect. 2.2) and the metal-dependent binding of sulfonamides (see Sect. 2.6).

The position of the Hg atom on the benzene ring of the acetoxy-mercurisulfonilamide is considerably farther away from the Zn(II) ion and would place the ring portion of the sulfonamide in the crevice

Figure 71   Polystyrene and molecular model of the complex of human carbonic anhydrase C with acetoxymercurisulfanilamide based on the electron density map at 5.5 Å resolution. From Fridborg et al. [77], by permission.

leading to the zinc ion [77]. The low dissociation constants observed for sulfonamide complexes of carbonic anhydrase in solution (10^{-5} to $10^{-9} M$) [345, 358–361] suggest bonding contacts of some sort between the ring portion of the molecule and the enzyme in addition to the metal ion coordination. Changes in absorption spectra and fluorescence of bound sulfonamides suggest that the binding cavity is hydrophobic [212, 334]. However, the base of the cavity around the Zn(II) ion may be quite hydrophilic.

A view of the electron density map at 2 Å resolution near the Zn(II) ion is shown in Figure 72a. Three groups from the protein appear to be coordinated to the metal ion. All three ligand electron densities can be fit adequately by histidyl residues as shown in Figure 72b overlaying some of the postulated structural features of the active center onto the electron density map. Since the amino acid sequence is not known, these identifications must be considered tentative. It is interesting that the two ligands on the left, tentatively identified as histidyl residues, come from the same section of the main chain, being separated by a single residue.

In addition to the three ligands described there is a fourth density approaching the Zn(II) from above and slightly to the right in Figure 72b. This is less dense than the protein side chains and probably belongs to the solution, possibly a water molecule. It is from the direction occupied by this density that the sulfonamide inhibitors bind. The position of the sulfonamide in the difference map is marked in Figure 72b by J_3. It appears as if the light density is displaced from the Zn(II) when the sulfonamide binds. A considerable amount of data on the enzyme in solution suggests that a coordinated H_2O or OH^- occupies this fourth position in the active enzyme (see below).

The precise geometry around the Zn(II) ion cannot be determined with certainty even from the electron density map at 2 Å resolution. Precise model building may improve the precision. While it would appear to approach a tetrahedral geometry, the spectrum of the Co(II) enzyme, for example (see p. 188), certainly distinguishes the d-orbital splitting from that observed in Co(II) carboxypeptides A, although the electron density maps indicate grossly similar geometries in the two enzymes (see Fig. 64). Data from spin-labeled sulfonamide complexes indicate that the conformations of the Zn(II) and Co(II) enzymes must not differ too greatly[209] (p. 236). Other features, such as the metal exchange reactions, also distinguish the complex at the active site of carbonic anhydrase from that at the active site of carboxypeptidase A (p. 178).

Some of the general features of the protein structure of carbonic anhydrase are evident from the 2 Å map to which a model of the polypeptide backbone has been fitted. Four small pieces of right-

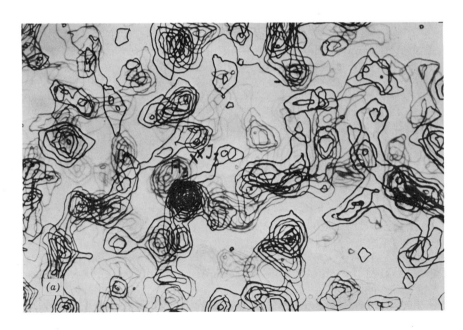

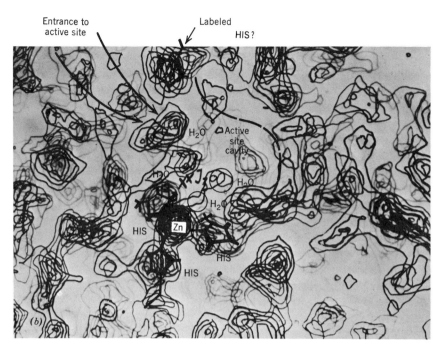

Figure 72(a) Electron density map at 2 Å resolution around the Zn(II) binding site in human carbonic anhydrase C. From Liljas et al.[73], by permission. (b) Electron density map of the active center of carbonic anhydrase with tentative identities of the densities indicated. Zinc is coordinated to three histidyl residues, in two cases through the 3-nitrogen and in one case through the 1-nitrogen. A water molecule occupies a fourth coordination site. A fourth histidine is located at the entrance to the active site cavity and may be the residue reported to be modified by bromopyruvate in the C isozyme[73]. The densities between this histidine and the zinc ion belong to the solvent and are apparently water molecules.

handed helix of about 2 turns each have been identified. Only one of these appears to be α-helix; the others are distorted and apparently closer to a 3_{10}-helix. This small amount of α-helix is of interest, since both the ORD and CD spectra of most isozyme and species variants of carbonic anhydrase have indicated very little α-helix content[25, 362–364]. Although both the primate C and bovine B isozymes have negative ellipticity in the region of 222 mμ, the major negative bands are located between 210 and 220 mμ[84]. The interpretations of these spectra have had to be qualified because of the large ellipticity bands associated with the aromatic chromophores[25, 362].

The predominant feature of protein secondary structure evident from the 2 Å map is a large region of β-structure extending through the center of the molecule. At least eight parts of this run mainly anti-parallel to each other and form a twisted pleated sheet. Thus the large hydrophobic core of the molecule is largely pleated sheet and bears a striking resemblance to the general structure observed in carboxy-peptidase A (see Sect. 3.1.1). The three "histidyl" ligands to the Zn(II) ion are from a portion of the main chain associated with the β-structure. It would appear that the major ellipticity bands in the ultraviolet CD must be associated with the β-structure.

Mechanism of Action. The pH–rate profile for carbonic anhydrase appears to be a single sigmoid curve describing the ionization of a group with a pK_a that varies between about 6.9 and 8.4, depending on the ionic environment, the buffers and particular substrate employed, and the species or isozyme variant of the enzyme. This pK_a is extremely sensitive to the anion environment (see Fig. 13), which may explain some of the observed variation. This same ionization appears to be the one affecting the transformation of the spectrum of the cobalt enzyme from its acid to alkaline form, as discussed in detail in Section 2.2.1. There is evidence from the complexometric titration data (see p. 172) that this pK_a represents that for a coordinated water molecule [37], and nmr data on the active Co(II) enzyme clearly show the presence of a coordinated H_2O molecule [228]. The observed pK_a has been considered somewhat low for a metal-coordinated water; however, the only models are the aquo species which do not correspond to a mixed ligand chelate complex with one open coordination site occupied by the water. Such a ligand is likely to lower the pK_a of the coordinated water.

A hydration mechanism involving the attack of a coordinated OH^- on the CO_2 molecule has been an attractive postulate [37, 82, 365, 366]. The minimal species involved in such a mechanism are indicated in equation 26.

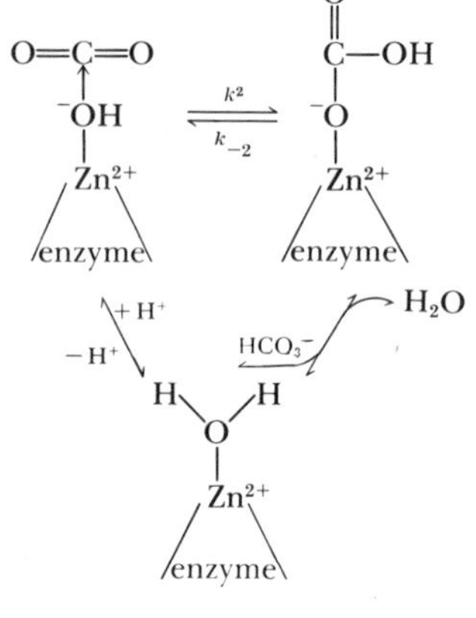

$$\text{(26)}$$

A mechanism as formulated in equation 26 would explain the lack of inhibition by weakly binding anions at high pH[81], the reversion at high pH of the spectra of the cobalt enzyme–anion or sulfonamide complexes to that typical of the alkaline form of the uninhibited enzyme [81], and the displacement of the pH–rate profile to higher pH in the presence of anions[81]. All can be related to competition with OH^-, which at high enough concentration displaces the anions and generates the active enzyme. The above mechanism is compatible with the kinetic evidence which shows the anion binding site to be coupled to a group, the basic form of which is essential for the hydration of CO_2 and the acidic form essential for the dehydration of bicarbonate[367, 368].

Infrared data on the carbonic anhydrase–CO_2 complex at pH 5.0[82] (see p. 206) show that CO_2 is bound in the hydrophobic cavity in an unstrained, hence, linear configuration[82]. The spectrum of the Co(II) enzyme at pH 8.2 in the presence of bicarbonate indicates that a bicarbonate complex is present in high equilibrium concentration at this pH[25, 84]. The first-order rate constant, k_2, for the enzyme-catalyzed hydration step is 4 to $6 \times 10^5 sec^{-1}$ at pH 7 25°[355]. As noted above this is seven orders of magnitude faster than the uncatalyzed hydration of CO_2, but only two orders of magnitude higher than the reaction of OH^- and CO_2 at pH > 10, where $k_2 = 8.5 \times 10^3 sec^{-1} M^{-1}$. Wang[369] has estimated the pseudo-first-order rate constant for a hypothetical system in which an OH^- ion is placed next to a CO_2 molecule and has derived a rate constant of $4 \times 10^5 sec^{-1}$, which is of the order of magnitude shown by the carbonic anhydrase reaction. The lowering of the K_a for H_2O from $10^{-15.7}$ to $10^{-7 \text{ to } -8}$ would, of course, meet the requirement that the OH^- and CO_2 reaction be the predominant one near neutral pH. One further conclusion based on the relative nucleophilicities of coordinated and free hydroxide ions has been pointed out by Wang[369]. While the ratio of nucleophilicities of the two species may not coincide with the ratio of their K_a values, one would not expect a difference of several orders of magnitude. This suggests that if the above features of the mechanism are correct, the nucleophilicity of the coordinated OH^- does not fall to the extent indicated by the fall in K_a from $10^{-15.6}$ to $10^{-7 \text{ or } 8}$, or else some other feature of the active center expidites the reaction.

While the above postulated features of the carbonic anhydrase reaction may seem self-consistent, some difficulties arise when the required proton transfers are considered. In the hydration reaction, the step from OH^-—Zn and CO_2 to the bicarbonate intermediate characterized by k_2 in equation 26 must be accompanied by proton transfer. This transfer between adjacent oxygen atoms may well be extremely rapid, although the precise mechanism, sequence, or nature of the transition state may be speculated upon [369].

A more serious difficulty arises when the other proton transfer is considered. At some point in the reaction a proton must be transferred to the solvent. In the formalism pictured in equation 26, this appears as the loss of a proton from a solvent or coordinated water molecule to regenerate the active $Zn-OH^-$. Based on model systems, the limiting value of the diffusion-controlled rate of proton transfer approaches $10^{10}-10^{11} M^{-1} sec^{-1}$ [370]. Since H_3O^+ is a strong acid and OH^- a strong base there is a gain in free energy when a proton transfer occurs under the conditions $pK_{H_3O^+} \ll pK_{HX}$ or $pK_{H_2O} \gg pK_{HX}$, where X is the acceptor. Under these conditions, the transfer approaches diffusion-controlled rates. If the pK_a of the acid formed is not higher than that of H_3O^+ then the proton transfer can no longer be diffusion controlled. Eigen has presented the dependence of $\log k$ (the rate constant for proton transfer) on ΔpK (pK-difference for donor and acceptor) [370].

The rate is diffusion controlled only if the acceptor has a pK_a 2–3 pH units higher than the donor. On the other hand, if the donor pK is above that of the acceptor the $\log k$ shows a linear decrease with increasing ΔpK. Using the lowest figure for the pK_a of the postulated coordinated water molecule in carbonic anhydrase, $pK_a = \sim 7$, donation of the proton to solvent H_2O involves an acceptor pK for H_3O^+ of ~ -1.7. Thus direct transfer of the proton from the coordinated H_2O to solvent involves a ΔpK of ~ -9 and suggests that even under the most favorable of circumstances k should not be much larger than 10^2 or $10^3 sec^{-1}$. However, the observed pseudo-first-order rate constant for the hydration reaction catalyzed by the C enzyme is $10^5 sec^{-1}$. Transfer of the proton to an adjacent imidazole nucleus to produce an imidazolium ion has been suggested by several investigators [329–333, 371]. This does not get around the dilemma appreciably, however, since the transfer of the imidazolium proton to solvent water also occurs with a rate constant of $\sim 10^3/sec$ [370]. There do appear to be histidyl residues near the active center, as is discussed below.

In association with their studies on the hydrolysis of the sulfonate ester (eq. 20), Kaiser and Lo [83] have proposed a cyclic mechanism involving the coordinated OH^- that avoids the transfer of a proton to the solvent. The mechanism is pictured in equation 27. Hydrolysis of this substrate by the bovine enzyme follows a sigmoid curve with a pK_a of 7.28 in a combination of buffers [83], while a pK_a of 7.8 has been observed for the human B isozyme in Tris–SO_4 buffer [335]. Kaiser and Lo in the mechanism pictured have interpreted this pK_a as that for a coordinated water [83]. An additional solvent water is included to

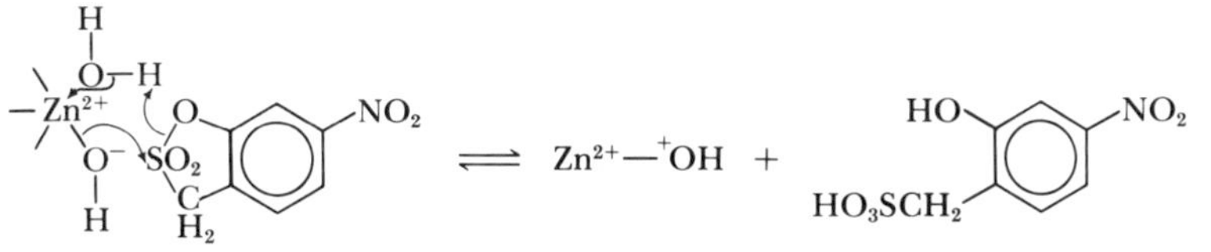

$$
\text{[structure]} \quad \rightleftharpoons \quad Zn^{2+}-{}^{+}OH \ + \quad \text{[structure]} \qquad (27)
$$

complete the cycle. Such a mechanism could also circumvent the proton transfer in the hydration reaction if the immediate product were carbonic acid rather than bicarbonate using solvent water as the source of the second hydrogen as indicated in equation 28. Since

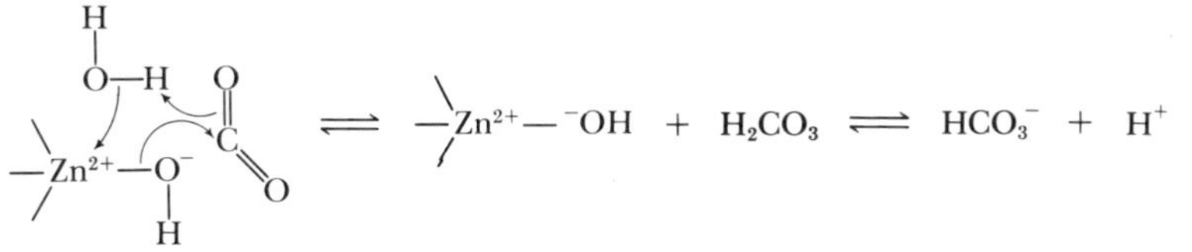

$$
\text{[structure]} \quad \rightleftharpoons \quad -\overset{|}{\underset{|}{Zn^{2+}}}-{}^{-}OH \ + \ H_2CO_3 \ \rightleftharpoons \ HCO_3^- \ + \ H^+ \qquad (28)
$$

H_2CO_3 has a pK of 3.6, the transfer to solvent presents no problem. Zn(II) coordination, of course, would help increase the acidity of the water protons.

The above type of mechanism has been criticized on the basis that the substrate in the reverse (dehydration) direction must be carbonic acid. Thus the rate of protonation of HCO_3^- would enter. Several arguments suggest that this protonation would not be fast enough at high enzyme concentrations for the reaction to be first order in enzyme [372]. Furthermore the rate of encounter of carbonic acid and enzyme, assuming the rate to be diffusion limited, is not fast enough to account for the enzymatic rate [368, 372] [hydration and dehydration by the enzyme are about equal at pH 7.05 [355, 368]. DeVoe and Kistiakowsky have previously presented kinetic arguments against H_2CO_3 as the substrate in the dehydration reaction [373].

Caplow has suggested a concerted reaction for the dehydration mechanism involving C—O bond cleavage [372] (eq. 29). C—O bond cleavage rather than ligand exchange has been observed in models involving pentamine cobaltic compounds [374] and has been mentioned

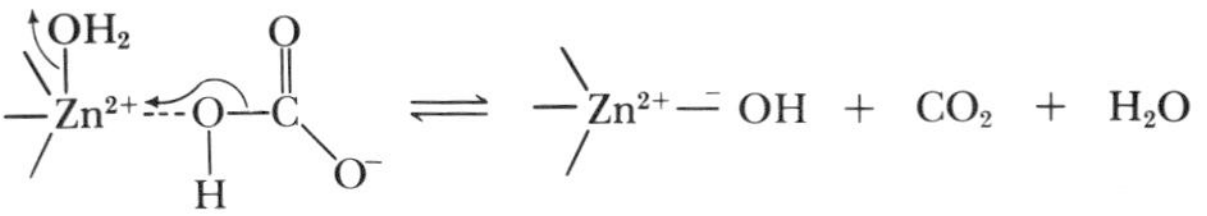

$$
\text{[structure]} \quad \rightleftharpoons \quad -\overset{|}{\underset{|}{Zn^{2+}}}-{}^{-} OH \ + \ CO_2 \ + \ H_2O \qquad (29)
$$

before in connection with possible carbonic anhydrase mechanisms [25]. Such a scheme, while attractive for the dehydration mechanism, does not circumvent the dilemma of the proton transfer in the hydration direction.

A mechanism has been proposed by Dennard and Williams[79] involving a nitrogen of the protein as the group attacking the CO_2 carbon with the intermediate formation of an unstable carbamate (16).

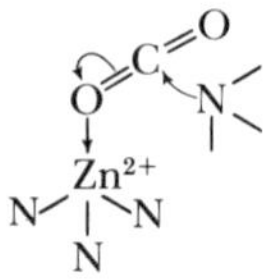

16

This has the drawback that known mechanisms of carbamate breakdown proceed by decarboxylation[375], hence a product bicarbonate would seem unlikely.

It does seem clear at the present time that the mechanism of proton transfer required in the hydration–dehydration reaction is a central problem in any mechanism. The rate of proton transfer may be the rate-limiting step in the hydration–dehydration reactions. The esterolytic reactions are considerably slower. Although they may proceed by the same mechanism, it is not a foregone conclusion that all features of the mechanism are the same.

The assumption has been made in the above discussion on the mechanism that proton transfers occur as observed in model systems. It should be kept in mind that structured or icelike water molecules near the protein surface might influence the rate of proton transfer. If some concerted or cooperative mechanism of proton transfer is involved, rates of transfer may be modified compared to proton transfer in simple model systems[370].

Amino Acid Residues Near the Active Site. Human carbonic anhydrase B is inactivated by iodoacetate to approximately 10–20% of its native activity[376, 377]. From this inactivated enzyme Bradbury has isolated a carboxymethylhistidine incorporated in the sequence Thr-His-Pro-Pro-Leu[378]. The irreversible reaction of iodoacetate with the enzyme first requires that it be bound in reversible form to the active center, a reaction that depends on the presence of the metal ion[376, 377]. The absorption and CD spectra of the Co(II) enzyme in the presence of acetate and iodoacetate suggest that both are bound in similar fashion and coordinate the Co(II) ion[379]. Iodoacetate then reacts irreversibly with the enzyme at a relatively slow rate[376–379]. Monkey carbonic anhydrase B requires up to 12 hr in the presence of $10^{-2} M$ iodoacetate to be completely irreversibly labeled with ^3H iodoacetate[379].

Once the enzyme is labeled, however, both CN^- and large sulfonamides can still be bound at the active site[379]. Cyanide reverts the

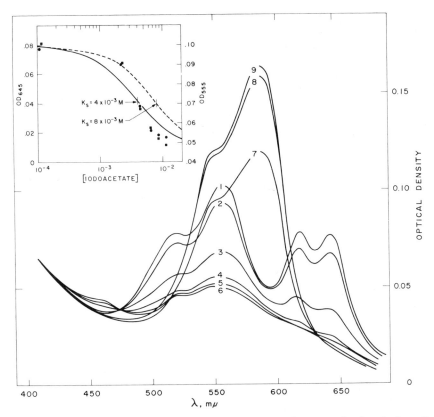

Figure 73   Effect of iodoacetate on the spectrum of Co(II) human carbonic anhydrase B at pH 9.0 (curves 1–6). Effect of CN⁻ on the spectrum of carboxymethylated Co(II) human carbonic anhydrase B (curves 7–9). Curves 1–6 are spectra taken with 0 to $2 \times 10^{-2} M$ iodoacetate as indicated in the insert (●, $O.D._{.555}$) (■, $O.D._{.645}$). K_s for iodoacetate is between 4 and $8 \times 10^{-3} M$. After 24 hrs the enzyme irreversibly reacted with iodoacetate was titrated with CN⁻ (7,1 equiv; 8, 2 equiv, and 9, excess CN⁻). Data from Coleman [379].

spectrum of the iodoacetate-Co(II) enzyme to that observed for the CN⁻ complex of the Co(II) native enzyme [379] (Fig. 73). Hence the coordination complex does not appear modified by the histidine modification. These findings have been synthesized in the model shown in Figure 74 and suggest that the carboxyl coordinates the metal ion and that the imidazole nitrogen is in a position to attack the iodinated carbon. This would place this nitrogen at least three bond distances from the metal. While this may be close enough to participate in proton transfers, this histidine has been shown to have a pK_a of 5.8 and only the basic form reacts with iodoacetate. This appears too low for the pK_a

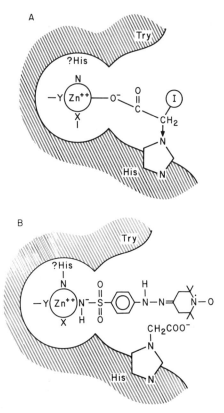

Fig. 74 (a) Model for the interaction of iodoacetate with the active center of human carbonic anhydrase B. (b). Model of interaction of a spin-labeled sulfonamide with carboxymethylated human carbonic anhydrase B.

involved in enzyme activity[377]. On the other hand, the reversible binding depends on a group with a pK_a estimated indirectly to be 7, suggested to be the pK_a of the coordinated water[377]. The complete carboxymethylated enzyme showing 10% of the activity continues to show a sigmoid pH–rate profile with a pK_a of 8.5. The latter also can be interpreted as the pK_a for a coordinated water on the modified enzyme [377, 378].

Sulfonamides, while still binding to the modified enzyme, have an altered less nonpolar environment, as shown by CD studies of a bound azosulfonamide[379] (see Figure 74B).

The reaction with iodoacetate shows considerable variation between isozyme and species variants of carbonic anhydrase, as shown in Figure 75 using ^{3}H-iodoacetate to label 4 species and isozyme variants of the

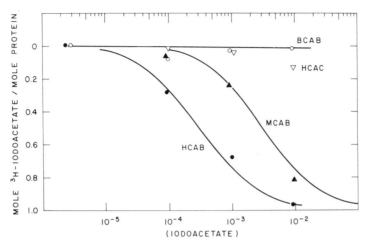

Figure 75   Reaction of species and isozyme variants of carbonic anhydrase with ³H-iodoacetate. Data from Coleman[379].

enzyme. At $10^{-2} M$ iodoacetate 1 mole of iodoacetate reacts irreversibly with the enzyme after 24 hr. The monkey B enzyme reacts almost as readily. No reaction with the bovine B enzyme is observed under these conditions and very little with the human C enzyme. Thus if this histidine is present in all the species and isozyme variants of the enzyme, it is much less reactive in some of them.

Sulfonamide inhibitors in which the nitrogen of the inhibitor is substituted are very poor inhibitors of carbonic anhydrase[345]. Whitney et al.[380] found, however, that while inhibitors of type (17) showed little if any binding, the chloroacetylation of a cyclothiazide such as (18)

$$CH_3\text{—}\bigcirc\text{—}SO_2\text{—}NH\text{—}CO\text{—}CH_2\text{—}Cl$$

17

$$\underset{R}{\overset{H}{\text{N}}}\overset{O_2}{\underset{N}{\text{S}}}\text{—}\bigcirc\text{—}SO_2\text{—}NH\text{—}CO\text{—}CH_2\text{—}Cl$$

18

resulted in an increase of K_i from $6 \times 10^{-7} M$ for the parent compound to $6.5 \times 10^{-5} M$ for the modified compound. Thus this type of compound still retains some affinity for the active site. Its binding is still metal-ion dependent[380], although the correspondence in mode of binding to unsubstituted sulfonamides cannot be judged. Bound compound **18**, however, leads to slow irreversible inactivation of human carbonic anhydrase B. The resultant inactivated enzyme contains a residue of carboxymethylated histidine. Peptide isolation and amino acid analysis suggest, however, that this is not the same histidine that reacts with iodoacetate [380].

Phosphorescence studies by Galley and Stryer [381] indicate that a tryptophan side chain must be located very close to the sulfonamide binding site in the bovine enzyme.

The two human isozymes and the bovine isozyme all have a methionyl residue at position 20 or 21 from the C-terminal end of the peptide chain [382–384]. Cyanogen bromide thus cleaves the proteins between residues 20 and 21 (or 19 and 20 [382–384] in the human B isozyme). The sequences of the resultant C-terminal peptides have been determined with the interesting result that the human C isozyme seems more closely related to the bovine enzyme than to the human isozyme B. The latter isozyme contains Phe as the C-terminal residue, while both the others contain C-terminal Lys, which appears to be an additional residue at the C-terminal.

3.1.4 Alkaline Phosphatase.

Alkaline phosphatase catalyzes the alkaline hydrolysis of phosphate monoesters, as illustrated in equation 30 by the hydrolysis of p-nitrophenyl phosphate. The enzyme is widely

$$NO_2-\langle\bigcirc\rangle-O-\overset{\overset{O}{\|}}{\underset{\underset{O_-}{|}}{P}}-O^- + H_2O \rightleftharpoons NO_2-\langle\bigcirc\rangle-O^- + HPO_4^{2-} \quad (30)$$

distributed in both animals and bacteria and is needed to release inorganic phosphate from organic phosphates when the former is required for other synthetic purposes, for example, bone formation in mammals.

The enzyme has been isolated in crystalline form from gram-negative (*E. coli*) and gram-positive (*B. Subtilis*) bacteria [385, 386]. It has also been partially purified from mammalian sources [387–389]. Substrates include phosphate monoesters of primary and secondary aliphatic alcohols, sugar alcohols, cyclic alcohols, phenols, and mono, di-, and triphosphomononucleotides. In addition, the enzyme hydrolyzes the terminal phosphate of oligonucleotides, the phosphate groups in phosvitin and casein, and phosphocellulose [254, 390]. It hydrolyzes pyrophosphates, short chain polyphosphates, and polymetaphosphate, and it very slowly hydrolyzes highly polymerized polymetaphosphate. Orthophosphate is also a substrate for the enzyme; the enzyme was shown to catalyze the hydrolysis of ^{18}O-orthophosphate, although the turnover is very slow compared to that observed for normal ester hydrolysis[269]. Alkaline phosphatase catalyzes the hydrolysis of S-substituted monoesters of phosphorothioic acid, for example, cysteamine S-phosphate to give orthophosphate and the corresponding thioalcohol[391]. It does not catalyze the hydrolysis of phosphodiesters. Thus the enzyme shows no specificity towards the organic ester moiety, but is quite specific in that it requires the divalent monosubstituted phosphate moiety[269].

The early studies carried out on intestinal enzyme showed that the P—OR bond is cleaved. The work demonstrated catalyzed exchange of ^{18}O between water and inorganic phosphate and demonstrated incorporation of ^{18}O into the phosphate moiety of glucose-1 phosphate, ATP, AMP, β-glycerol-phosphate, or phenylphosphate, although exchange for inorganic phosphate was relatively slow[268, 392]. The evidence is consistent with a Walden inversion type displacement in which the P—OR bond is broken simultaneously with the formation of a P—OH or P— enzyme bond. For the E. coli enzyme a similar cleavage is suggested and the mechanism proposed includes the formation of a phosphoryl enzyme intermediate.

Enzymatic synthesis of phosphomonoesters in the presence of inorganic phosphate and alcohol was demonstrated by Meyerhof and Green[393] in 1949 using the impure intestinal enzyme. Studies later demonstrated phosphotransferase activity by the purified intestinal enzyme which is carried out much more rapidly than direct synthesis [394]. Transphosphorylation catalyzed by E. coli alkaline phosphatase has been verified[395, 396]. A good acceptor must contain a hydroxyl group and a second hydroxyl group or amino group; a quaternary amino group will not serve, but a tertiary amine is functional. The two groups must not be separated by more than one carbon atom, thus good acceptors are Tris, ethanolamine, and glucose, but 4-amino butanol-1 is inactive. Active donors include p-nitrophenyl-phosphate, pyrophosphate, mono-, di-, and triphosphomononucleotides, and other sugar phosphates. The biological significance of the trans-phosphorylation phenomenon is questionable. Very high concentrations of acceptor are necessary to induce this activity and still the ratio of phosphotransferase to phosphohydrolase activity is only about 0.10.

Ionic substances increase the rate of hydrolysis of phosphomonoesters by alkaline phosphatase[396]. Magnesium salts, manganese salts, and sodium salts activate to about the same degree when present at similar ionic strengths. Ammonium salts activate at concentrations under 1 M, but at higher concentration produce inhibition. There is some indication that this is true of Tris also, although since Tris is an acceptor in the transphorylation reaction this must be carefully interpreted [396].

The effect of magnesium as activator is of particular interest since it has been suggested that at millimolar concentrations of magnesium the substrate for the enzyme is the magnesium complex of the phosphate ester [387]. Mg^{2+} also has a very definite effect in stabilizing the enzyme against heat denaturation [254]. In view of the fact that any ionic substance has an effect on the rate, it seems likely that Mg^{2+} also operates by an ionic strength mechanism.

The pH–rate profiles which have been determined for the enzyme under various conditions of ionic strength and buffer species show sigmoid-shaped profiles with inflection points between 7 and 8. pH–rate profiles have been given for several substrates, including p-nitrophenylphosphate [254, 391, 395], β-glycerol phosphate [269], cysteamine S-phosphate [391], and pyrophosphate [396], and for transphosphorylation activity from pyrophosphate to glucose [396]. All display an inflection point between pH 7 and 8. In the cases where Tris is used as a buffer and the ionic strength is not maintained the activity reaches a maximum near pH 8 and then decreases. This decrease may be due to loss of the activating effect of Tris upon dissociation of its cationic species, thus obscuring the true pH dependence.

Alkaline phosphatase from *E. coli* has been the most extensively studied highly purified alkaline phosphatase. The enzyme is located between the cell wall and cell membrane of these bacteria and is released into the medium when the cells are subject to osmotic shock [385]. This gives rise to a rapid and convenient method of preparation [385]. The enzyme was discovered to be a Zn(II) metalloenzyme by Plocke et al. in 1962 [255]. The first physicochemical studies of the *E. coli* enzyme were performed by Garen and Levinthal [254], who showed the active enzyme to be a dimer of molecular weight approximately 80,000 at alkaline pH [254]. Between pH 4 and 2 the dimer dissociates into monomers of molecular weight $\sim 40,000$ [254]. Peptide maps of the trytic peptides from digests of the protein suggest that the monomers are identical in primary structure [397]. There appears to be one structural gene for the enzyme. Recent X-ray data (see below) also suggest that the subunits are identical [398]. Monomers prepared at pH 2 can be reassociated into active dimers by readjusting the solution to neutral pH [258, 399], even though ORD and CD studies show the monomer to be extensively unfolded at pH 2. The native conformation appears to be completely restored at neutral pH [26, 256].

Two Zn(II) ions are associated with the active dimer [26]. As many as eight metal ions can be bound by the protein [26], but there appear to be only two metal binding sites of an unusual and distorted coordination geometry, as outlined in connection with the discussion of Co(II) spectra in Section 2.2. Monomers which bind the metal ion at neutral pH have not been prepared. Several mutant enzymes are monomeric at neutral pH, but both activity[400–401] and metal binding are defective[51, 402]. It may be that the coordination site is fully formed only in the dimer.

Extensive studies of the substrate specificity of the enzyme have revealed that there is little specificity for the R group on the alcohol portion of the ester[271]. A great variety of phosphate monoesters with widely different R groups, such as p-chlorophenyl phosphate and β-naphthylphosphate, are hydrolyzed with the same maximum velocity and have the same K_m values[271].

This finding suggests that the rate-limiting step in the hydrolysis reaction involves an intermediate which does not contain R[271]. The transient formation of a phosphate ester with a group on the enzyme has been an attractive postulate to account for the above findings[271]. If such a phosphoryl enzyme is formed, and its subsequent hydrolysis is the rate-limiting step in the reaction, then its initial formation should be accompanied by a burst of the alcohol. A burst is observed, but only at pH 6 and below where the enzyme is relatively inactive[264–266].

Incubation of the enzyme with ^{32}P-orthophosphate or ^{32}P-labeled substrates at low pH results in the covalent labeling of the enzyme by phosphate with the uptake of approximately 1 mole of phosphate per mole of enzyme[260–263]. The ^{32}phosphorus can be isolated from the protein as ^{32}P-phosphoserine after partial acid hydrolysis. This phosphoserine is contained in the following peptide sequence[270]:

Thr–Gly–Lys–Pro–Asp–Tyr–Val–Thr–Asp–SerP–Ala–Ala–Ser–Ala

The isolation of this phosphoryl enzyme suggested that a phosphoryl enzyme, specifically a phosphoserine, might be an intermediate on the reaction path. Proof that this particular phosphoryl enzyme is a kinetically significant intermediate is the object of continuing work in a number of laboratories.

Present information does suggest some of the mechanistic features, but leaves many interesting questions unanswered. The phosphoserine that is formed on the protein at low pH is stable to perchloric acid and is 10^5 times as stable as O-phosphoserine[260, 269, 271]. The appearance of a phosphoprotein might represent a transfer of phosphate to a suitably located serine hydroxyl[271].

The extraordinary stability of the phosphoryl enzyme is also puzzling if it represents a kinetically significant intermediate. Wilson and co-workers[271–273] have studied this problem using both kinetic analysis of phosphatase activity and labeling of the protein with ^{32}P-phosphate. Phosphate forms a stable complex with the enzyme from pH 5 to 9, and this complex is responsible for the potent inhibition of alkaline phosphatase by inorganic phosphate which has received extensive kinetic analysis. At alkaline pH the phosphate complex is entirely non-covalent[269]. Ried et al.[273] have measured both the equilibrium constant for the noncovalent Michaelis complex and the equilibrium constant for the hydrolysis of the covalent phosphoprotein as a function of pH by their ^{32}P-labeling technique. The constants derived from these measurements agree closely with the values of the same constants determined by kinetic means. These authors conclude that the phosphoserine is a kinetically significant intermediate[273].

The phosphoprotein is stable relative to the free enzyme and inorganic phosphate at all pH values, however, no phosphoryl enzyme is formed at alkaline pH. The findings of Ried et al.[273] suggest that the reason for this is that while the phosphoryl enzyme is very stable at alkaline pH, the Michaelis complex is even more stable, hence the equilibrium concentration of phosphoryl enzyme becomes negligible. This pH function is shown on page 261.

The Michaelis complex with phosphate is observed to alter the stereochemistry around the metal-binding site[52, 86, 88] (see p. 194) and alkaline phosphatase also catalyzes the exchange of [18]O from the solvent into inorganic phosphate[269]. Thus phosphate can be considered in a sense the simplest substrate for the enzyme. Catalysis of this exchange is more rapid at pH 5 than at pH 7, as would be predicted if the hydrolysis of a phosphoprotein with a greater equilibrium concentration at low pH is involved in the exchange mechanism[269].

On the basis of the information discussed above a scheme for the reaction pathway which includes a minimal number of intermediates can be formulated as in equation 31. $E \cdot ROP$ is the Michaelis complex with substrate, $E \cdot P$ the Michaelis complex with phosphate, while $E{-}P$

$$E + ROP \underset{k_{-1}}{\overset{k_1}{\rightleftharpoons}} E{\cdot}ROP \overset{k_3}{\longrightarrow} E-P \underset{k_{-2}}{\overset{k_2}{\rightleftharpoons}} E{\cdot}P \rightleftharpoons E + P_i \qquad (31)$$

$$R'OH$$
$$\Updownarrow$$
$$R'OP$$

is the phosphoryl enzyme. $R'OH$ is an acceptor alcohol. The reactions with inorganic phosphate can be formulated as in equation 32.

$$E + HOPO_3^= \underset{k_{-1}}{\overset{k_1}{\rightleftharpoons}} E \cdot HOPO_3^{2-} \underset{k_{-2}}{\overset{k_2}{\rightleftharpoons}} E-OPO_3^2 + H_2O \qquad (32)$$

Many of the features indicated by this formalism appear to be supported by the presently available evidence. At pH values near 5 a burst of the alcohol product is observed [264–266]. Barrett et al.[271] have shown that at alkaline pH, the hydrolysis of nine phosphate esters in the presence of Tris as the acceptor results in a constant ratio for two of the products, phenol and phosphate. The phenol/phosphate ratio was 2.37 to 2.42 for all nine substrates [271]. Phosphorylated Tris accounts for the remainder of the phosphate. The constancy of this ratio and the constancy of k_{cat} implies that the rate-limiting step involves the transfer of phosphate from an intermediate that does not contain R. The phosphoryl enzyme described above would appear to meet the requirement for such an intermediate. However, the ratio of phosphoryl enzyme (E—P) to the Michaelis complex (E · P) is given by k_2/k_{-2} (eq. 32), and this ratio varies according to the pH function shown in Figure 59. At low pH E—P is the predominant species, thus k_2 must be greater than k_{-2}. As the pH rises, significant equilibrium concentrations of E—P disappear and thus the ratio k_2/k_{-2} must be decreasing. The kinetic measurements of Fernley and Walker [266] suggest that k_{-2} becomes larger than k_2 at alkaline pH. The burst phenomenon also disappears at high pH. Thus it is not clear that dephosphorylation of the phosphoryl enzyme remains the rate-limiting step at high pH, although the rate of formation of E—P from substrate k_3 could be an order of magnitude larger than in the case of inorganic phosphate, still making k_{-2} the rate-limiting step.

The question of what is the rate-limiting step at alkaline pH is still an important one. In the case of the active Co(II) enzyme, it has been shown that at low temperature, a burst continues to be observed throughout the pH range, suggesting that dephosphorylation of a phosphoryl enzyme is rate limiting [403]. It is not certain, however, if all features of the mechanism are the same in the Co(II) and Zn(II) enzymes.

Recently, Halford et al.[404] have studied the interaction of benzyl phosphonates with the active center. They found that 2-hydroxy-5-nitrobenzyl-phosphonate changes its spectrum on binding to the enzyme. Kinetics of the reaction followed by stopped-flow and temperature-jump techniques show that binding of this inhibitor occurs in two successive and reversible steps: enzyme–phosphonate complex formation followed by rearrangement of the complex. The spectral change is associated with the rearrangement. At pH 8 in 1 M NaCl at 22° the rate constant for this rearrangement is 167 sec^{-1} and it is 18 sec^{-1} for the reverse process [404]. These authors suggest that this rearrangement may be a relatively slow conformational change in the protein which could also accompany phosphate ester binding and be a required step prior to the transfer of the phosphate from substrate to the enzyme [404]. Trentham and Gutfreund [265] have proposed a mechanism in which a conformational change in the enzyme–substrate complex rather than dephosphorylation is the rate-limiting step in the mechanism at alkaline pH.

A puzzling feature related to the mechanism of action of alkaline phosphatase is the presence of some uncertainty as to how many active sites are present in the active dimer. Kinetic studies of the burst phenomenon have uniformly revealed that approximately one site per enzyme dimer is phosphorylated during this process[264–266, 269]. Equilibrium dialysis studies with ^{32}P-orthophosphate at alkaline pH and covalent labeling of the enzyme at low pH have both generally shown one phosphate binding site per dimer[52, 273]. A recent study shows that two sites can be covalently labeled at pH 2–3 in the presence of very high phosphate concentrations if a rapid labeling technique is employed[405]. This involves a pH region where the enzyme is rapidly denatured, and the same study shows that a single high-affinity site is present at alkaline pH[405]. Binding studies by Neumann with a ^{14}C-labeled oxygen ester of phosphorothioic acid have also shown the binding of 1 mole of the nonhydrolyzed ester per enzyme dimer at alkaline pH[406].

All present evidence seems to indicate that the dimer has twofold symmetry. The enzyme crystallizes in the space group $P3_121$ with the monomer as the asymmetric unit[398]. Each unit cell contains three dimers and the unit cell dimensions are: $a = 70.50$ Å, $b = 70.5$ Å, $c = 155.6$ Å, $\beta = 120°$. It can be speculated that the twofold axes between monomer units relate the monomer units of a functional dimer. Hence identical structural features must exist in each half. The spectral titrations (p. 194) already suggests that two identical metal binding sites exist[86].

Several alternative explanations for the apparent presence of a single active site are discussed below. (1) The active center could be located on a twofold axis between monomer units and thus explain its unique nature in an otherwise symmetrical environment[398]. There is some evidence from work on *in vitro* complementation that parts of each of the two monomers are required for complete formation of the active center. Schlesinger and Levinthal[274] showed that activity could be generated by mixing two inactive mutant alkaline phosphatases. Electrophoresis showed that activity was associated only with the hybrid dimer[274]. This suggests that different parts of the active center are defective in each of the two mutant enzymes and that each monomer in the hybrid dimer contributes part of the intact active center.

The changes in the absorption and CD of the d-d absorption bands of Co(II) alkaline phosphatase induced by phosphate and arsenate suggest that one anion does modify both metal-binding sites. It is possible that both metal ions are involved in a single active center. (2) Two sites could be located such that direct interference with the second site is produced by occupancy of the first site such that only one of two alternate sites is effectively operating. (3) Negative cooperativity between more distantly separated identical sites, as postulated by Koshland [407, 408] could also explain the observation of only one site with high binding affinity for phosphate, perhaps mediated by a conformational change induced by phosphate binding.

The detailed function of the metal ion in this enzyme is less clear than in the two enzymes just discussed. Zn(II) functions in inducing phosphate binding [52] (p. 260). Whether it has a similar role in phosphate ester binding is not clear as yet. Since the enzyme has practically no specificity for the R group on the ester and K_m values for the esters are about equal to the equilibrium constants for inorganic phosphate binding, $10^{-6} M$, one of the chief features of the active center must be a binding site of high affinity for phosphate, probably as the dianion. The binding of both metal ions is required to induce the binding of one phosphate, which suggests that the phosphate binding site is not completely constituted except in the dimer containing two metal ions [52].

The metal ions do influence the monomer–dimer equilibrium, as discussed in Section 2.1, but the metal protein bonds are not the major factor in dimerization, since the apoenzyme is 75% dimer at neutral pH [26]. There is also evidence that bound metal ions in excess of the two bound at highly specific sites may affect function [26, 88, 409]. The mechanistic significance of this is unclear as yet.

In model systems where the metal ion catalyzed hydrolysis of phosphate esters has been observed, there is not a great deal of specificity for a particular species of metal ion [23]. This may arise from the fact that phosphate oxygens are relatively poor donors and, hence, not much ligand field stabilization is involved [3]. However, a considerable amount of information on the metal ion catalysis of the hydrolysis of phosphate esters and phosphoric acid anhydrides, including pyrophosphate derivatives and ATP, has now accumulated and some complex effects are observed. One general observation can be made, however, and that is that the catalytic effects are disappointingly small.

Hydroxides of lanthanum, cerium, and thorium have been shown to promote the hydrolysis of α-glycerol phosphate in the region from pH 7 to 10 [410]. The enhancement of the rate by these hydroxides is $\sim 10^3$ and is by far the greatest enhancement observed in any of the model systems. Intermediates of the type **19** and **20** have been proposed. It has been suggested that these might serve as models for the metal ion promoted alkaline phosphatases [411]. Here, however, the primary

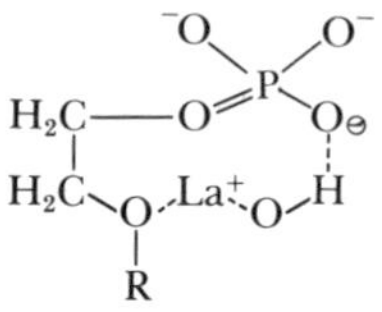

19 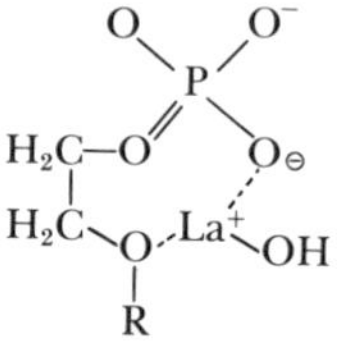   20

interaction is electrostatic or via the coordinated hydroxide. While the latter feature could be present in Zn(II) alkaline phosphatase, the effects of metal ion species on the hydrolysis of the phosphoryl enzyme (p. 261) suggest that more complex features of transition and IIB metal coordination are involved.

While specific structural features of the coordination chemistry in alkaline phosphatase are not available and all we can say about general structure is that a dimer of identical monomers containing two Zn(II) ions is necessary for the catalysis, it is useful to discuss some facets of nonenzymatic phosphate ester hydrolysis.

The hydrolysis of the monoanions of phosphate monoesters presumably involves protonation of the leaving alcohol and its expulsion from a dipolar intermediate [412] 21. Hydrolysis of the dianion species involves

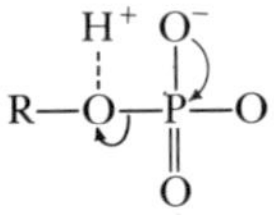

21 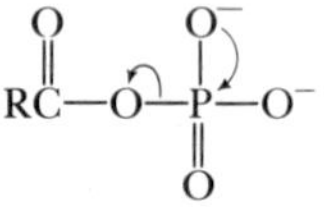   22

direct expulsion of the oxygen anion and is important in the case of good leaving groups, such as the carboxylate group of acyl phosphates 22 [412]. It would appear likely that the dianion species of phosphate monoesters and pyrophosphate derivatives are the substrates for alkaline phosphatase in view of the pH range of maximum hydrolysis. The latter may be regarded as phosphate monoesters where the "alcohol" components are alkyl phosphates or dialkyl phosphates (see below). It appears unlikely that alkaline phosphatase catalyzes the formation of a metaphosphate intermediate, since structural changes in the leaving group do not produce changes in the rates of phosphorylation.

There are three fundamental aspects of phosphate ester hydrolysis that should be mentioned. It is generally agreed that under many circumstances this reaction proceeds by the elimination of a metaphosphate intermediate. This requires either the zwitterionic form (21) or the dianion (22). Thus anything that neutralizes these charges (protonation of the phosphate oxygens or metal ion coordination) would be expected to inhibit this mechanism (see below). Secondly, the nature of the leaving group has a major effect on the rate of hydrolysis. A very poor leaving group, especially a negatively charged one as in pyrophosphate, slows the reaction (see below). Hence, hydrolysis proceeds from the dianion form (22) only when excellent leaving groups are involved (groups with relatively low pK_a's)[412]. Thirdly, the hydrolysis may proceed by nucleophilic attack on the phosphorus. This mechanism is relatively poor in the case of phosphate esters because of the high electron density contributed by the phosphate oxygens. Thus only good nucleophiles would be expected to catalyze this hydrolysis. Evidence suggests that the dianion form is the substrate for alkaline phosphatase [413]. This might suggest a mechanism in which a metaphosphate intermediate occurs. However, if the negative charge density on the terminal phosphate is neutralized by charge on the protein or coordination to the metal ion, then nucleophilic attack is more likely. The presence of a serine hydroxyl which participates by the formation of a phosphoryl enzyme also suggests nucleophilic attack. The metal may assist by acting as an electrophile to reduce the negative charge density around the phosphorus.

The unsymmetrical pyrophosphate diester (eq. 33) has recently been synthesized and very rapid hydrolysis is observed in the presence of the neutral leaving group[414]. Calculations suggest also that the observed

$$\begin{array}{c} C_2H_5O \\ C_2H_5O \end{array} \!\!\! \underset{\underset{O}{\overset{\parallel}{P}}}{\overset{O}{\parallel}} \!\!\! -O- \!\!\! \underset{\overset{\parallel}{O}}{\overset{O^-}{\underset{\parallel}{P}}} \!\!\! -O^- + H_2O \; \rightleftharpoons \; \begin{array}{c} C_2H_5O \\ C_2H_5O \end{array} \!\!\! \overset{O}{\underset{\parallel}{P}} \!\!\! -O^- + H_2PO_4^{2-} \quad (33)$$

hydrolysis of pyrophosphate in water occurs exclusively via the protonated form (23)[414]. In the case of the unsymmetrical pyrophosphate

$$\begin{array}{c} HO \\ HO \end{array} \!\!\! \underset{\underset{}{\overset{\parallel}{P}}}{\overset{O}{\parallel}} \!\!\! -O- \!\!\! \underset{\overset{\parallel}{O}}{\overset{O^-}{\underset{\parallel}{P}}} \!\!\! -O^-$$

23

diester, pH studies show that the dianion is rapidly hydrolyzed, while the protonated form is not. Thus the hydrolysis resembles that of phosphate monoesters and is consistent with the generally accepted mechanism of a metaphosphate intermediate (24).

$$\left[\begin{array}{c} O \underset{\diagdown}{} \overset{\diagup}{} O \\ P \\ \underset{\parallel}{} \\ O \end{array} \right]^-$$

24

It has thus been natural to propose that one function of the metal ion might be an electrostatic or charge neutralization function. However, those situations involving chelate formation with the negatively charged

$$\text{Me}^{2+} \cdots \begin{matrix} \text{-O}^- \\ \text{O}^- \end{matrix} P \begin{matrix} \text{O} \\ \text{OH} \end{matrix}$$

25   (R)

oxygens (25) have been uniformly disappointing as models of catalysis, although the situation might be quite different if nucleophilic attack on the phosphorus by a powerful nucleophile is also involved.

Metal ion catalyzed hydrolysis of pyrophosphate derivatives such as ATP has received considerable attention[415–417]. One of the most complete studies shows that at pH 5, the order of effectiveness of a series of metal ions in the catalysis of the hydrolysis of ATP is Cu(II) > Zn(II) > Cd(II) > Mn(II) > Be(II)[415].

An interesting finding is that Ba(II) and Mg(II) do not catalyze the reaction[415]. At pH 9.0 the order was observed to be Ca(II) > Mn(II) > Cu(II) > Cd(II) > Zn(II) > Co(II), although the maximum rate of hydrolysis obtained in the presence of Ca(II) ions was only 1/28 of the maximum rate observed at pH 5 in the presence of Cu(II) ions[415].

The uncatalyzed hydrolysis of ATP proceeds with hydrolysis of either the terminal or the internal pyrophosphate bond. The ratio ADP/AMP is 10/1 in the uncatalyzed reaction. In the case of the metal-catalyzed reaction, this ratio varies. For example, Cd(II) favors the terminal bond hydrolysis, while Cu(II) relatively favors the formation of AMP[415]. This finding suggests the participation of both types of chelates shown in 26. Cohn and Hughes have studied the formation of these chelates extensively with nmr techniques[418]. They have shown that both types form and that Mg(II), Ca(II), and Zn(II) bind predominantly to the β and γ phosphates of ATP, while Cu(II) binds predominantly to the α and β-phosphates. Mn(II) can bind to all three[418]. The mechanisms proposed for the nucleophilic attack of water on phosphorous in these chelates is indicated in 26.

The problem with this system is that the hydrolysis is increased only 10–60-fold over the noncatalyzed system and the rate is completely out of the range of enzyme-catalyzed hydrolysis. It may be that the function of Mg(II) or Mn(II) in the enzymes such as kinases may be primarily to function in binding the nucleotide to the protein (see Sect. 2.4). The chelates appear to be the actual substrates.

Some ambiguity resides in the study of the metal ion catalysis of pyrophosphate esters, since several coordination complexes are possible

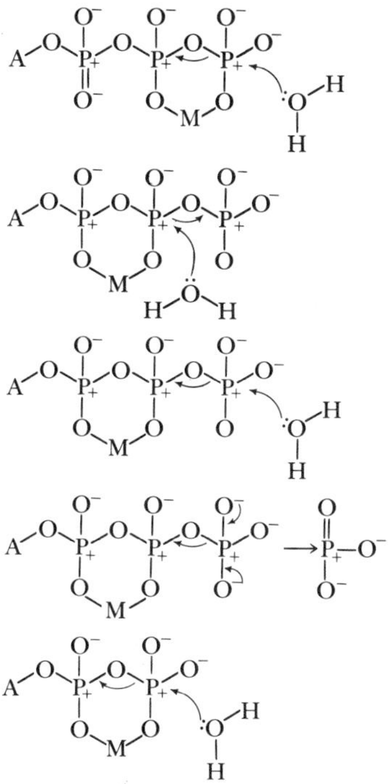

Mechanism that may be involved in the hydrolysis of ATP and ADP catalyzed by divalent metal ions.

26

(**26**). On the basis of the observed rapid hydrolysis of the unsymmetrical diester of pyrophosphate [414], it has been suggested that if exclusive internal coordination could be induced, leaving the terminal dianion free, the catalysis might be enhanced [417]. In order to test this hypothesis Cooperman synthesized pyrophosphate derivatives of the type shown in **27** (tetraaminophosphonyl pyrophosphate) and **28**([7-(2-hydroxyphenyl)-3,6-diazaheptyl] phosphonyl phosphate). The substituents would be expected to direct coordination as shown in **27** and **28**. The metals Cu(II), Ni(II), Zn(II), Co(II)-Co(III), Mn(II), Pd(II), Hg(II), Mg(II), Ca(II), and Fe(III) were all tested and no significant catalysis was found [417]. Thus the failure of the model suggests that some special features must be present in the protein–metal–substrate complexes.

Possible features of an alkaline phosphatase mechanism involving the metal as an electrophile are suggested in Figure 76. The studies with phosphate binding suggests that the phosphate dianion is somehow

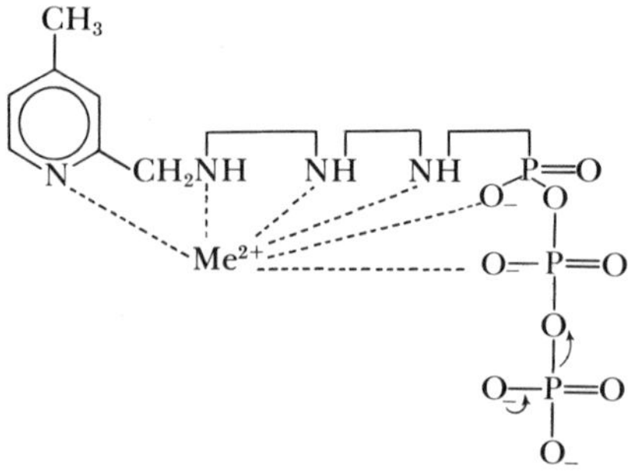

27

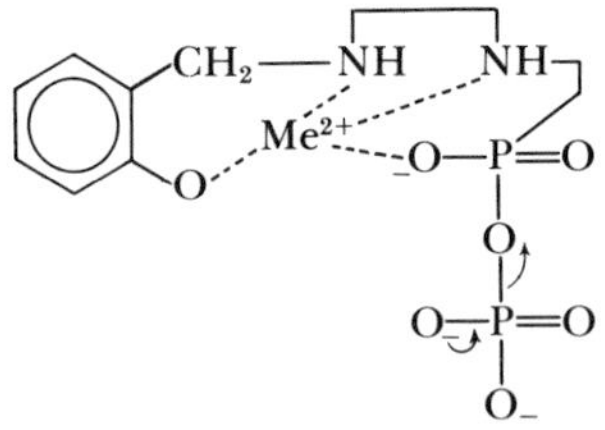

28

involved with the metal ion, perhaps through coordination of one or two of the oxygens [52]. The studies of the phosphoryl enzyme (see p. 261) show that the metal ion is intimately involved with the pH stability of this intermediate [52], hence catalysis of the breakdown of this intermediate could also involve the metal ion.

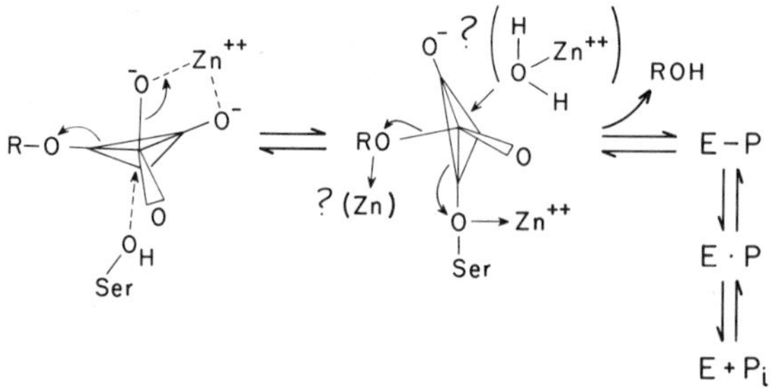

Fig. 76   Possible modes of action of Zn(II) in the catalysis of the hydrolysis of phosphate monoesters by *E. coli* alkaline phosphatase. The positions and functions of the Zn ions pictured in this scheme are only suggested as possibilities. From left to right: direct binding of Zn^{2+} to the phosphate oxygens; interaction with the ester oxygen of the serine phosphate or alternatively (in parenthesis) the ester oxygen of the substrate; either group might leave as R-$\bar{O}$-Zn. A zinc coordinated H_2O or OH^- might serve as a nucleophile attacking the phosphorus.

Since the metal ion species so drastically affects the pH stability of this intermediate and this stability appears to be related to activity (see p. 261), there is the strong suggestion that this intermediate is involved in the mechanism. The Co(II) enzyme is also reported to have lost transferase activity suggesting that the metal ion may alter the step after the phosphoryl intermediate[419]. The question of whether one or two metal ions participate in the hydrolysis of one phosphate ester or the reactive intermediates can only be a matter of speculation, and the interactions in Figure 76 are meant only to convey the possibilities. Attack on the phosphorus nucleus by a coordinated H_2O or OH^- at some point in the mechanism might also be a possibility (Fig. 76).

There are several other aspects of phosphate ester hydrolysis that are of potential interest in the consideration of the enzymatic catalysis. The hydrolysis of ethylene phosphate at alkaline pH occurs 10^8 times as fast as that of the non-cyclic compound dimethylphosphate[420–422]. It is assumed that the rate difference must reflect some distortion or destabilization brought about by the ring structure which forces the starting material part way towards the structure of the transition state. It is also observed that there is rapid ^{18}O exchange from solvent water into the oxygens of the unreacted ethylene phosphate[423, 424]. Jencks has pointed out that if an enzyme could induce a similar strain in the substrate relative to the transition state for its reaction, a large fraction of the rate acceleration brought about by the enzyme would be explained [412].

This brings up the other interesting feature observed in the hydrolysis of cyclic phosphate esters. The hydrolysis of methylethylene phosphate which is analogous to oxygen exchange in hydrogen ethylene phosphate is accompanied by the hydrolytic cleavage of the methoxyl group[422, 423, 425], as shown in equation 34. This occurs one million times faster than the analogous hydrolysis of trimethyl phosphate[423, 425]. While rapid opening of the ring might be explained by ring strain, enormous enhancement of the hydrolysis of the ester external to the ring is difficult to explain.

Westheimer has proposed that this can be explained if the external

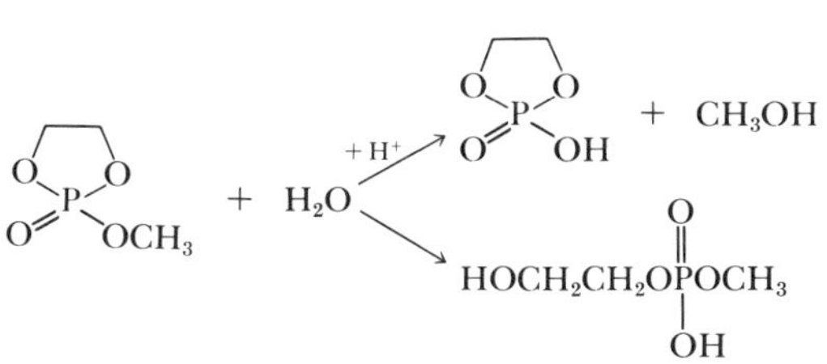

$$\tag{34}$$

hydrolysis proceeds with "pseudo-rotation" between trigonal–bipyramidal intermediates, as shown in equation 35. In equation 35 the 5-membered ring occupies one apical and one equatorial position.

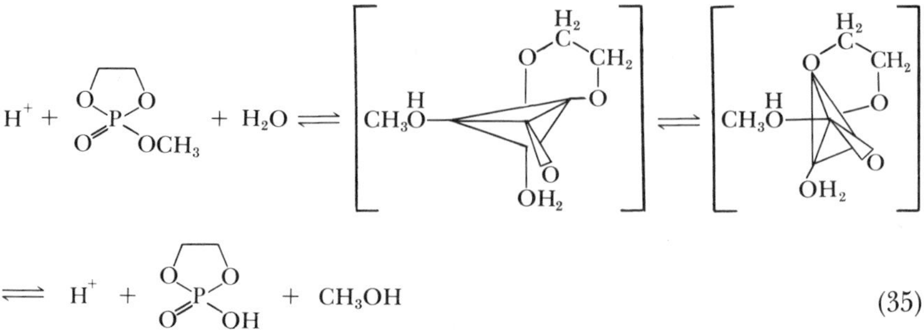

$$
H^+ + \text{(cyclic phosphate)} + H_2O \rightleftharpoons \left[\text{intermediate 1}\right] \rightleftharpoons \left[\text{intermediate 2}\right]
$$

$$
\rightleftharpoons H^+ + \text{(cyclic phosphate)} + CH_3OH \tag{35}
$$

Pseudo-rotation is a distortion about the phosphoryl oxygen such that it remains equatorial, while the substituents that were equatorial in the first intermediate move forward to become apical and the apical substituents move backward to become equatorial. A great deal of evidence has been obtained for such a mechanism and has been reviewed by Westheimer[423].

Original references should be consulted for the details of this mechanism, but it essentially provides a pathway where the ring oxygens always occupy one apical and one equatorial position and induces a 90° ring angle which reduces the strain for intermediates leading either to ring opening or hydrolysis external to the ring. The latter would explain how ring strain can induce rapid reaction external to the ring. It also provides equivalent apical positions for the entering and leaving groups in the intermediates consistent with microscopic reversibility.

Whether any such mechanism could operate in the case of an alkaline phosphatase intermediate is pure conjecture. The tremendous enhancement in rate is intriguing and the formation of an enzyme–substrate complex of such low dissociation constant as observed might result in a structure equivalent in many ways to cyclic ester, perhaps after initial reaction of the serine hydroxyl. It is also intriguing to think of such mechanisms in terms of possible conformational changes in the dimer during the hydrolysis, as suggested by Trentham and Gutfreund[265].

3.2 Oxidative Metalloenzymes

Knowledge of the physicochemistry and mechanistic features of oxidative metalloenzymes is not nearly so far advanced as in the case of the enzymes described in Section 3.1. Except for the ferredoxins and the Azurins, these enzymes are large proteins, MW 60,000–1,000,000, and have complex subunit structure. Much of the detailed information has come from the application of specific spectroscopic techniques and this information has been largely covered in Section 2. This section briefly summarizes the characteristics of the enzymes in this category, with emphasis on the present information relating the metal ions to the reactions catalyzed.

3.2.1 Zinc Dehydrogenases.

3.2.1 Zinc Dehydrogenases. Dehydrogenases utilizing the oxidized and reduced forms of NAD and NADP in the oxidation (dehydrogenation) of a great variety of biological substrates have received a great deal of attention[426]. Among this group the alcohol dehydrogenases from a number of species and glutamic dehydrogenase from liver contain zinc (Table 13). Less extensive studies have indicated that Zn(II) is contained in L-lactate dehydrogenase and L-malate dehydrogenase (Table 13).

The mechanism of NAD- and NADP-dependent dehydrogenases is accepted to proceed by addition or subtraction of a hydride ion from the substrate at the α- or β-position of C_4 of the nicotinamide ring. The other proton appears in or is extracted from the solvent. For a given enzyme the hydride ion transfer is stereospecific, either the α- or β-position. Both α- and β-transferring dehydrogenases exist. The Zn-containing alcohol dehydrogenases transfer to the α-position[436]. An enormous amount of data are available on the formation of binary ADH–NADH and ternary ADH–NADH–substrate (inhibitor) complexes as measured by absorption, optical activity, and fluorescence of the coenzyme chromophores. This information has been summarized by Sund and Theorell[437].

The specific part that zinc ions play in this general mechanism for some of these dehydrogenases is not entirely clear. The metal ion has variously been postulated to coordinate the adenine portion of the coenzyme and thereby participate in binding the coenzyme[437–439], to coordinate the oxygen of the alcoholic OH of the substrate (in the case of the alcohol dehydrogenases)[438, 439], and to function in maintaining the tertiary and quaternary structure of the protein[24, 440]. There is good evidence for the last in the case of the yeast[24] and liver enzymes[440]. Horse liver alcohol dehydrogenase is the only zinc dehydrogenase for which extensive information relating the metal ion to function is available.

The extensive studies of Vallee and co-workers on the liver alcohol dehydrogenase have given the most information directly concerning the possible function of the metal ion[24, 53, 54, 428–431, 440–447].

TABLE 13 ZINC DEHYDROGENASES

Enzyme	Source	MW	Cofactor Stoichiometry, mole/mole[a]	Reaction	References
Alcohol dehydrogenase	Horse liver	84,000	2–4Zn(II) 2NAD (tightly bound) nNAD (loosely bound)	Alcohols + NAD $\rightleftharpoons$ aldehydes + NADH	427–429
Alcohol dehydrogenase	Human liver	87,000	2Zn(II) 2NAD	Alcohols + NAD $\rightleftharpoons$ aldehydes + NADH	430
Alcohol dehydrogenase	Yeast	150,000	4Zn(II) 4NAD	Alcohols + NAD $\rightleftharpoons$ aldehydes + NADH	431
Lactate dehydrogenase	Rabbit muscle		nZn nNAD	L-Lactate + NAD $\rightleftharpoons$ pyruvate + NADH	432
Malate dehydrogenase	Porcine heart	40,000	1–2Zn(II) nNAD	L-Malate + NAD $\rightleftharpoons$ oxaloacetate + NADH	433
Glutamate dehydrogenase	Bovine liver	1,000,000	2–4Zn(II) nNAD(P)	L-Glutamate + H_2O + NAD(P) $\rightleftharpoons$ α-Ketoglutarate + NH_3 + NAD(P)H	434, 435

[a] n is used when exact stoichiometry is not known.

Inhibition studies with chelating agents have shown that inhibition involves the formation of mixed enzyme–Zn–inhibitor complexes [24, 53, 54, 444]. These complexes, that with 1,10-phenanthroline for example, occupy highly dissymmetric environments, as is shown by the fact that large optical activity is generated in the absorption bands of the bound inhibitor [444, 445]. Cotton effects are also observed in the adsorption bands of the bound coenzyme molecules [445, 446]. These optical properties can be used to show that the formation of the mixed complexes with chelating inhibitors competes with coenzyme binding [445]. It is also clear that the metal ion does maintain the quaternary structure of the protein as outlined for the yeast alcohol dehydrogenase in Section 2.1. The liver enzyme is also irreversibly denatured by removal of the zinc [440].

Liver alcohol dehydrogenase contains between two and four zinc ions [429, 448, 449]. It has two tight binding sites for NAD [437] and two of the zinc ions appear to be directly involved in activity (see below). The additional zinc ions may participate in maintaining the structural integrity of the protein [429]. The most interesting studies concerning the function of these zinc ions have been carried out by observing the exchange of the intrinsic zinc with ^{65}Zn in the medium [440, 447]. While the zinc does not exchange at neutral pH, it will exchange at pH 5.5 in $0.1 M$ acetate. Under these conditions two zinc atoms are exchangeable [440, 447]. Among a variety of chelating agents tested, diethydithiocarbamate was found to specifically remove these two freely exchangeable zinc ions [440]. Enzyme activity was observed to be directly proportional to the content of this "freely" exchangeable zinc, but not to total zinc. Additional zinc found in the native enzyme is not exchangeable in acetate and appears to be "buried," perhaps participating in structural stabilization. Both the freely exchangeable and buried zinc will exchange in $0.1 M$ phosphate at pH 5.5 [440]. This dependence of exchange on the buffer species present has no counterpart in model systems.

It would appear that the two "freely" exchangeable zinc ions are those directly participating in the mechanism and closely associated with the coenzyme molecules (as well as forming mixed complexes with the chelating agents). NADH or coenzyme analogs do not retard the exchange of the two free zinc ions [440, 447]. If substrate or substrate homologs are added in the presence of coenzyme, however, the exchange is retarded, suggesting that it is only the ternary complex that has a major effect on the zinc coordination site.

Studies with a spin-labeled analog of NAD [215, 216] (see p. 237) have shown that the binding of two tightly bound NAD molecules is not metal

316

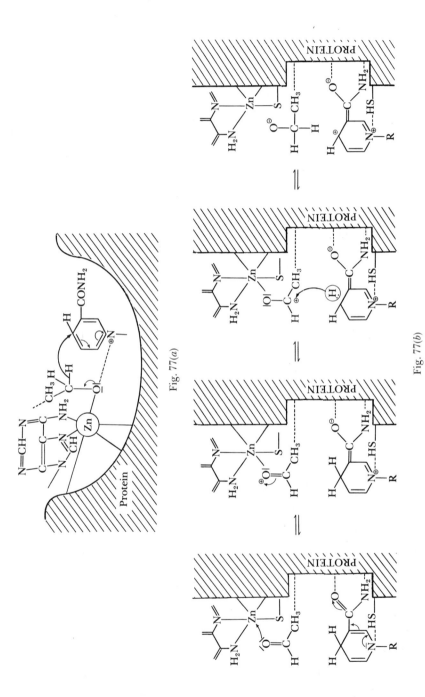

Fig. 77(a)

Fig. 77(b)

Figure 77 (a) Complex between alcohol dehydrogenase, NAD, Zn(II), and ethanol proposed by Theorell and McKinley McKay for the active site of alcohol dehydrogenase [488]. (b) Mechanism for alcohol dehydrogenase proposed by Wallenfels and Sund [439].

ion dependent. Sulfhydryl groups may be involved more directly in coenzyme binding, since NADH protects two sulfhydryl groups per molecule from modification by iodoacetate[450]. These are contained in the peptide sequence: Met–Val–Ala–Thr–Gly–Ileu–S–carboxy-methylcysteine – Arg. Earlier studies with p-chloromercuribenzoate had also implicated—SH groups in coenzyme binding[437, 451].

Theorell, McKinley McKee[438] and Wallenfels and Sund[439] have postulated certain mixed complexes between coenzyme and substrate and enzyme zinc as participating in the reaction of alcohol dehydrogenase. The features of this complex and the proposed mechanism are pictured in Figure 77. If significant coordination of the adenine to the zinc occurs, it is surprising that it does not retard zinc exchange. The precise nature of the zinc participation in mixed active complex formation must await further work.

3.2.2 Metalloflavoenzymes. All these molecules are large complex proteins, MW 100,000–300,000 and contain molybdenum and nonheme iron, as well as the flavin coenzyme (Table 14). The most direct work on the mechanism of action has been done with electron spin resonance, which has determined many features of the electron transport sequence and the species involved[147, 148, 199, 452]. This work has been reviewed in Section 2.3. On the basis of this work the skeleton reactions for three of these enzymes, xanthine oxidase, aldehyde oxidase, and dihydroorotate dehydrogenase, can be described as shown in equations 36–38.

Recent work by Nelson and Handler has shown that while subunit structure is present in these enzymes, dissociation appears to be associated with loss of function[453]. The best molecular weight for xanthine oxidase is ~300,000[453], and the protein dissociates to a species of MW 150,000 in acid or guanidine, but apparently cannot be dissociated further[453]. The 300,000 unit in the native state shows multiple cotton effects from 300 to 600 mμ associated with the flavin chromophores [453]. These all disappear when the protein is dissociated, suggesting that the highly specific dissymmetric environment of the flavin is no longer present. Aldehyde oxidase has a molecular weight of 275,000 and also can be dissociated into units of about half this size[453]. The best molecular weight for dihydroorotate dehydrogenase is 112,000–124,000 and it will dissociate in guanidine into four identical subunits[453].

The esr results on xanthine oxidase have been reviewed (see p. 230) and a similar series of results have recently been obtained on aldehyde oxidase[452]. Qualitatively similar esr data are obtained on the two enzymes; signals appear on reduction corresponding to a flavin free radical and reduced forms of both Fe and Mo[452]. There are some

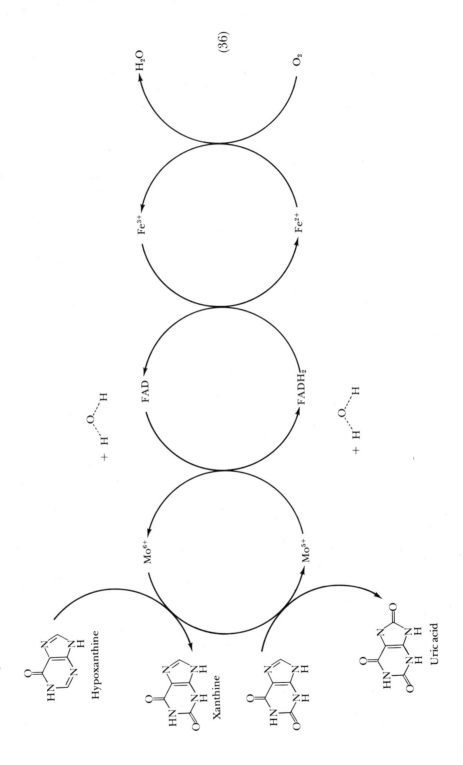

(36)

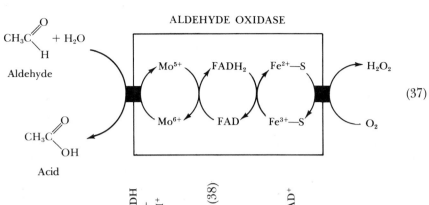

$$(37)$$

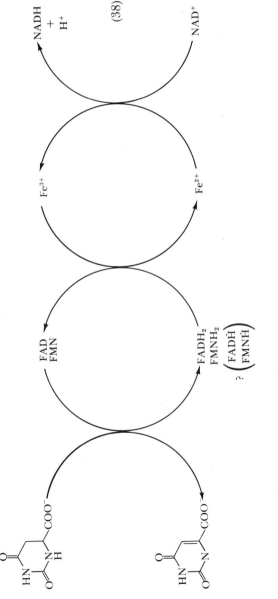

$$(38)$$

TABLE 14 METALLOFLAVOENZYMES

Enzyme	Source	MW	Cofactor Stoichiometry, mole/mole[a]	Reaction	References
Xanthine oxidase	Milk	300,000	8Fe 2Mo 2FAD	Xanthine + O_2 → urate + H_2O_2	147, 148, 195, 197, 452, 454
Xanthine oxidase	Bird liver	—	8Fe 1Mo 1FAD	Xanthine + O_2 → urate + H_2O_2	195, 454
Xanthine oxidase	Mammalian liver	—	4Fe (calculated per 1Mo mole of flavin) 1FAD	Xanthine + O_2 → urate + H_2O_2	195
Aldehyde oxidase	Porcine liver	300,000	8Fe 2Mo 2Fad 1–2Coenzyme Q_{10}	Aldehyde + O_2 → acid + H_2O_2	196, 198, 452
Nitrate reductase	Neurospora crassa Soybean leaves	Partially purified	nMo nFAD	Nitrate + NADPH ⇌ nitrite + $NADP^+$ + H_2O	455
Succinate dehydrogenase	Bovine heart	200,000	4Fe 1Flavin	Succinate + certain dyes ⇌ Fumarate + reduced dyes	123, 153, 456
Succinate dehydrogenase	Yeast		4Fe 1Flavin		123
Dihydroorotate dehydrogenase	Zymobacterium oroticum	112,000–124,000	2Fe 1FAD 1FMN	L-4,5-Dihydroorotate + O_2 → orotate + H_2O_2 (NAD can replace O_2; NADH can replace dihydroorotate)	453, 457, 458

[a] n is used when stoichiometry is not known.

differences in detail in the kinetic behavior of the esr signals in the two enzymes. In contrast to xanthine oxidase, there is some paramagnetic molybdenum present in resting aldehyde oxidase. Under conditions where the maximal esr signals were obtained, 24% of the flavin was in the semiquinone form ($g = 2.00$), 25% of the iron was in the reduced form responsible for the signal at $g = 1.94$, and 15–20% of the molybdenum was in the Mo(V) form. The kinetics of the esr signals indicate that all three species are reduced and oxidized rapidly enough to participate in the reaction. The interesting findings in the steady state are that the Mo is partly reduced, whereas the Fe and flavin components are largely oxidized. Hence, the reduction of Mo and the oxidation of flavin and iron appear to be the rapid reactions in the sequence [452].

The actual sequence of the electron transfers could not be determined unequivocally, but most of the data are compatible with an electron transfer sequence of substrate $\rightarrow$ Mo $\rightarrow$ FAD $\rightarrow$ Fe $\rightarrow$ O$_2$ as described in the case of xanthine oxidase [148]. Studies of electron relaxation indicate that there are subtle interactions among the electron carriers which depend on the time of reaction, the substrate, and the degree of reduction [452]. This suggests that all the paramagnetic species are within 10–20 Å of each other, a not unexpected feature if sequential electron transport is involved. It is impossible at the present time to suggest any more detailed mechanisms for these complex and important enzymes.

3.2.3 Copper Oxidases.

Copper oxidases that have been well characterized are listed in Table 15 along with a brief description of their physicochemistry and the reaction catalyzed. They can be further classified on the basis of the type of copper complex that appears to be present. One class appears to contain exclusively Cu(I) in the active form, and no evidence for a Cu(II) state during the reaction has been obtained. The tyrosinases (phenolases) are of this class [61].

A majority of the copper oxidases are blue copper proteins, the blue chromophore being clearly associated with a Cu(II) complex in the oxidized state of these proteins. A large proportion of the work on this group of copper oxidases has centered on the visible absorption and electron spin resonance characteristics of these highly unusual copper chromophores. This work has been discussed in Section 2.3.4. While the unusual geometry of these copper–protein sites must be related to the activity, it is too early to even attempt to picture the relationship among the substrates, O$_2$, and the metal ion in these instances. In practically all cases of the blue proteins, the oxidation of the substrate is accompanied by acceptance of an electron by the Cu(II), as has been documented by the esr techniques (see Sects. 2.2 and 2.3).

322

TABLE 15 COPPER OXIDASES

Enzyme	Source	MW	Subunits MW	Cofactor Stoichiometry, mole/mole	Reaction	References
Tyrosinase	Mushroom	120,000	4 30,000	4Cu(I)	Oxidation of monohydric phenols (cresol) to the o-dihydric compound and oxidation of the o-dihydric phenols (catechol) to the o-quinone stage	61, 459
Tyrosinase	*Neurospora crassa*	30,000– 120,000	1–4 30,000	1Cu(I)/subunit	Oxidation of monohydric phenols (cresol) to the o-dihydric compound and oxidation of the o-dihydric phenols (catechol) to the o-quinone stage	460
Tyrosinase	Mammalian melanoma	64,000		Cu(I)	Oxidation of monohydric phenols (cresol) to the o-dihydric compound and oxidation of the o-dihydric phenols (catechol) to the o-quinone stage	461a, 461b, 478, 479
Laccase (p-diphenol oxidase)	Latex of the lac tree					101
	1. *Rhus vernicifera*	120,000		4Cu [both Cu(II) and Cu(I)]	p-Diphenols + O_2 → p-quinones + H_2O	462, 189
	2. *Rhus succedanea*	120,000		4Cu [both Cu(II) and Cu(I)]	p-Diphenols + O_2 → p-quinones + H_2O	92, 171
Laccase	*Polyporus versicolor*	62,000		4Cu	p-Diphenols + O_2 → p-quinones + H_2O	463, 464
Laccase	Indochinese lac trees	141,000 130,000		4.5–6Cu	p-Diphenols + O_2 → p-quinones + H_2O (more recent preparations than 1 and 2)	189, 462
Ascorbic acid oxidase	Squash, cucumber	146,000		6Cu	L-Ascorbate + O_2 → dehydroascorbate ± H_2O	55

Enzyme	Source	Molecular weight	Metal content	Reaction	Reference
Ceruloplasmin	Human plasma	151,000	8Cu	Oxidizes p-phenylenediamine, ascorbic acid, and some o- and p-dihydroxyphenols	
Urate oxidase	Mammalian liver	120,000	1Cu	Urate $+ O_2 \rightarrow$ Unstable products (not allantoin)	465
D-Galactose oxidase	$Polyporus\ circinatus$ (Dactylium dendroides)	75,000	1Cu	D-Galactose $+ O_2 \rightarrow$ D-galactohexodialdose $+ H_2O_2$	466
Monoamine oxidase	Bovine plasma	255,000	4Cu	$RCH_2NH_2 + O_2 + H_2O \rightarrow RCHO + NH_3 + H_2O_2$	467, 468
Monoamine oxidase	Bovine liver mitochondria	200,000–300,000	nCu ?FMN	$RCH_2NH_2 + O_2 + H_2O \rightarrow RCHO + NH_3$ H_2O_2	467
Diamine oxidase	Pea seedlings	96,000	1Cu Pyridoxal-P	A diamine $+ H_2O + O_2 \rightarrow$ an amino-aldehyde $+ NH_3 + H_2O$ (also oxidizes some monoamines)	469
Cytochrome oxidase	Bovine heart	200,000–10^6 polymolecular aggregate	1Cu/heme	Reduced cytochrome c $+ O_2 \rightarrow$ oxidized cytochrome c $+ H_2O$	106
Azurins (Pseudomonas blue proteins)					470–472
1. $Pseudomonas$ $aeruginosa$		16,400	1Cu(II)	Functions as a respiratory chain protein, an intermediate between bacterial cytochromes and cytochrome oxidase	
2. $Pseudomonas$ $denitrificans$		16,500	1Cu(II)	Functions as a respiratory chain protein, an intermediate between bacterial cytochromes and cytochrome oxidase	
3. $Pseudomonas$ $fluorescens$		14,010[a]	1Cu(II)		472

[a] Determined from amino acid sequence.

323

A third class of copper proteins is represented by the amine oxidases. These proteins appear yellow or red, perhaps due to charge transfer bands from the copper complex[23]. The copper appears to be present as Cu(II)[467] and no evidence has been found for reduction of Cu(II) by substrate[467]. The situation is complicated by the presence of cofactors in at least two of these enzymes, pyridoxal phosphate in the pea seedling enzyme[469] and the apparent presence of FMN in the enzyme from bovine mitochondria[467]. No mechanism of action can be proposed as yet.

Some information is available bearing on the mechanism of action of the phenolases and the blue copper oxidases and the comparison is rather interesting. Tyrosinase catalyzes two ostensibly different reactions, the *ortho* hydroxylation of phenols to their dihydric form and the oxidation of the diphenols to the corresponding quinones (eq. 39). A very

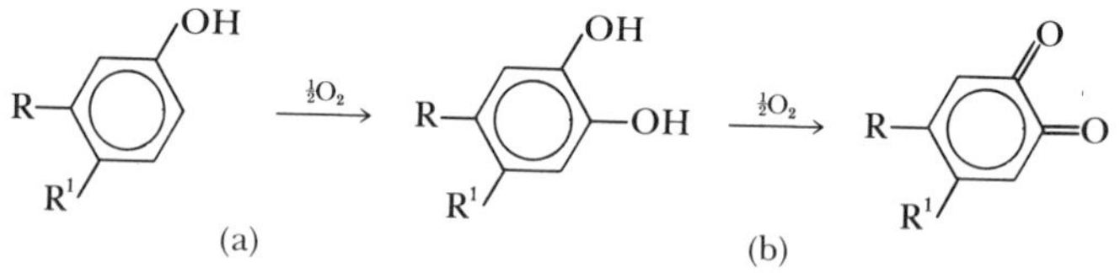

(39)

significant feature of this reaction was revealed when Mason et al. [473] demonstrated that the oxygen incorporated into the phenol in reaction (a) came from molecular oxygen. Their ^{18}O experiments on the oxidation of 3,4-dimethyl phenol gave the results shown in equation 40.

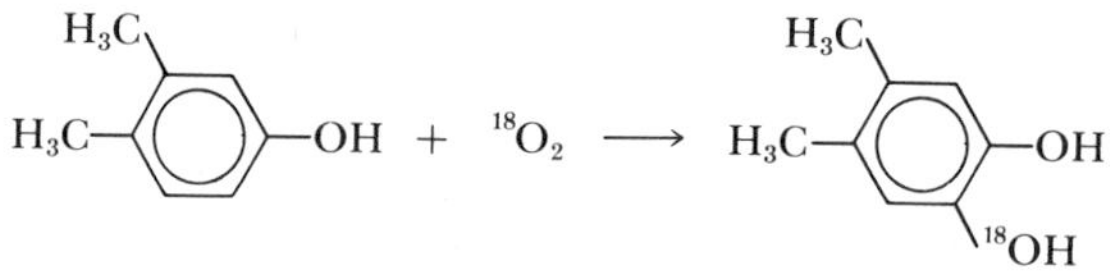

(40)

Determination of K_m and k_{cat} for the oxidation of a homologous series of p-substituted catechols by tyrosinase shows that K_m and k_{cat} decrease in the order $H > SCN > COCH_3 > CHO > CN > NO_2$, referring to the p-substituents. Both kinetic constants obey satisfactory Hammett relationships, suggesting that each is oxidized by the same mechanism [474]. Catechol is oxidized 10^4 times as rapidly as nitrocatechol, hence a density of readily removed electrons must be maintained around the phenolic hydroxyls.

Since several of these enzymes appear to be tetramers (Table 15), the number of active sites or number of Cu(I) ions per site is unclear. Presently available evidence would suggest that there is at least one site per dimer and perhaps one per monomer[474]. The aggregation and dissociation phenomena make it difficult to know just what quaternary form is active.

The nature of the oxygen complex formed with the enzyme is unclear. The information that is available on the oxygen complex shows oxygen to be very firmly bound, $K_s \sim 10^{-5} M$ [69, 474]. The kinetically determined binding constants appear to vary with the catechol used, suggesting that a tertiary complex involving enzyme–substrate–oxygen is formed [474]. Benzoic acid is a powerful inhibitor of tyrosinase and its inhibition is competitive with catechol, but noncompetitive with O_2 [474]. On the other hand, CN^- is competitive with O_2, but noncompetitive with catechol. Thus two substrate sites appear to be present, one for catechol (and perhaps phenols) and one for O_2. The latter presumably involves the metal ion. ^{14}C-Benzoic acid binding, however, is dependent on the presence of copper [474], hence both substrate sites, although kinetically distinguishable, may be related to the metal ion.

It has never been clear whether both catechol and phenol oxidation are catalyzed by the same or different sites. Enzymes (isozymes) have been isolated that differ in the ratio of the two activities, suggesting that two sites might be involved [459]. Recent data on end groups and peptide analysis indicate that these isozymes are products of combinations of the same basic subunits [475]. One molecule of oxygen does oxidize two molecules of catechol in reactions catalyzed by tyrosinase [61], and it is necessary to include a second catechol molecule in any mechanistic scheme [474]. Furthermore, search by esr for a one electron intermediate has failed to reveal one in contrast to the laccase mechanism [170] (see below). Catechols are activators of the phenolase reaction [461a, 476, 477], and the enzyme appears to have a very tight binding site for an activator catechol $(K_m = 10^{-6} M)$. This site has a much lower affinity for phenols [474]. Present information can be synthesized in a unified mechanism by proposing that a quaternary complex of enzyme, oxygen, and two catechols (or a phenol and catechol) is formed during oxidations catalyzed by tyrosinase [474]. Proof of such a scheme must await further investigation.

In contrast to tyrosinase oxidations, a clear demonstration of a catechol semiquinone has been demonstrated in the oxidation of this substrate by laccase [170]. In addition, the reduction of Cu(II) to Cu(I) during this oxidation is well documented. While laccases share the catecholase specificity with tyrosinase, they also oxidize such substrates as p-phenylenediamine. Thus some features of the mechanism must be

$$\text{hydroquinone} \xrightarrow{\text{enzymatic}} \text{semiquinone} \rightarrow 2 \text{ semiquinone}$$
$$\xrightarrow{\text{nonenzymatic}} \text{quinone} + \text{hydroquinone} \quad (41)$$

different in the laccase enzymes. The sequence shown in equation 41 has been proposed[101]. There is also evidence for heterogeniety of the copper in laccases[186] and all of it may not be Cu(II). Electron spin resonance data show that CN does combine with some of the Cu(II) in *polyporus* laccase[186]. The extensive esr results on the blue copper oxidases have been discussed in Section 2.3.4.

The azurins are the only small molecular weight proteins in this group, and the amino acid sequence of the protein from *Pseudomonas fluorescens* has been determined[472]. These enzymes are interesting in that they apparently resume their native structure after severe acid and alkali denaturation, even to the extent of regenerating the precise geometry at the copper binding site, as judged by the spectral characteristics[92, 93].

3.2.4 Non-heme Iron Enzymes. Since the early 1960's it has become clear that an iron–sulfur complex involving both cysteine sulfur and a type of labile sulfur (a bound form of inorganic sulfide) as donors forms a part of the active centers of many important enzymes participating in oxidation–reduction reactions related to photosynthesis, anaerobic bacterial metabolism, and mitochondrial oxidations. This particular type of non-heme iron has been most extensively studied in a series of small molecular weight proteins from plants and bacteria. Many of these appear to be rather similar molecules from both structural and functional characteristics. These have been classed together as the ferredoxins. The nature of the iron complex in these proteins has been discussed extensively in Sections 2.2, 2.3, and 2.5.

Several of the large metalloflavoenzymes discussed in Section 3.2.2 appear to contain a similar type of iron complex, although extensive information is not available and these will not be discussed further. Table 16 summarizes the properties of the first class of non-heme iron enzymes (the small molecular weight enzymes), including the ferredoxins. A number of excellent recent reviews are available on these enzymes[63, 156, 480–483], and only a few of the most significant functional and physicochemical features will be mentioned here.

From the data in Table 16 it is clear that the iron content and molecular weight vary considerably, although the spectral, esr, and magnetic properties indicate that the iron complex is similar in all of them. The model of the active center of *C. pasteurianum* was shown in Figure 33 involving and linear series of Fe—S_4 complexes. The model has considerable chemical evidence to support it. All the sulfur in the molecule reacts with PCMB[65]. Approximately 24 moles of PCMB react with one mole of ferredoxin[65]. Since two moles of PCMB are expected to

TABLE 16 SMALL MOLECULAR WEIGHT NON-HEME IRON PROTEINS

Enzyme	Source	MW	Oxidation–Reduction Potential, mV	Iron moles/mole	Sulfide moles/mole	References
Ferredoxin	*Clostridium pasteurianum*	5,600–6,000	−418	7–8	8	65, 484–486
	Clostridium acidi-urici	6,000		5–6	4	65
	Clostridium tetanomorphum	6,000		5–6	5	65
	Clostridium cylindrosporum	6,000		5	4	65
	Clostridium butyricum	6,000		5	5	65
Ferredoxin	Chromatium	5,600	−490	3	3	125, 487
Ferredoxin	Spinach	13,000	−432	2	2	484, 488, 489
Rubredoxin	*Clostridium pasteurianum*	6,000	−57	1	0	490, 491
High-potential iron protein	Chromatium	9,500	+350	4	4	158, 492
Adrenodoxin	Adrenal cortex	22,000	+150	2	2	493, 494
Protein component of coenzyme a-cytochrome c reductase	Bovine heart mitochondria	26,000	+220	2	2	495, 496

react with sodium sulfide and there are eight cysteine residues in the molecule, this finding is compatible with eight sulfides and eight cysteines per molecule. In addition, it is found that the sulfhydryl groups in the native protein do not react with N-ethylmaleimide or iodoacetate. They can be alkylated with iodoacetate after treatment of the protein with 2-mercaptoethanol and 8 M urea [65].

While the chemical evidence is compatible with the linear model, several arrangements of the Fe-sulfide and cysteine SH are possible. Persulfide structures such as **29** have been proposed [497]. The varying

$$
\begin{array}{c}
\text{Protein}-\text{Fe}-\text{S} \\
\diagdown \\
\text{S} \\
\diagup \\
-\text{CH}_2-\text{CH}_2 \\
\textbf{29}
\end{array}
$$

stoichiometry of iron and sulfide among the various species of ferredoxin suggest that there must be some variation in precise structure at these active centers. The amino acid sequences have been determined for the ferredoxin from *C. pasteurianum* [498] and *C. butyricum* [499], and the alternating sequence of cysteinyl residues does accommodate the model (Fig. 33).

Some doubt about the linear Fe—S_4 model has been raised by the X-ray work on the crystalline structure of these proteins. X-ray analysis on the ferredoxins from *M. aerogenes* [500] and *C. acidi-urici* [501] have progressed to 5 and 3 Å resolution, respectively, and neither structure reveals a linear array of dense Fe and S atoms which should be visible at this resolution. High resolution X-ray work has actually been hampered somewhat because of the large number of electron dense atoms.

In the case of the clostridial ferredoxins the density of these components was illustrated in striking fashion by the results of molecular weight determinations. The molecular weight was originally reported to be 12,000 [484]. The usual protein partial specific volume of 0.73, however, is not applicable to these molecules, since such a high percentage of the atoms are iron and sulfur. An accurate determination of the ferredoxin specific volume by sedimentation equilibrium in H_2O and D_2O revealed $V = 0.61$ [502]. Thus the true molecular weight of the clostridial ferredoxin is around 6000 (Table 16).

The chloroplast-type ferredoxins seem to share many of the properties of the bacterial ferredoxins, since they are frequently interchangeable in a number of ferredoxin-dependent enzymatic reactions [481]. The most striking physico-chemical finding with implications as to function is the low oxidation–reduction potential of the proteins of the ferredoxin class (Table 16), ~100 mV lower than the NAD–NADH couple. Thus ferredoxin of the chloroplast-type is able to function as an electron acceptor between the photoactivated chlorophyl molecule and the enzymes reducing NADP[156, 503], as shown in equation 42. In

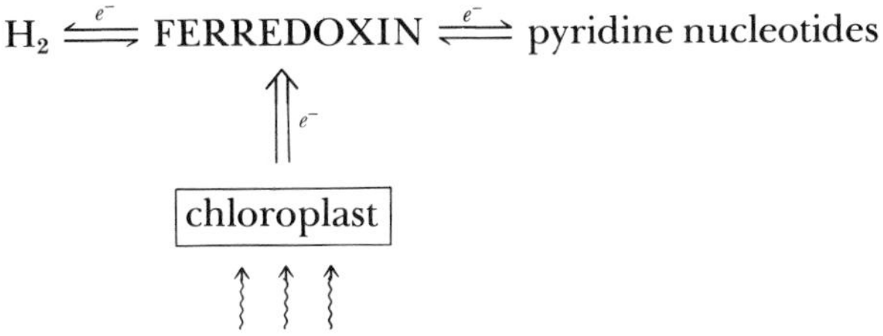

$$H_2 \underset{\longleftarrow}{\overset{e^-}{\longrightarrow}} \text{FERREDOXIN} \underset{\longleftarrow}{\overset{e^-}{\longrightarrow}} \text{pyridine nucleotides}$$

$$\uparrow \Big| e^-$$

$$\boxed{\text{chloroplast}}$$

$$\uparrow \uparrow \uparrow$$

$$(42)$$

anaerobic bacteria, particularly hydrogen-producing Clostridial species, the protein also functions in the conversion of pyruvate to acetyl phosphate, the reduction of sulfite, the production of hydrogen from hydrosulfite, the reduction of hydroxylamine to ammonia, and the reduction of N_2 to ammonia[156, 481–484]. The specific features of these reactions in the anaerobic fermentative bacteria have been reviewed by Buchanan [63]. The details of one of these reactions are shown in equations 43. Pyruvate supplies electrons for the final production of acetyl-CoA. There is some conflicting evidence on whether ferredoxin functions as a one- or two-electron carrier[481].

$$\text{Pyruvate} \xrightarrow{\text{thiamine-PP}} CO_2 + C_2 - \text{``acetaldehyde''} \qquad (43a)$$

$$C_2 - \text{``acetaldehyde''} + \text{ferredoxin}_{ox} + \text{CoA} \rightarrow \text{ferredoxin}_{red} + \text{acetyCoA}$$
$$(43b)$$

$$\text{Ferredoxin}_{red} \xleftarrow{\text{enzymes}} H_2 + \text{ferredoxin}_{ox} \qquad (43c)$$

Total reaction =

$$\text{Pyruvate} + \text{CoA} \xrightarrow[\text{ferredoxin}]{\text{thiamine-PP}} \text{acetylCoA} + H_2 + CO_2 \qquad (43d)$$

Since discovery of ferredoxin, a number of similar small molecular weight non-heme iron proteins have been isolated from various sources. All of them, however, do not have the low oxidation–reduction potential of the ferredoxins (Table 16). One of them, adrenodoxin, isolated from bovine[494] and porcine[493] adrenal cortex, is one of the protein components that participate in the 11β-hydroxylation of desoxycorticosterone [493, 494]. In many respects the iron–sulfide complex at the active center is very similar to ferredoxin, judging from esr and CD data[127], yet the protein environment clearly shifts the oxidation–reduction potential by over 500 mV to +150 mV.

4 CONCLUSION

This chapter has attempted to summarize the present state of knowledge concerning the mechanisms by which metals participate in catalysis by protein molecules by selecting for detailed discussion those enzyme systems that best illustrate the various physicochemical approaches to the study of metal enzymes. Reactions catalyzed by metalloenzymes or enzymes activated by metal ions span a remarkably broad spectrum of reaction types. Practically every class of enzymes includes at least one, if not several, metal enzymes. Those metalloenzymes for which detailed three-dimensional structural information is available from X-ray analysis of the crystalline state are clearly those about which the most detailed statements concerning mechanism can be made. At the same time it is obvious that the successful approach to the understanding of these mechanisms will come from multiple approaches involving solution chemistry complemented by the data on crystal structure.

Striking examples of the need for these multiple approaches are provided by the elegant demonstrations of the presence of coordinated water molecules in certain metalloenzymes by proton magnetic resonance and the demonstration of particular features of electron transport in several of the oxidative metalloenzymes by electron spin resonance. Neither type of information is readily accessible to present X-ray techniques. It is worth reemphasizing that the presence of the metal ion in these macromolecular catalysts provides the opportunity to apply certain powerful physicochemical techniques that depend on the chemistry of the metal. This distinctive chemistry is maintained in the protein environment and thus allows the metal ions to serve as powerful probes of the structure of these enzymes.

The combined study of solution chemistry and X-ray analysis of structure has been most successful in the case of the smaller monomeric metalloenzymes, such as carboxypeptidase and carbonic anhydrase. With the rapid advance in techniques such efforts applied to the more complex multimeric metalloenzymes containing several metal ions should reveal additional features of metal-assisted enzyme mechanisms.

ACKNOWLEDGMENTS

I thank Dr. Anders Liljas, Dr. Bror Strandberg, and Dr. Mary Ellen Riepe (Fabry) for supplying their data prior to publication. I thank my colleagues Dr. Meredithe Applebury, Dr. Michael Caplow, and Dr. June Taylor for many helpful discussions and suggestions. Original research covered in this chapter was supported by grants AM-09070 from the National Institutes of Health and GB 13344 from the National Science Foundation.

REFERENCES

1. O. Warburg, *Heavy Metal Prosthetic Groups and Enzyme Action*, London, Oxford University Press, 1949.
2. B. L. Vallee, *Enzymes*, **3**, 225 (1959).
3. R. J. P. Williams, *Enzymes*, **1**, 391 (1959).
4. B. L. Vallee, *Physiolog. Rev.*, **39**, 443 (1959).
5. B. L. Vallee and J. F. Riordan, in "Structure, Function, and Evolution in Proteins," *Brookhaven Symp. Biol.*, **21**, 91 (1968).
6. P. D. Boyer, H. Lardy, and K. Myrback, *Enzymes*, **6**, (1962).
7. P. Desnuelle, *Enzymes*, **4**, 119 (1960).
8. D. F. Waugh and D. J. Baughman, *Enzymes*, **4**, 215 (1960).
9. A. J. Lotka, *Elements of Mathematical Biology*, Dover Publications, New York, 1956, p. 194.
10. A. E. Martell and M. Calvin, *Chemistry of the Metal Chelate Compounds*, Prentice-Hall, Englewood Cliffs, New Jersey, 1952.
11. L. E. Orgel, *An Introduction to Transition-Metal Chemistry Ligand-Field Theory*, Methuen, London, 1960.
12. F. Basolo and R. C. Johnson, *Coordination Chemistry*, Benjamin, New York, 1964.
13. F. A. Cotton and F. R. S. Wilkinson, *Advanced Inorganic Chemistry*, 2nd ed., Interscience, New York, 1966.
14. J. P. Hunt, *Metal Ions in Aqueous Solution*, Benjamin, New York, 1965.
15. L. G. Sillén and A. E. Martell, *Chem. Soc. Spec. Publ.* **17**, (1964).
16. J. T. Edsall and J. Wyman, *Biophysical Chemistry*, Academic Press, New York, 1958, p. 477.
17. L. M. Riddiford, *J. Biol. Chem.*, **239**, 1079 (1964).
18. L. M. Riddiford, R. H. Stellwagen, S. Mehta, and J. T. Edsall, *J. Biol. Chem.*, **240**, 3305 (1965).
19. C. Tanford and J. Epstein, *J. Amer. Chem. Soc.*, **76**, 2163, 2170 (1954).
20. K. Marcker, *Acta Chem. Scand.*, **14**, 2071 (1960).
21. D. C. Hodgkin, "Three Dimensional Structure of Insulin by X-ray Crystallography," 8th International Congress of Crystallography, New York, September, 1969.
22. J. E. Coleman and B. L. Vallee, *J. Biol. Chem.*, **235**, 390 (1960).
23. B. L. Vallee and J. E. Coleman, *Comprehensive Biochem.*, **12**, 165 (1964).
24. J. H. R. Kägi and B. L. Vallee, *J. Biol. Chem.*, **235**, 3188 (1960).
25. J. E. Coleman, *Biochemistry*, **4**, 2644 (1965).
26. M. L. Applebury and J. E. Coleman, *J. Biol. Chem.*, **244**, 308 (1969).
27. B. L. Vallee, R. J. P. Williams and J. E. Coleman, *Nature*, **190**, 633 (1961).
28. J. E. Coleman and B. L. Vallee, *J. Biol. Chem.*, **236**, 2244 (1961).
29. J. E. Coleman, Metallocarboxypeptidases, Physicochemical and Enzymatic Properties, Doctoral Dissertation, Massachusetts Institute of Technology, 1963.
30. J. A. Rupley and H. Neurath, *J. Biol. Chem.*, **235**, 609 (1960).

31. R. P. Davis, *Enzymes*, **5**, 545 (1961).
32. R. W. Henkens and J. M. Sturtevant, *J. Amer. Chem. Soc.*, **90**, 2669 (1968).
33. S. Lindskog and B. G. Malmstrom, *J. Biol. Chem.*, **237**, 1129 (1962).
34. E. E. Rickli and J. T. Edsall, *J. Biol. Chem.*, **237**, PC258 (1961).
35. S. Lindskog and P. O. Nyman, *Biochim. Biophys. Acta*, **85**, 462 (1964).
36. S. R. Cohen and I. B. Wilson, *Biochemistry*, **5**, 904 (1966).
37. J. E. Coleman, *J. Biol. Chem.*, **242**, 5212 (1967).
38. D. R. Stranks, in *Modern Coordination Chemistry*, J. Lewis and R. G. Wilkins, Eds., Interscience, New York, 1960, p. 78.
39. M. Eigen and R. G. Wilkins, *Advan. Chem. Ser.*, **49**, 55 (1965).
40. R. H. Holyer, C. D. Hubbard, S. F. A. Kettle, and R. G. Wilkins, *Inorg. Chem.*, **4**, 929 (1965).
41. R. H. Holyer, C. D. Hubbard, S. F. A. Kettle, and R. G. Wilkins, *Inorg. Chem.*, **5**, 622 (1966).
42. D. H. Busch, in *Cobalt, Its Chemistry, Metallurgy, and Uses*, R. S. Young, Ed., Rheinhold, New York, 1960, p. 88.
43. F. J. C. Rossotti, in *Modern Coordination Chemistry*, J. Lewis and R. G. Wilkins, Eds., Interscience, New York, 1960, p. 1.
44. J. E. Coleman and B. L. Vallee, *Biochemistry*, **3**, 1874 (1964).
45. J. E. Coleman and B. L. Vallee, *J. Biol. Chem.*, **237**, 3430 (1962).
46. J. E. Coleman and B. L. Vallee, *Biochemistry*, **1**, 1083 (1962).
47. H. Dressler and C. R. Dawson, *Biochim. Biophys. Acta*, **45**, 508, 515 (1960).
48. B. Tilander, B. Strandberg, and K. Fridborg, *J. Mol. Biol.*, **12**, 740 (1965).
49. D. J. Plocke and B. L. Vallee, *Biochemistry*, **1**, 1039 (1962).
50. H. Csopak, *Europ. J. Biochem.*, **7**, 186 (1969).
51. M. I. Harris and J. E. Coleman, *J. Biol. Chem.*, **243**, 5063 (1968).
52. M. L. Applebury, B. P. Johnson, and J. E. Coleman, *J. Biol. Chem.*, **245**, 4968 (1970).
53. F. L. Hoch and B. L. Vallee, *J. Biol. Chem.*, **221**, 491 (1956).
54. B. L. Vallee and T. L. Coombs, *J. Biol. Chem.*, **234**, 2615 (1959).
55. C. R. Dawson, in *The Biochemistry of Copper*, J. Peisach, P. Aisen, and W. E. Blumberg, Eds., Academc Press, New York, 1966, p. 305.
56. Z. Penton, Ph.D. Thesis, Columbia University, 1964.
57. A. Tissieres, *Nature*, **162**, 340 (1948).
58. F. Kubowitz, *Biochem. Z.*, **299**, 32 (1938).
59. D. Kertész, in *The Biochemistry of Copper*, J. Peisach, P. Aisen, and W. E. Blumberg, Eds., Academic Press, New York, 1966, p. 359.
60. Z. G. Penton and C. R. Dawson, in *Oxidases and Related Redox Systems*, T. E. King, H. S. Mason, and M. Morrison, Eds., Wiley, New York, 1965, p. 222.
61. D. W. Brooks and C. R. Dawson, *The Biochemistry of Copper*, J. Peisach, P. Aisen, and W. E. Blumberg, Eds., Academic Press, New York, 1966, p. 343.
62. B. G. Malmström, B. Reinhammar, and T. Vänngärd, *Biochim. Biophys. Acta*, **156**, 67 (1968).
63. B. B. Buchannan, *Struct. Bonding*, **1**, 109 (1966).
64. K. T. Fry and A. San Pietro, *Biochem. Biophys. Res. Commun.*, **9**, 218 (1962).
65. W. Lovenberg, B. B. Buchannan, and J. C. Rabinowitz, *J. Biol. Chem.*, **238**, 3899 (1963).
66. T. M. Dunn, in *Modern Coordination Chemistry*, J. Lewis and R. G. Wilkins, Eds., Interscience, New York, 1960, p. 229.
67. E. B. Sandell, *Colorimetric Determinations of Traces of Metals*, 3rd ed., Interscience, New York, 1959.
68. J. Peisach, P. Aisen, and W. E. Blumberg, *The Biochemistry of Copper*, Academic Press, New York, 1966.

69. E. Frieden, S. Osaki, and H. Kobayashi, *Oxygen*, Little Brown, Boston, 1965, p. 213.

70. W. E. Blumberg, in *The Biochemistry of Copper*, J. Peisach, P. Aisen, and W. E. Blumberg, Eds., Academic Press, New York, 1966, p. 49.

71. A. S. Brill, G. F. Bryce, and H. J. Maria, *Biochim. Biophys. Acta*, **154**, 342 (1968).

72. A. S. Brill and C. F. Bryce, *J. Chem. Phys.*, **48**, 4398 (1968).

73. A. Liljas, P. C. Bergstén, U. Carlbom, K. Fridborg, L. Järup, K. K. Kannan, S. Lövgren, M. Petef, B. Strandberg, and I. Weapa, in preparation.

74. S. Lindskog, *J. Biol. Chem.*, **238**, 945 (1963).

75. J. E. Coleman, *Nature*, **214**, 193 (1967).

76. J. E. Coleman, *J. Amer. Chem. Soc.*, **89**, 6757 (1967).

77. K. Fridborg, K. K. Kannan, A. Liljas, J. Lundin, B. Strandberg, R. Strandberg, B. Tilander, and G. Wiren, *J. Mol. Biol.*, **25**, 505 (1967).

78. J. E. Coleman, *Proc. Natl. Acad. Sci. U.S.*, **59**, 123 (1968).

79, A. C. Dennard and R. J. P. Williams, *Trans. Metal Chem.*, **2**, 115 (1966).

80. S. Lindskog and A. Ehrenberg, *J. Mol. Biol.*, **24**, 133 (1967).

81. S. Lindskog, *Biochemistry*, **5**, 2641 (1966).

82. M. E. Riepe and J. H. Wang, *J. Biol. Chem.*, **243**, 2779 (1968).

83. E. T. Kaiser and K-W. Lo, *J. Amer. Chem. Soc.*, **91**, 4912 (1969).

84. J. E. Coleman, in *Symposium on the Chemical, Biochemical and Physiological Aspects of CO_2*, F. J. W. Roughton, J. T. Edsall, A. B. Otis, and R. E. Forster, Eds., NASA Publication, SP-188, 1970, p. 141.

85. J. A. Schellman, *J. Chem. Phys.*, **44**, 55 (1966).

86. M. L. Applebury and J. E. Coleman, *J. Biol. Chem.*, **244**, 709 (1969).

87. B. L. Vallee and R. J. P. Williams, *Chem. Britain*, **4**, 397 (1968).

88. R. T. Simpson and B. L. Vallee, *Biochemistry*, **7**, 4343 (1968).

89. H. C. Freeman, *Advan. Protein Chem.*, **22**, 257 (1967).

90. S. A. Latt and B. L. Vallee, *Federation Proc.* **28**, 534 (1969).

91. T. A. Duff and J. E. Coleman, *Biochemistry*, **5**, 2009 (1966).

92. S-P.W. Tang, J. E. Coleman, and Y. P. Myer, *J. Biol. Chem.*, **243**, 4286 (1968).

93. H. J. Maria, *Nature*, **209**, 1023 (1966).

94. C. R. Dawson, *Ann, N.Y. Acad. Sci.*, **88**, 353 (1960).

95. W. E. Blumberg, J. Wisinger, P. Aisen, A. G. Morell, and I. H. Scheinberg, *J. Biol. Chem.*, **238**, 1675 (1963).

96. W. L. Koltun, R. H. Roth, and F. R. N. Gurd, *J. Biol. Chem.*, **238**, 124 (1963).

97. G. F. Bryce, J. M. H. Pinkerton, L. K. Steinrauf, and F. R. N. Gurd, *J. Biol. Chem.*, **240**, 3829 (1965).

98. F. R. N. Gurd and G. F. Bryce, in *The Biochemistry of Copper*, J. Peisach, P. Aisen, and W. E. Blumberg, Eds., Academic Press, New York, 1966, p. 115.

99. R. P. Ambler, *Biochem. J.*, **89**, 341 (1963).

100. H. S. Mason, *Biochem. Biophys. Res. Commun.*, **10**, 11 (1963).

101. W. G. Levine, in *The Biochemistry of Copper*, J. Peisach, P. Aisen, and W. E. Blumberg, Eds., Academic Press, New York, 1966, p. 371.

102. C. G. Holmberg and C. B. Laurell, *Acta Chem. Scand.*, **2**, 550 (1948).

103. F. J. Dunn and C. R. Dawson, *J. Biol. Chem.*, **189**, 485 (1951).

104. T. Nakamura, *Biochim. Biophys. Acta*, **30**, 44 (1958).

105. H. Beinert, D. E. Griffiths, D. C. Wharton, and R. H. Sands, *J. Biol. Chem.*, **237**, 2337 (1962).

106. H. Beinert, in *The Biochemistry of Copper*, J. Peisach, P. Aisen, and W. E. Blumberg Eds., Academic Press, New York, p. 213.

107. D. C. Wharton and Q. H. Gibson, in *The Biochemistry of Copper*, J. Peisach, P. Aisen, and W. E. Blumberg, Eds., Academic Press, New York, 1966, p. 235.

108. B. F. van Gelder and E. C. Slater, in *The Biochemistry of Copper*, J. Peisach, P. Aisen, and W. E. Blumberg, Eds., Academic Press, New York, 1966, p. 245.

109. H. Dobbie and W. O. Kermack, *Biochem. J.*, **59**, 246 (1955).

110. S. P. Datta and B. R. Rabin, *Trans. Faraday Soc.*, **52**, 1123 (1956).

111. M. K, Kim and A. E. Martell, *Biochemistry*, **3**, 1169 (1964).

112. W. L. Koltun, M. Fried, and F. R. N. Gurd, *J. Amer. Chem. Soc.*, **82**, 233 (1960).

113. T. Yasui, J. Hidaka, and Y. Shimura, *J. Amer. Chem. Soc.*, **87**, 2762 (1965).

114. B. H. Campbell, F. S. Shu, and S. Hubbard, *Biochemistry*, **2**, 764 (1963).

115. J. E. Coleman, *Biochem. Biophys. Res. Commun.*, **24**, 208 (1966).

116. G. Nilsson, *Z. Anal. Chem.*, **153**, 161 (1956).

117. M. Ciampolini, *Struct. Bonding*, **6**, 52 (1969).

118. L. Sacconi, P. L. Orioli, and M. DiVaira, *J. Amer. Chem. Soc.*, **87**, 2059 (1965).

119. L. Sacconi, P. Nannelli, N. Nardi, and U. Campigli, *Inorg. Chem.*, **4**, 943 (1965).

120. Z. Dori and H. B. Gray, *J. Amer. Chem. Soc.*, **88**, 1394 (1966).

121. A. D. Liehr, *J. Phys. Chem.*, **68**, 665 (1964).

122. P. Handler, K. V. Rajagopalan, and V. Aleman, *Federation Proc.*, **23**, 30 (1964).

123. T. P. Singer and B. Kearney, *Enzymes*, **7**, 383 (1963).

124. R. Hill and A. San Pietro, *Z. Naturforsch.*, **B18**, 677 (1963).

125. D. I. Arnon, *Science*, **149**, 1460 (1965).

126. H. E. Davenport, *in Non-heme Iron Proteins: Role in Energy Conversion*, A. San Pietro, Ed.), Antioch Press, Yellow Springs, Ohio, 1965, p. 115.

127. G. Palmer, H. Brintzinger, R. Estabrook, and R. H. Sands, in *Magnetic Resonance in Biological Systems*, A. Ehrenberg, B. G. Malmstrom and T. Vanngard, Eds., Pergamon Press, New York, 1967, p. 159.

128. F. A. Cotton, in *Modern Coordination Chemistry*, J. Lewis and R. G. Wilkins, Eds., Interscience, New York, 1960, p. 301.

129. B. N. Figgis and J. Lewis, in *Modern Coordination Chemistry*, J. Lewis and R. G. Wilkins, Eds., Interscience, New York, 1960, p. 400.

130. F. A. Cotton and G. Wilkinson, *Advances Inorganic Chemistry*, 2nd ed. Interscience, New York, 1966, p. 695.

131. R. G. Shulman, S. Ogawa, K. Wuthrich, T. Yamane, J. Peisach, and W. E. Blumberg, *Science*, (1969) 251.

132. L. Pauling and C. D. Coryell, *Proc. Natl. Acad. Sci. U.S.*, **22**, 159, 210 (1936).

133. W. D. Phillips, E. Knight, and D. C. Blomstrom, in *Non-heme Iron Proteins*, A. San Pietro, Ed., Antioch Press, Yellow Springs, Ohio, 1965, p. 69.

134. A. Ehrenberg and R. C. Bray, *Arch. Biochem. Biophys.*, **109**, 199 (1965).

135. A. Hudson and G. R. Luckhurst, *Chem. Rev.*, **69**, 191 (1969).

136. A. Ehrenberg, B. G. Malmstrom, and T. Vänngård, *Magnetic Resonance in Biological Systems*, Pergamon Press, New York, 1967.

137. S. I. Chan, B. M. Fung, and H. Lutje, *J. Chem. Phys.*, **47**, 2121 (1967).

138. H. Levanon and Z. Luz, *J. Chem. Phys.*, **49**, 2031 (1968).

139. B. G. Malmström, T. Vänngård, and M. Larsson, *Biochim. Biophys. Acta*, **30**, 1 (1958).

140. A. S. Mildvan and M. Cohn, *Biochem.*, **2**, 910 (1963).

141. W. J. O'Sullivan and M. Cohn, *J. Biol. Chem.*, **241**, 3104 (1966).

142. G. Reed and M. Cohn, *J. Biol. Chem.*, **245**, 662 (1970).

143. D. Lopiekes and S. Liebman, as reported in G. Reed and M. Cohn, *J. Biol. Chem.*, **245**, 662 (1970).

144. B. Bleaney and R. S. Rubins, *Proc. Phys. Soc.*, **77**, 103 (1961).

145. H. Beinert and R. H. Sands, *Biochem. Biophys. Res. Commun.*, **3**, 41 (1960).

146. K. V. Rajagopalan, V. Aleman, P. Handler, W. Heinen, G. Palmer, and H. Beinert, *Biochem. Biophys. Res. Commun.*, **8**, 220 (1962).

147. R. C. Bray, G. Palmer, and H. Beinert, *J. Biol. Chem.*, **239**, 2667 (1964).
148. G. Palmer, R. C. Bray and H. Beinert, *J. Biol. Chem.*, **239**, 2657 (1964).
149. H. Beinert, in *Non-heme Iron Proteins*, A. San Pietro, Ed., Antioch Press, Yellow Springs, Ohio, 1965, p. 23.
150. G. Palmer, H. Brintzinger, R. W. Estabrook, and R. H. Sands, in *Magnetic Resonance in Biological Systems*. A. Ehrenberg, B. G. Malmström and T. Vanngrad, Eds., Pergamon Press, New York, 1967, p. 159.
151. Y. I. Shethna, P. W. Wilson, R. E. Hansen, and H. Beinert, *Proc. Natl. Acad. Sci. U.S.*, **52**, 1263 (1964).
151a. J. C. M. Tsibris, R. L. Tsai, I. C. Gunsalus, W. H. Orme-Johnson, R. E. Hansen, and H. Beinert, *Proc. Natl. Acad. Sci. U.S.*, **59**, 959 (1968).
152. K. V. Rajagopalan and P. Handler, *J. Biol. Chem.*, **239**, 1509 (1964).
153. T. P. Singer, in *Non-heme Iron Proteins*, A. San Pietro, Ed., Antioch Press, Yellow Springs, Ohio, 1965, p. 349.
154. G. Palmer and R. H. Sands, *J. Biol. Chem.*, **241**, 253 (1966).
155. G. Palmer, R. H. Sands, and L. E. Mortensen, *Biochem. Biophys. Res. Commun.*, **23**, 357 (1966).
156. D. I. Arnon, in *Non-heme Iron Proteins*, A. San Pietro, Ed., Antioch Press, Yellow Springs, Ohio, 1965, p. 137.
157. B. E. Sobel and W. Lovenberg, *Biochemistry*, **5**, 6 (1966).
158. A. J. Bearden and T. H. Moss, in *Magnetic Resonance in Biological Systems*, Pergamon Press, New York, 1967, p. 391.
159. D. C. Blomstrom, E. Knight, W. D. Phillips, and J. F. Weicher, *Proc. Natl. Acad. Sci. U.S.*, **51**, 1085 (1964).
160. M. Tanaka, A. M. Benson, H. F. Mower, and K. T. Yasunobu, in *Non-heme Iron Proteins*, A. San Pietro, Ed., Antioch Press, Yellow Springs, Ohio, 1965, p. 221.
161. T. C. Hollocher, F. Solomon, and T. E. Ragland, *J. Biol. Chem.*, **241**, 3452 (1966).
162. J. M. Assour, *J. Amer. Chem. Soc.*, **87**, 4701 (1965).
163. G. N. Schrauzer and L. P. Lee, *J. Amer. Chem. Soc.*, **90**, 6541 (1968).
164. H. Kon and N. E. Sharpless, *Spectr. Letters*, **49** (1968).
165. J. H. Bayston, N. K. King, F. D. Looney, and M. E. Winfield, *J. Amer. Chem. Soc.*, **91**, 2775 (1969).
166. B. M. Hoffman, D. L. Diemente, and F. Basolo, *J. Amer. Chem. Soc.*, **92**, 61 (1970).
167. Y. Ting and P. Williams, *Phys. Rev.*, **82**, 507 (1951).
168. J. H. E. Griffiths and J. Owen, *Proc. Roy. Soc. (London)*, **A213**, 459 (1952); **A213**, 951 (1952); **A226**, 96 (1954).
169. A. K. Wiersema and J. J. Windle, *J. Phys. Chem.*, **68**, 2316 (1964).
170. T. Nakamura and Y. Ogura, in *Magnetic Resonance in Biological Systems*, A. Ehrenberg, B. G. Malmström, and T. Vänngård, Eds., Pergamon Press, New York, 1967, p. 205.
171. B. G. Malmström and T. Vänngård, *J. Mol. Biol.*, **2**, 118 (1960).
172. A. H. Maki and B. R. McGarvey, *J. Phys. Chem.*, **29**, 35 (1958).
173. E. M. Roberts and W. S. Koski, *J. Amer. Chem. Soc.*, **82**, 3006 (1960); **83**, 1865 (1961).
174. D. Kivelson and R. Neiman, *J. Chem. Phys.*, **35**, 149 (1961); **35**, 156 (1961).
175. H. R. Gersmann and J. D. Swalen, *J. Chem. Phys.*, **36**, 3221 (1962).
176. R. Aasa, B. G. Malmström, P. Saltman and T. Vänngård, *Biochim. Biophys. Acta*, **75**, 203 (1963).
177. J. J. Windle, A. K. Wiersema, J. R. Clark, and R. E. Feeney, *Biochemistry*, **2**, 1341 (1963).
178. G. F. Bryce, *J. Phys. Chem.*, **70**, 3549 (1966).
179. J. H. Venable and A. S. Brill, personal communication.
180. A. S. Brill and J. H. Venable, in *The Biochemistry of Copper* (J. Peisach, P. Aisen, and

W. E. Blumberg, Eds. Academic Press, New York, 1966, p. 67.

181. S. Bouchilloux, P. McMahill, and H. S. Mason, *J. Biol. Chem.*, **238**, 1699 (1963).

182. M. Fling, N. H. Horowitz, and S. E. Heinemann, *J. Biol. Chem.*, **238**, 2045 (1963).

183. T. Nakamura, S. Sho, and Y. Ogura, *J. Biochem. (Tokyo)*, **59**, 481 (1966).

184. W. E. Blumberg, J. Eisinger, P. Aisen, A. G. Morell, and I. H. Scheinberg, *J. Biol. Chem.*, **238**, 1675 (1963).

185. T. Vänngard, in *Magnetic Resonance in Biological Systems*, A. Ehrenberg, B. G. Malmström, and T. Vänngard, Eds., Pergamon Press, New York, 1967, p. 213.

186. B. G. Malmström, B. Reinhammar, and T. Vänngard, *Biochim. Biophys. Acta*, **156**, 67 (1965).

187. A. Ehrenberg and T. Yonetani, *Acta Chem. Scand.*, **15**, 1071 (1961).

188. I. Yamazaki and L. H. Piette, *Biochim. Biophys. Acta*, **50**, 62 (1961).

189. T. Nakamura and Y. Ogura, in *The Biochemistry of Copper*, J. Peisach, P. Aisen and W. E. Blumberg, Eds., Academic Press, New York, 1966, p. 389.

190. W. E. Blumberg, W. G. Levine, S. Margolis, and J. Peisach, *Biochem. Biophys. Res. Commun.*, **15**, 277 (1964).

191. W. E. Blumberg, in *The Biochemistry of Copper*, J. Peisach, P. Aisen, and W. E. Blumberg, Eds. Academic Press, New York, 1966, p. 399.

192. H. Beinert, in *The Biochemistry of Copper*, J. Peisach, P. Aisen and W. E. Blumberg, Eds), Academic Press, New York, 1966, p. 436.

193. M. Goldstein, in *The Biochemistry of Copper*, J. Peisach, P. Aisen and W. E. Blumberg, Eds., Academic Press, New York, 1966, p. 443.

194. C. Remy, D. A. Richert, R. H. Doisy, I. C. Wells, and W. W. Westerfeld, *J. Biol. Chem.*, **217**, 293 (1955).

195. E. C. DeResnzo, *Advances in Enzymology*, Vol. 17, F. F. Nord, Ed., Interscience, New York, 1956, p. 293.

196. I. Fridovich and P. Handler, *J. Biol. Chem.*, **231**, 899, 1581 (1958).

197. P. G. Avis, F. Bergel, and R. C. Bray, *J. Chem. Soc.*, **1955**, 1100; **1956**, 1219.

198. K. V. Rajagopalan, I. Fridovich, and P. Handler, *J. Biol. Chem.*, **237**, 922 (1962).

199. R. C. Bray, P. F. Knowles, and L. S. Meriwether, in *Magnetic Resonance in Biological Systems*, A. Ehrenberg, B. G. Malmstrom, and T. Vangard, Eds., Pergamon Press, New York, 1967, p. 249.

200. R. J. P. Williams, in *Advances in the Chemistry of the Coordination Compounds*, S. Kirschner, Ed., MacMillan, New York, 1961, p. 65.

201. C. L. Hamilton and H. M. McConnell, in *Structural Chemistry and Molecular Biology*, A. Rich and N. Davidson, Eds., Freeman, San Francisco, 1968, p. 115.

202. O. H. Griffith and A. S. Waggoner, *Accounts Chem. Res.*, **2**, 17, (1969).

203. T. J. Stone, T. Buckman, P. L. Nordio, and H. M. McConnell, *Proc. Natl. Acad. Sci. U.S.*, **54**, 1010 (1965).

204. H. M. McConnell and C. L. Hamilton, *Proc. Natl. Acad. Sci. U.S.*, **60**, 776 (1968).

205. W. L. Hubbell and H. M. McConnell, *Proc. Natl. Acad. Sci. U.S.*, **63**, 16 (1969).

206. E. G. Rozantzev and M. B. Neiman, *Tetrahedron*, **20**, 131 (1964).

207. E. G. Rozantzev and L. A. Krinitzkaya, *Tetrahedron*, **21**, 491 (1965).

208. A. K. Hoffman and A. T. Henderson, *J. Amer. Chem. Soc.*, **83**, 4671 (1961).

209. J. S. Taylor, P. Mushak, and J. E. Coleman, *Proc. Nat. Acad. Sci. U.S.*, **67**, 1410 (1970).

210. S. Lindskog and A. Thorslund, *Europ. J. Biochem.* **3**, 453 (1968).

211. M. S. Itzkowitz, Ph.D. Thesis, California Institute of Technology, 1966.

212. R. F. Chen and J. C. Kernohan, *J. Biol. Chem.*, **242**, 5813 (1967).

213. J. S. Taylor, J. S. Leigh and M. Cohn, *Proc. Natl. Acad. Sci. U.S.*, **64**, 219 (1969).

214. D. C. Watts and B. R. Rabin, *Biochem. J.*, **85**, 507 (1962).

215. H. Weiner, *Biochemistry*, **8**, 526 (1969).

216. A. S. Mildvan and H. Weiner, *Biochemistry*, **8**, 552 (1969).
217. D. H. Meadows, J. L. Markley, J. S. Cohen, and O. Jardetzky, *Proc. Natl. Acad. Sci. U.S.*, **58**, 1307 (1967).
218. A. S. Mildvan and M. Cohn, *Advances in Enzymology*, **33**, 1 (1970).
219. O. Jardetzky, *Avdn. Chem. Phys.*, **7**, 499 (1964).
220. I. Solomon,.*Phys. Rev.*, **99**, 559 (1955).
221. I. Solomon and N. Bloembergen, *J. Chem. Phys.*, **25**, 261 (1956).
222. N. Bloembergen, *J. Chem. Phys.*, **27**, 572 (1957).
223. R. G. Shulman, G. Navon, B. J. Wyluda, D. C. Douglass, and T. Yamane, *Proc. Natl. Acad. Sci. U.S.*, **56**, 39 (1966).
224. G. Navon, R. G. Shulman, B. J. Wyluda, and T. Yamane, *Proc. Natl. Acad. Sci. U.S.*, **60**, 86 (1968).
225. W. J. O'Sullivan, R. Virden, and S. Blethen, *Europ. J. Biochem.*, **8**, 562 (1969).
226. W. J. O'Sullivan and L. Noda, *J. Biol. Chem.*, **243**, 1424 (1968).
227. R. D. Kobes, A. S. Mildvan, and W. J. Rutter, Abstracts, 158th Meeting, American Chemical Society, New York, 1969.
228. M. E. (Riepe) Fabry, S. H. Koenig, and W. E. Schillinger, *J. Biol. Chem.*, **245**, 4256 (1970).
229. R. L. Ward and P. A. Srere, *Biochem. Biophys. Acta*, **99**, 270 (1965).
230. M. Cohn, in *Magnetic Resonance in Biological Systems*, A. Ehrenberg, B. G. Malmström, and T. Vanngard, Eds., Pergamon Press, New York, 1967, p. 101.
231. M. Cohn and J. S. Leigh, Jr., *Nature*, **193**, 1037 (1962).
232. R. S. Miller, A. S. Mildvan, H. C. Chang, R. L. Easterday, H. Maruyama, and M. D. Lane, *J. Biol. Chem.*, **243**, 6030 (1968).
233. A. S. Mildvan, M. C. Scrutton, and M. F. Utter, *J. Biol. Chem.*, **241**, 3488 (1966).
234. A. S. Mildvan, R. W. Estabrook, and G. Palmer, in *Magnetic Resonance in Biological Systems*, A. Ehrenberg, B. G. Malmström, and T. Vanguard, Eds., Pergamon Press, New York, 1967, p. 175.
235. M. C. Scrutton and A. S. Mildvan, in *Symposium on CO_2: Chemical Biochemical and Physiological Aspects*, J. T. Edsall, R. E. Forster, A. B. Otis, and F. J. W. Roughton, Eds., NASA publication, SP-188 (1969), p. 207.
236. M. Eigen and K. Z. Tamm, *Elektrochem.*, **66**, 107 (1962).
237. M. C. Scrutton and A. S. Mildvan, *Biochemistry*, **7**, 1490 (1968).
238. A. S. Mildvan and M. C. Scrutton, *Biochemistry*, **6**, 2978 (1967).
239. G. K. Wertheim, *Mössbauer Effect: Principle and Applications*, Academic Press, New York, 1964.
240. C. E. Johnson, R. C. Bray, and P. F. Knowles, in *Magnetic Resonance in Biological Systems*, A. Ehrenberg, B. G. Malmström and T. Vänngård, Eds., Pergamon Press, New York, 1967, p. 417.
241. A. Werner, *Neurere Anschauungen auf dem Gebiete der anorganischen Chemie*, 4th ed., Friedrich Vieweg and Son, Brunswick, 1920, p. 44.
242. Werner, A., *Chem. Ber,*, **34**, 2584 (1901).
243. L. Pauling, *Nature of the Chemical Bond*, 2nd ed., Cornell University Press, Ithaca, New York, 1962.
244. K. S. V. S. Kumar, K. A. Walsh, J. P. Bargetzi, and H. Neurath, *Biochemistry*, **2**, 1475 (1963).
245. W. N. Lipscomb, J. A. Hartsuck, G. N. Reeke, F. A. Quiocho, P. H. Bethge, M. L. Ludwig, T. A. Steitz, H. Muirhead, and J. C. Coppola, in *Structure, Function, and Evolution of Proteins*, Brookhaven Symposium in Biology, No. 21, 1969, p. 24.
245a. W. N. Lipscomb, J. A. Hartsuck, F. A. Quiocho, and G. N. Reeke, *Proc. Natl. Acad. Sci. U.S.* **64**, 28 (1969).

246. K. A. Walsh, L. H. Ericsson, R. A. Bradshaw, and H. Neurath, *Biochemistry*, **9**, 219 (1970).
247. J. J. Hoppe and J. E. Prue, *J. Chem. Soc.*, **1957**, 1775.
248. J. E. Prue, *J. Chem. Soc.*, **1952**, 2331.
249. R. Steinberger and F. H. Westheimer, *J. Amer. Chem. Soc.*, **73**, 429 (1951).
250. J. F. Speck, *J. Biol. Chem.*, **178**, 315 (1949).
251. C. Lazdunski, C. Petitclere, and M. Lazdunski, *Europ. J. Biochem.*, **8**, 510 (1969).
252. K. M. Wilbur and N. G. Anderson, *J. Biol. Chem.*, **176**, 147 (1948).
253. B. L. Vallee, J. F. Riordan, and J. E. Coleman, *Proc. Natl. Acad. Sci. U.S.*, **49**, 109 (1963).
254. A. Garen and C. Levinthal, *Biochim. Biophys. Acta*, **38**, 470 (1960).
255. D. J. Plocke, C. Levinthal, and B. L. Vallee, *Biochemistry*, **1**, 373 (1962).
256. J. A. Reynolds and M. J. Schlesinger, *Biochemistry*, **8**, 588 (1969).
257. J. A. Reynolds and M. J. Schlesinger, *Biochemistry*, **6**, 3552 (1967).
258. M. J. Schlesinger and K. Barrett, *J. Biol. Chem.*, **240**, 4284 (1965).
259. R. T. Simpson, B. L. Vallee, and G. H. Tait, *Biochemistry*, **7**, 4336 (1968).
260. J. H. Schwartz and F. Lipmann, *Proc. Natl. Acad. Sci. U.S.*, **47**, 1996 (1961).
261. L. Engström, *Arkiv Kimi*, **19**, 129 (1962).
262. L. Engström, *Biochim. Biophy. Acta*, **56**, 606 (1962).
263. M. M. Pigretti and C. Milstein, *Biochem. J.*, **94**, 106 (1965).
264. S. H. D. Ko and F. J. Kezdy, *J. Amer. Chem. Soc.*, **89**, 7139 (1967).
265. D. R. Trentham and H. Gutfreund, *Biochem. J.*, **106**, 455 (1968).
266. H. N. Fernley and P. G. Walker, *Biochem. J.*, **111**, 187 (1969).
267. C. Lazdunski, C. Petitclerc, D. Chappelet, and M. Lazduncki, *Biochem. Biophys. Res. Commun.*, **37**, 744 (1969).
268. S. S. Stein and D. E. Koshland, *Arch. Biochem. Biophys.*, **39**, 229 (1952).
269. James H. Schwartz, *Proc. Natl. Acad. Sci. U.S.*, **49**, 871 (1963).
270. James H. Schwartz, A. M. Crestfield, and F. Lipmann, *Proc. Natl. Acad. Sci. U.S.*, **49**, 722 (1963).
271. H. Barrett, R. Butler, and I. B. Wilson, *Biochemistry*, **8**, 1042 (1969).
272. D. Levine, T. W. Reid, and I. B. Wilson, *Biochemistry*, **9**, 2374 (1969).
273. T. W. Reid, M. Pavlic, D. J. Sullivan, and I. B. Wilson, *Biochemistry*, **8**, 3184 (1969).
274. M. J. Schlesinger and C. J. Levinthal, *Mol. Biol.*, **7**, 1 (1963).
275. M. S. Mohamed and D. M. Greenberg, *Arch. Biochem.*, **8**, 349 (1945).
276. D. M. Greenberg, *Enzymes* **4**, 257 (1960).
277. S. Udenfriend, C. T. Clark, J. Axelrod, and B. B. Brodie, *J. Biol. Chem.*, **208**, 731 (1954).
278. H. S. Mason, *Ann. Rev. Biochem.*, **34**, 595 (1965).
279. R. O. C. Norman and G. K. Radda, *Proc. Chem. Soc.*, **138** (1962).
280. G. A. Hamilton, R. J. Workman, and L. Woo, *J. Amer. Chem. Soc.*, **86**, 3390 (1964).
281. G. A. Hamilton, J. P. Friedman, and P. M. Campbell, *J. Amer. Chem. Soc.*, **88**, 5266 (1966).
282. G. A. Hamilton, J. W. Hanifin, and J. P. Friedman, *J. Amer. Chem. Soc.*, **88**, 5269 (1966).
283. G. A. Hamilton, *J. Amer. Chem. Soc.*, **86**, 3391 (1964).
284. H. Neurath, *Enzymes*, **4**, 11 (1960).
285. M. L. Anson, *J. Gen. Physiol.*, **20**, 663, 777 (1937).
286. L. J. Greene, C. H. W. Hirs, and G. E. Palade, *J. Biol. Chem.*, **238**, 2054 (1963).
287. P. J. Keller, E. Cohen, and H. Neurath, *J. Biol. Chem.*, **223**, 457 (1956); **230**, 905 (1958).
288. J. R. Brown, D. J. Cox, R. N. Greenshields, K. A. Walsh, M. Yamasaki, and H. Neurath, *Proc. Natl. Acad. Sci. U.S.*, **47**, 1554 (1961).

289. M. Yamasaki, J. R. Brown, D. J. Cox, R. N. Greenshields, R. D. Wade, and H. Neurath, *Biochemistry*, **2**, 859 (1963).

290. J. R. Brown, R. N. Greenshields, M. Yamasaki, and H. Neurath, *Biochemistry*, **2**, 867 (1963).

291. J. R. Brown, K. Yamasaka, and H. Neurath, *Biochemistry*, **2**, 877 (1963).

292. R. Piras and B. L. Vallee, *Biochemistry*, **6**, 348 (1967).

293. M. Bergmann, *Harvey Lectures*, **31**, 37 (1935–36).

294. E. L. Smith and R. Lumry, *Cold Spring Harbor Symp. Quant. Biol.*, 14, 168 (1950).

295. H. Neurath and G. W. Schwert, *Chem. Rev.*, **46**, 69 (1950).

296. E. L. Smith, *Advances in Enzymology*, Vol. 12, F. F. Nord, Ed., Interscience, New York, 1951, p. 191.

297. M. Bergmann and J. S. Fruton, *J. Biol. Chem.*, **145**, 247 (1952).

298. E. L. Smith, *Proc. Natl. Acad. Sci. U.S.*, **35**, 80 (1949).

299. B. L. Vallee and H. Neurath, *J. Amer. Chem. Soc.*, **76**, 5006 (1964).

300. B. L. Vallee, J. A. Rupley, T. L. Coombs, and H. Neurath, *J. Amer. Chem. Soc.*, **80**, 4750 (1958).

301. H. Neurath, R. A. Bradshaw, L. H. Ericsson, D. R. Babin, P. H. Petra, and K. A. Walsh, in *"Structure, Function, and Evolution in Proteins,"* *Brookhaven Symp. Biol.* **21**, 1 (1969).

302. S. Yanari and M. A. Mitz, *J. Amer. Chem. Soc.*, **79**, 1150, 1154 (1957).

303. J. P. Felber, T. L. Coombs, and B. L. Vallee, *Biochemistry*, **1**, 231 (1962).

304. N. Abramowitz, I. Schechter, and A. Berger, *Biochem. Biophys. Res. Commun.*, **29**, 862 (1967).

305. R. T. Simpson, J. F. Riordan, and B. L. Vallee, *Biochemistry*, **2**, 616 (1963).

305a. O. A. Roholt and D. Pressman, *Proc. Natl. Acad. Sci. U.S.*, **58**, 280 (1967).

306. J. F. Riordan, B. L. Vallee, and D. M. Saunders, *Biochemistry*, **2**, 1460 (1963).

307. R. T. Simpson and B. L. Vallee, *Biochemistry*, **5**, 1760 (1966).

308. J. E. Coleman, P. Pulido, and B. L. Vallee, *Biochemistry*, **5**, 2019 (1966).

308a. M. L. Bender, J. R. Whitaker, and F. Menger, *Proc. Natl. Acad. Sci. U.S.*, **53**, 711 (1965).

309. B. L. Vallee, *Federation Proc.*, **23**,,8 (1964).

310. J. F. Riordan, M. Sokolovsky, and B. L. Vallee, *Biochemistry*, **6**, 358 (1967).

311. M. Sokolovsky and B. L. Vallee, *Biochemistry*, **6**, 700 (1967).

312. F. A. Quiocho and F. M. Richards, *Biochemistry*, **5**, 4062 (1966).

313. W. H. Bishop, F. T. Quiocho, and F. M. Richards, *Biochemistry*, **5**, 4077 (1966).

314. G. Nemethy, D. C. Phillips, S. J. Leach, and H. A. Scheraga, *Nature*, **214**, 363 (1967).

315. H. Fujioka and K. Imahori, *J. Biol. Chem.*, **237**, 2804 (1962).

316. D. S. Auld, and B. L. Vallee, *Biochemistry*, **9**, 602 (1970).

317. B. L. Vallee, J. F. Riordan, J. L. Bethune, T. L. Coombs, D. S. Auld and M. Sokolovsky, *Biochemistry*, **7**, 3547 (1968).

317a. E. T. Kaiser and F. W. Carson, *Biochem. Biophys. Res. Commun.*, **18**, 457 (1965).

318. I. Schecter and A. Berger, *Biochemistry*, **5**, 3371 (1966).

319. T. C. Bruice and S. J. Benkovic, *Bioorganic Mechanisms*, Benjamin, New York, 1966, p 27.

320. F. W. Carson and E. T. Kaiser, *J. Amer. Chem. Soc.*, **88**, 1212 (1966).

321. R. C. Davies, J. F. Riordan, D. S. Auld, and B. L. Vallee, *Biochemistry*, **7**, 1090 (1968).

321a. J. E. Folk, E. C. Wolff, and E. W. Schirmer, *J. Biol. Chem.*, **237**, 3100 (1962).

322. E. Wintersberger, D. J. Cox, and H. Neurath, *Biochemistry*, **1**, 1069, 1078 (1962).

323. E. Wintersberger, H. Neurath, T. L. Coombs, and B. L. Vallee, *Biochemistry*, **4**, 1526 (1965).

324. D. Keilin and T. Mann, *Biochem. J.*, **34**, 1163 (1940).

325. D. Keilin and T. Mann, *Nature*, **148**, 493 (1941).
326. R. E. Tashian, C. C. Plato, and T. B. Shows, *Science*, **140**, 53 (1963).
327. Y. Pocker and J. T. Stone, *J. Amer. Chem. Soc.*, **87**, 5497 (1965).
328. K-W. Lo and E. T. Kaiser, *Chem. Commun.*, **1966**, 834.
329. Y. Pocker and J. E. Meany, *Biochemistry*, **4**, 2535 (1965).
330. Y. Pocker and J. E. Meany, *Biochemistry*, **6**, 239 (1967).
331. Y. Pocker and D. G. Dickerson, *Biochemistry*, **7**, 1995 (1968).
332. Y. Pocker and J. T. Stone, *Biochemistry*, **7**, 3021 (1968).
333. Y. Pocker and D. R. Storm, *Biochemistry*, **7**, 1202 (1968).
334. J. E. Coleman, *J. Biol. Chem.*, **243**, 4574 (1968).
335. J. Maynard and J. E. Coleman, *J. Biol. Chem.* (in press, 1971).
336. M. Eigen, K. Kustin, and G. Moss, *Z. Physik. Chem.*, **30**, 130 (1961).
337. D. M. Kern, *J. Chem. Educ.*, **37**, 14 (1960).
338. C. Ho and J. M. Sturtevant, *J. Biol. Chem.*, **238**, 3499 (1963).
339. B. H. Gibbons and J. T. Edsall, *J. Biol. Chem.*, **238**, 3502 (1963).
340. T. H. Maren, *J. Pharmacol. Exptl. Therap.*, **139**, 140 (1963).
341. F. J. W. Roughton, *Physiol. Rev.*, **15**, 241 (1935).
342. F. J. W. Roughton, *Harvey Lectures*, **39**, 96 (1943).
343. R. E. Davies, *Biol. Rev. Cambridge Phil. Soc.*, **26**, 87 (1951).
344. S. M. Cain and A. B. Otis, *J. Appl. Physiol.*, **16**, 1023 (1961).
345. T. H. Maren, *Physiol. Rev.*, **47**, 595 (1967).
346. E. E. Rickli, S. A. S. Ghazanfar, B. H. Gibbons, and J. T. Edsall, *J. Biol. Chem.*, **239**, 1065 (1964).
347. G. Laurent, C. Marriq, D. Nahon, M. Charrel, and Y. Derrien, *Compt. Rend., Soc. Biol.*, **156**, 1456 (1962).
348. G. Laurent, M. Charrel, F. Luccioni, M. F. Autran, and Y. Derrien, *Bull Soc. Chim. Biol.*, **47**, 1101 (1965).
349. S. Lindskog, *Biochim. Biophys. Acta*, **39**, 218 (1960).
350. P. O. Nyman, *Biochim. Biophys. Acta*, **52**, 1 (1961).
351. J. T. Edsall, *Harvey Lectures*, **62**, 191 (1968).
352. J. T. Edsall, *Ann. N.Y. Acad. Sci.*, **151**, 41 (1968).
353. A. J. Furth, *J. Biol. Chem.*, **243**, 4832 (1968).
354. P. Byvolt and A. Gotti, *Mol. Pharmacol.*, **3**, 142 (1967).
355. B. H. Gibbons and J. T. Edsall, *J. Biol. Chem.*, **239**, 2539 (1964).
356. E. E. Rickli and J. T. Edsall, *J. Biol. Chem.*, **237**, PC 258 (1962).
357. B. Strandberg, B. Tilander, K. Fridborg, S. Lindskog, and P. O. Nyman, *J. Mol. Biol.*, **5**, 583 (1962).
358. T. H. Maren, B. Robinson, R. F. Palmer, and M. E. Griffith, *Biochem. Pharmacol.* **6**, 21 (1960).
359. T. H. Maren, A. L. Parcell, and M. N. Malik, *J. Pharmacol. Exptl. Therap.*, **130**, 389 (1960).
360. T. H. Maren, *J. Pharmacol. Exptl. Therap.*, **139**, 129 (1963).
361. J. C. Kernohan, *Biochim. Biophys. Acta*, **118**, 405 (1966).
362. D. V. Myers and J. T. Edsall, *Proc. Natl. Acad. Sci. U.S.*, **53**, 169 (1965).
363. S. Beychok, J. M. Armstrong, C. Lindblow, and J. T. Edsall, *J. Biol. Chem.*, **241**, 5150 (1966).
364. A. Rosenberg, *J. Biol. Chem.*, **241**, 5126 (1966).
365. R. P. Davis, *J. Amer. Chem. Soc.*, **60**, 5209 (1958).
366. R. P. Davis, *The Enzymes*, **5**, 545 (1961).
367. J. C. Kernohan, *Biochim. Biophys. Acta*, **81**, 346 (1964).

368. J. C. Kernohan, *Biochim. Biophys. Acta*, **96**, 304 (1965).

369. J. H. Wang, *Science*, **161**, 328 (1968).

370. M. Eigen, *Discussions Faraday Soc.*, **39**, 7 (1965).

371. Y. Pocker and J. T. Stone, *Biochemistry*, **7**, 2936 (1968).

372. M. Caplow, *J. Amer. Chem. Soc.*, **93**, 230 (1971).

373. H. DeVoe and G. B. Kistiakowsky, *J. Amer. Chem. Soc.*, **83**, 274 (1961).

374. A. M. Sargeson, in *Chelating Agents and Metal Chelates*, F. P. Dwyer and D. P. Mellor, Eds. Academic Press, New York, 1964, p. 183.

375. M. Caplow, *J. Amer. Chem. Soc.*, **90**, 6795 (1968).

376. P. L. Whitney, P. O. Nyman, and B. G. Malmstrom, *J. Biol. Chem.*, **242**, 4212 (1967).

377. S. L. Bradbury, *J. Biol. Chem.*, **244**, 2002 (1969).

378. S. L. Bradbury, *J. Biol. Chem.*, **244**, 2010 (1969).

379. J. E. Coleman, unpublished observation.

380. P. L. Whitney, G. Folsch, P. O. Nyman, and B. G. Malmstrom, *J. Biol. Chem.*, **242**, 4206 (1967).

381. W. C. Galley and L. Stryer, *Proc. Natl. Acad. Sci. U.S.*, **60**, 108 (1968).

382. P. O. Nyman, L. Strid, and G. Westermark, *Biochim. Biophys. Acta*, **122**, 554 (1966).

383. P. O. Nyman, L. Strid, and G. Westermark, *Europ. J. Biochem.*, **6**, 72 (1968).

384. B. Andersson, P. O. Gothe, T. Nilsson, P. O. Nyman, and L. Strid, *Europ. J. Biochem.*, **6**, 190 (1968).

385. M. H. Malamy and B. L. Horecker, *Biochemistry*, **3**, 1893 (1964).

386. K. Takeda and A. Tsugita, *J. Biochem. (Tokyo)*, **61**, 231 (1967).

387. J. C. Mathies, *J. Biol. Chem.*, **233**, 1121 (1958).

388. S. Trubowitz, D. Feldman, S. W. Morgenstern, and V. M. Hunt, *Biochem. J.*, **80**, 369 (1961).

389. L. Engström, *Biochim. Biophys. Acta*, **52**, 36 (1961).

390. L. A. Heppel, D. R. Harkness, and R. J. Hilmoe, *J. Biol. Chem.*, **237**, 841 (1962).

391. H. Neumann, L. Boross, and E. Katchalski, *J. Biol. Chem.*, **242**, 3142 (1967).

392. M. Cohn, *J. Biol. Chem.*, **180**, 771 (1949).

393. O. Meyerhof and H. Green, *Biochem. J.*, **178**, 655 (1949).

394. R. K. Morton, *Biochem. J.*, **70**, 139 (1958).

395. I. B. Wilson, J. Dayan, and K. Cyr, *J. Biol. Chem.*, **239**, 4182 (1964).

396. W. B. Anderson and R. C. Nordlie, *J. Biol. Chem.*, **242**, 114 (1967).

397. F. Rothman and R. Byrne, *J. Mol. Biol.*, **6**, 330 (1963).

398. A. W. Hanson, M. L. Applebury, J. E. Coleman, and H. W. Wyckoff, *J. Biol. Chem.*, **245**, 4975 (1970).

399. M. J. Schlesinger, *J. Biol. Chem.*, **240**, 4293 (1965).

400. A. Garen, C. Levinthal, and F. Rothman, *J. Chim. Phys.*, **58**, 1068 (1961).

401. A. Garen and S. Garen, *J. Mol. Biol.*, **7**, 13 (1963).

402. M. J. Schlesinger, *J. Biol. Chem.*, **241**, 3181 (1966); **242**, 1604 (1967).

403. M. Gottesman, R. T. Simpson, and B. L. Vallee, *Biochemistry*, **8**, 3776 (1969).

404. S. E. Halford, N. G. Bennett, D. R. Trentham, and H. Gutfreund, *Biochem. J.*, **114**, 243 (1969).

405. C. Lazdunski, C. Petitclerc, D. Chappelet, M. Lazdunski, *Biochem. Biophys. Res. Commun.* **37**, 744 (1969).

406. H. Neuman, *J. Biol. Chem.*, **243**, 4671 (1968).

407. D. E. Koshland, Jr., and K. E. Neet, *Ann. Rev. Biochem.*, **37**, 359 (1968).

408. A. Levitzki and D. E. Koshland, Jr., *Proc. Natl. Acad. Sci. U.S.*, **62**, 359 (1968).

408. A. Levitzki and D. E. Koshland, Jr., *Proc. Natl. Acad. Sci. U.S.* **62**, 1121 (1969).

409. R. T. Simpson, B. L. Vallee, and G. H. Tait, *Biochemistry*, **7**, 4336 (1968).

410. W. W. Butcher and F. H. Westheimer, *J. Amer. Chem. Soc.*, **77**, 2420 (1955).
411. M. L. Bender and R. Breslow, *Comprehensive Biochem.* **2**, 1 (1962).
412. W. P. Jencks, *Catalysis in Chemistry and Enzymology*, McGraw-Hill, New York, 1969, p. 97, 233, 498.
413. C. Lazdunski and M. Lazdunski, *Biochim. Biophys. Acta*, **113**, 551 (1966).
414. D. L. Miller and T. Ukena, *J. Amer. Chem. Soc.*, **91**, 11 (1969).
415. M. Tetas and J. M. Lowenstein, *Biochemistry*, **2**, 350 (1963).
416. D. L. Miller, G. J. Krol, and U. P. Strauss, *J. Amer. Chem. Soc.*, **91**, 6882 (1969).
417. B. S. Cooperman, *Biochemistry*, **8**, 5005 (1969).
418. M. Cohn and T. R. Hughes, *J. Biol. Chem.*, **237**, 176 (1962).
419. G. H. Tait and B. L. Vallee, *Proc. Nat. Acad. Sci., U.S.*, **56**, 1247 (1966).
420. J. Kumamoto, J. R. Cox, and F. H. Westheimer, *J. Amer. Chem. Soc.*, **78**, 4858 (1956).
421. F. Covitz and F. H. Westheimer, *J. Amer. Chem. Soc.*, **85**, 1773 (1963).
422. J. R. Cox and B. Ramsay, *Chem. Rev.*, **64**, 317 (1964).
423. F. H. Westheimer, *Accounts Chem. Res.*, **1**, 70 (1968).
424. P. C. Haake and F. H. Westheimer, *J. Amer. Chem. Soc.*, **83**, 1102 (1961).
425. E. A. Dennis and F. H. Westheimer, *J. Amer. Chem. Soc.*, **88**, 3432 (1966).
426. P. D. Boyer, H. Lardy, and K. Myrbäck, *Enzymes*, **7** (Part A) (1963).
427. H. Theorell, A. P. Nygaard, and R. K. Bonnicksen, *Acta Chem. Scand.*, **5**, 1105 (1955).
428. B. L. Vallee and F. L. Hoch, *J. Biol. Chem.*, **225**, 185 (1957).
429. D. E. Drum, T-K Li, and B. L. Vallee, *Biochemistry*, **8**, 3783 (1969).
430. J. P. von Wartburg, J. L. Bethune, and B. L. Vallee, *Biochemistry*, **3**, 1775 (1964).
431. B. L. Vallee and F. L. Hoch, *Proc. Natl. Acad. Sci. U.S.*, **41**, 327 (1955).
432. B. L. Vallee and W. E. C. Wacker, *J. Amer. Chem. Soc.*, **78**, 1771 (1956).
433. J. H. Harrison, *Federation Proc.*, **22**, 493 (1963).
434. B. L. Vallee, S. J. Adelstein, J. A. Olson, *J. Amer. Chem. Soc.*, **77**, 5196 (1955).
435. S. J. Adelstein and B. L. Vallee, *J. Biol. Chem.*, **233**, 589 (1958).
436. H. R. Levy and B. Vennesland, *J. Biol. Chem.*, **228**, 85 (1957).
437. H. Sund and H. Theorell, *Enzymes*, **7**, 25 (1963).
438. H. Theorell and J. S. McKinley McKee, *Acta Chem. Scand.*, **15**, 1834 (1961).
439. K. Wallenfels and H. Sund, *Biochem. Z.*, **329**, 59 (1957).
440. D. E. Drum, T-K Li, and B. L. Vallee, *Biochemistry*, **8**, 3792 (1969).
441. F. L. Hoch, R. J. P. Williams, and B. L. Vallee, *J. Biol. Chem.*, **232**, 453 (1958).
442. R. J. P. Williams, F. L. Hoch, and B. L. Vallee, *J. Biol. Chem.*, **232**, 465 (1958).
443. B. L. Vallee, R. J. P. Williams, and F. L. Hoch, *J. Biol. Chem.*, **234**, 2621 (1959).
444. D. D. Ulmer and B. L. Vallee, *J. Biol. Chem.*, **236**, 730 (1961).
445. D. D. Ulmer, T-K Li, and B. L. Vallee, *Proc. Natl. Acad. Sci. U.S.*, **47**, 1155 (1961).
446. T-K Li, D. D. Ulmer, and B. L. Vallee, *Biochemistry*, **1**, 114 (1962).
447. R. Druyan and B. L. Vallee, *Biochemistry*, **3**, 944 (1964).
448. A. Åkeson, *Biochem. Biophys. Res. Commun.*, **17**, 211 (1964).
449. H. L. Oppenheimer, R. W. Green, and R. H. McKay, *Arch. Biochem. Biophys.*, **119**, 552 (1967).
450. T-K Li and B. L. Vallee, *Biochemistry*, **3**, 869 (1964).
451. H. Theorell and R. Bonnichsen, *Acta Chem. Scand.*, **5**, 329, 1105 (1951).
452. K. V. Rajagopalan, P. Handler, G. Palmer, and H. Beinert, *J. Biol. Chem.*, **243**, 3784, 3797 (1968).
453. C. A. Nelson and P. Handler, *J. Biol. Chem.*, **243**, 5368 (1968).
454. C. Remy, D. A. Richert, R. H. Doisy, I. C. Wells, and W. W. Westerfeld, *J. Biol. Chem.*, **217**, 293 (1955).
455. D. J. D. Nicholas and A. Nason, *J. Biol. Chem.*, **207**, 352 (1954); **211**, 183 (1954).

456. T. P. Singer, E. B. Kearney, and V. Massey, *Advances in Enzymology*, Vol. 18 F. F. Nord, Ed., Interscience, New York, 1957, p. 65.
457. P. Handler, K. V. Rajagopalan, and V. Aleman, *Federation Proc.*, **23**, 30 (1964).
458. H. C. Friedman and B. Vennesland, *J. Biol. Chem.*, **235**, 1526 (1960).
459. S. Bouchilloux, P. McMahill, and H. S. Mason, *J. Biol. Chem.*, **238**, 1699 (1963).
460. M. Fling, N. H. Horowitz, and S. F. Heineman, *J. Biol. Chem.*, **238**, 2045 (1963).
461a. A. B. Lerner, T. B. Fitzpatrick, E. Calkins, and W. H. Summerson, *J. Biol. Chem.*, **178**, 185 (1949); **187**, 793 (1950); **191**, 799 (1951).
461b. Burnett, personal communication.
462. T. Omura, *J. Biochem. (Tokyo)*, **50**, 264 (1961).
463. R. Mosbach, *Biochim. Biophys. Acta*, **73**, 204 (1963).
464. G. Fåhraeus and B. Reinhammar, *Acta Chem. Scand.*, **21**, 2367 (1967).
465. H. R. Mahler, in *Trace Elements*, C. A. Lamb, O. G. Bentley and J. M. Beattie, Eds., Academic Press, New York, 1958, p. 311.
466. D. Amaral, L. Bernstein, D. Morse, and B. L. Horecker, *J. Biol. Chem.*, **238**, 2281 (1963).
467. S. Nara and K. T. Yasunobu, in *The Biochemistry of Copper*, J. Peisach, P. Aisen and W. E. Blumberg, Eds., Academic Press, New York, 1966, p. 423.
468. H. Yamada and Y. T. Yasunobu, *J. Biol. Chem.*, **237**, 1511, 3077 (1962).
469. J. M. Hill and P. J. G. Mann, *Biochem. J.*, **91**, 171 (1964).
470. R. P. Ambler, *Biochem. J.*, **89**, 341 (1963).
471. H. Suzuki and H. Iwasaki, *J. Biochem. (Tokyo)*, **52**, 193 (1962).
472. R. P. Ambler and L. H. Brown, *J. Mol. Biol.*, **9**, 825 (1964).
473. H. S. Mason, W. L. Fowlks, and E. Peterson, *J. Amer. Chem. Soc.*, **77**, 2914 (1955).
474. H. W. Duckworth and J. E. Coleman, *J. Biol. Chem.*, **245** (1970).
475. R. L. Jolley, D. A. Robb, and H. S. Mason, *J. Biol. Chem.*, **244**, 1593 (1969).
476. S. Osaki, *Arch. Biochem. Biophys.*, **100**, 378 (1963).
477. M. Sato, *Phytochemistry*, **8**, 353 (1969).
478. S. H. Pomerantz, *J. Biol. Chem.*, **238**, 2351 (1963).
479. S. H. Pomerantz and M. C. Warner, *J. Biol. Chem.*, **242**, 5308 (1967).
480. A. San Pietro, Ed., *Non-heme Iron Proteins: Role in Energy Conversion*, Antioch Press, Yellow Springs, Ohio, 1965.
481. R. Malkin and J. C. Rabinowitz, *Ann. Rev. Biochem.*, **36**, 113 (1967).
482. R. C. Valentine, *Bacteriol. Rev.*, **28**, 497 (1964).
483. T. Kimura, *Struct. Bonding*, **5**, 1 (1968).
484. K. Tagawa and D. I. Arnon, *Nature*, **195**, 537 (1962).
485. L. E. Mortenson, *Biochim. Biophys. Acta*, **81**, 71 (1964).
486. B. E. Sobel and W. Lovenberg, *Biochemistry*, **5**, 6 (1966).
487. R. Bachofen and D. I. Arnon, *Biochim. Biophys. Acta*, **120**, 259 (1966).
488. A. San Pietro and H. M. Lang, *J. Biol. Chem.*, **231**, 211 (1958).
489. P. Böger, C. C. Black, and A. San Pietro, *Arch. Biochem. Biophys.*, **115**, 35 (1966).
490. W. Lovenberg and B. E. Sobel, *Proc. Natl. Acad. Sci. U.S.*, **54**, 193 (1965).
491. W. Lovenberg and B. E. Sobel, *Federation Proc.*, **24**, 233 (1965).
492. R. G. Bartsch, in *Bacterial Photosynthesis*, H. Gest, A. San Pietro and L. P. Vernon, Eds., Antioch Press, Yellow Springs, Ohio, 1963, p. 315.
493. K. Suzuki and T. Kimura, *Biochem. Biophys. Res. Commun.*, **19**, 340 (1965).
494. T. Omura, E. Sanders, D. Y. Cooper, O. Rosenthal, and R. W. Estabrook, in *Non-heme Iron Proteins*, A. San Pietro, Ed., Antioch Press, Yellow Springs, Ohio, 1965, p. 401.
495. R. Coleman, J. S. Rieske, and D. Wharton, *Biochem. Biophys. Res. Commun.*, **15**, 345 (1964).

496. J. S. Rieske, in *Non-heme Iron Proteins*, A. San Pietro, Ed., Antioch Press, Yellow Springs, Ohio, 1965, p. 461.
497. R. W. Miller and J. Massey, *J. Biol. Chem.*, **240**, 1453 (1965).
498. M. Tanaka, T. Nakashima, A. Benson, H. Mower, and K. T. Yasunobu, *Biochemistry*, **5**, 1666 (1966).
499. A. M. Benson, H. F. Mower, and K. T. Yasunobu, *Proc. Natl. Acad. Sci. U.S.*, **55**, 1532 (1966).
500. R. D. Gillard, E. D. McKenzie, R. Mason, S. G. Mayhew, J. L. Peel, and J. E. Stangroom, *Nature*, **208**, 769 (1965).
501. L. C. Sieker and L. H. Jensen, *Biochem. Biophys. Res. Commun.*, **20**, 33 (1965).
502. S. J. Edelstein and H. J. Schachman, *J. Biol. Chem.*, **242**, 306 (1967).
503. A. San Pietro and C. C. Black, *Ann. Rev. Plant Physiol.*, **16**, 155 (1965).

AUTHOR INDEX

Numbers in parentheses are reference numbers and show that an author's work is referred to although his name is not mentioned in the text. Numbers in *italics* indicate the pages on which the full references appear.

SUBJECT INDEX